Advances in Soft Computing 51

Editor-in-Chief: J. Kacprzyk

Advances in Soft Computing

Editor-in-Chief

Prof. Janusz Kacprzyk
Systems Research Institute
Polish Academy of Sciences
ul. Newelska 6
01-447 Warsaw
Poland
E-mail: kacprzyk@ibspan.waw.pl

Further volumes of this series can be found on our homepage: springer.com

Mieczyslaw A. Klopotek, Slawomir T. Wierzchon, Kryzysztof Trojanowski (Eds.)
Intelligent Information Processing and Web Mining, 2006
ISBN 978-3-540-33520-7

Ashutosh Tiwari, Joshua Knowles, Erel Avineri, Keshav Dahal, Rajkumar Roy (Eds.)
Applications and Soft Computing, 2006
ISBN 978-3-540-29123-7

Bernd Reusch, (Ed.)
Computational Intelligence, Theory and Applications, 2006
ISBN 978-3-540-34780-4

Jonathan Lawry, Enrique Miranda, Alberto Bugarín Shoumei Li, María Á. Gil, Przemysław Grzegorzewski, Olgierd Hryniewicz,
Soft Methods for Integrated Uncertainty Modelling, 2006
ISBN 978-3-540-34776-7

Ashraf Saad, Erel Avineri, Keshav Dahal, Muhammad Sarfraz, Rajkumar Roy (Eds.)
Soft Computing in Industrial Applications, 2007
ISBN 978-3-540-70704-2

Bing-Yuan Cao (Ed.)
Fuzzy Information and Engineering, 2007
ISBN 978-3-540-71440-8

Patricia Melin, Oscar Castillo, Eduardo Gómez Ramírez, Janusz Kacprzyk, Witold Pedrycz (Eds.)
Analysis and Design of Intelligent Systems Using Soft Computing Techniques, 2007
ISBN 978-3-540-72431-5

Oscar Castillo, Patricia Melin, Oscar Montiel Ross, Roberto Sepúlveda Cruz, Witold Pedrycz, Janusz Kacprzyk (Eds.)
Theoretical Advances and Applications of Fuzzy Logic and Soft Computing, 2007
ISBN 978-3-540-72433-9

Katarzyna M. Węgrzyn-Wolska, Piotr S. Szczepaniak (Eds.)
Advances in Intelligent Web Mastering, 2007
ISBN 978-3-540-72574-9

Emilio Corchado, Juan M. Corchado, Ajith Abraham (Eds.)
Innovations in Hybrid Intelligent Systems, 2007
ISBN 978-3-540-74971-4

Marek Kurzynski, Edward Puchala, Michal Wozniak, Andrzej Zolnierek (Eds.)
Computer Recognition Systems 2, 2007
ISBN 978-3-540-75174-8

Van-Nam Huynh, Yoshiteru Nakamori, Hiroakira Ono, Jonathan Lawry, Vladik Kreinovich, Hung T. Nguyen (Eds.)
Interval / Probabilistic Uncertainty and Non-classical Logics, 2008
ISBN 978-3-540-77663-5

Ewa Pietka, Jacek Kawa (Eds.)
Information Technologies in Biomedicine, 2008
ISBN 978-3-540-68167-0

Didier Dubois, M. Asunción Lubiano, Henri Prade, María Ángeles Gil, Przemysław Grzegorzewski, Olgierd Hryniewicz (Eds.)
Soft Methods for Handling Variability and Imprecision, 2008
ISBN 978-3-540-85026-7

Juan M. Corchado, Francisco de Paz, Miguel P. Rocha, Florentino Fernández Riverola (Eds.)
2nd International Workshop on Practical Applications of Computational Biology and Bioinformatics (IWPACBB 2008), 2009
ISBN 978-3-540-85860-7

Juan M. Corchado, Sara Rodríguez, James Llinas, José M. Molina (Eds.)
International Symposium on Distributed Computing and Artificial Intelligence 2008 (DCAI 2008), 2009
ISBN 978-3-540-85862-1

Juan M. Corchado, Dante I. Tapia, José Bravo (Eds.)
3rd Symposium of Ubiquitous Computing and Ambient Intelligence 2008, 2009
ISBN 978-3-540-85866-9

Juan M. Corchado, Dante I. Tapia,
José Bravo (Eds.)

3rd Symposium of Ubiquitous Computing and Ambient Intelligence 2008

Editors

Juan M. Corchado
Departamento de Informática y Automática
Facultad de Ciencias
Universidad de Salamanca
Plaza de la Merced S/N
37008, Salamanca
Spain
E-mail: corchado@usal.es

José Bravo
Universidad de Castilla-La Mancha
Departamento de Tecnologías y Sistemas de Información (UCLM)
Escuela Superior de Informática
Paseo de la Universidad, s/n
13071, Ciudad Real
Spain
E-mail: Jose.Bravo@uclm.es

Dante I. Tapia
Departamento de Informática y Automática
Facultad de Ciencias
Universidad de Salamanca
Plaza de la Merced S/N
37008, Salamanca
Spain
E-mail: dantetapia@usal.es

ISBN 978-3-540-85866-9 e-ISBN 978-3-540-85867-6

DOI 10.1007/978-3-540-85867-6

Advances in Soft Computing ISSN 1615-3871

Library of Congress Control Number: 2008933603

Typeset & Cover Design: Scientific Publishing Services Pvt. Ltd., Chennai, India.

Printed in acid-free paper

5 4 3 2 1 0

springer.com

Preface

The Symposium on Ubiquitous Computing and Ambient Intelligence (UCAmI) began as a workshop held in 2003 in San Sebastián (Spain) under the Spanish Artificial Intelligence Conference. This event gathered 32 attendees and 18 papers were presented. The second edition, already as a Symposium, took place in Granada (Spain) under the first Spanish Computer Science Conference (CEDI). Later, in 2006, a second workshop was celebrated in Ciudad Real and, in 2007; the second Symposium was organized in Zaragoza by the CEDI conference. Now we continue to work on the organization of this event in Salamanca, a beautiful Spanish city.

The European Community and the Sixth and Seventh Framework Programs encourage researchers to explore the generic scope of the AmI vision. In fact, some researchers have a crucial role in this vision. Emile Aarts from Philips describes Ambient Intelligence as "the integration of technology into our environment, so that people can freely and interactively utilize it". This idea agrees with the proposal of Mark Weiser regarding the Ubiquitous Computing paradigm.

The UCAmI community tries to join experts around the world in order to promote collaborations and to put into practice studies for involving people into intelligent environments so that the "Everyday Computing" concept can be a reality. The UCAmI technical program includes 40 papers (31 long paper, 6 short paper and 3 doctoral consortium) selected from a submission pool of 56 papers, from 11 different countries. We thank the excellent work of the local organization members and also from the members of the Program Committee for their excellent reviewing work.

October 2008

Juan M. Corchado
José Bravo
Dante I. Tapia

Organization

General Co-chairs

Juan M. Corchado – University of Salamanca (Spain)
José Bravo – University of Castilla-La Mancha (Spain)
Dante I. Tapia – University of Salamanca (Spain)

Program Committee

José Bravo (Chairman) – University of Castilla-La Mancha (Spain)
Xavier Alamán – Autonomous University of Madrid (Spain)
Emilio S. Corchado – University of Burgos (Spain)
Mariano Alcañiz – Polytechnic University of Valencia (Spain)
Carlos Carrascosa – Polytechnic University of Valencia (Spain)
Cecilio Angulo – Polytechnic University of Catalonia (Spain)
Francisco J. Garijo – Telefónica I+D (Spain)
José M. Molina – University Carlos III of Madrid (Spain)
Francisco J. Ballesteros – Rey Juan Carlos University (Spain)
Yang Cai – Carnegie Mellon University (USA)
Rafael Corchuelo – University of Sevilla (Spain)
Jesús Favela – CICESE (Mexico)
Álvaro Herrero – University of Burgos (Spain)
Marcela Rodríguez – Autonomous University of Baja California (Mexico)
Pablo Vidales – Deutsche Telekom Laboratories (Germany)
Lidia Fuentes – University of Malaga (Spain)
Nuria Oliver – Microsoft Research (USA)
Juan A. Botía – University of Murcia (Spain)
Iñaki Vazquez – University of Deusto (Spain)
Carlos García – University Carlos III of Madrid (Spain)
Victor M. Gonzalez – University of Manchester (UK)
Cristina Pelayo –University of Oviedo (Spain)
Hani Hagras – University of Essex (UK)
Günter Haring – University of Vienna (Austria)
Lourdes Borrajo – University of Vigo (Spain)
Antonio F. Gómez Skarmeta – University of Murcia (Spain)
Carlos Juiz – University of the Balearic Islands (Spain)
Soraya Kouadri Mostéfaoui – The Open University (UK)

Diego Gachet – European University of Madrid (Spain)
Isidro Laso – D.G. Information Society and Media (European Commission)
Diego López de Ipiña – University of Deusto (Spain)
Javier Jaen – Polytechnic University of Valencia (Spain)
Juan Carlos López – University of Castilla-La Mancha (Spain)
Ricardo J. Machado – University of Minho (Portugal)
Juan Pavón – Complutense University of Madrid (Spain)
Óscar Mayora – Create-Net (Italy)
René Meier – Trinity College Dublin (Ireland)
Bruno Baruque – University of Burgos (Spain)
Sergio Ochoa – University of Chile (Chile)
Francisco Moya – University of Castilla-La Mancha (Spain)

Organising Committee

Juan M. Corchado (Chairman) – University of Salamanca (Spain)
Dante I. Tapia (Cochairman) – University of Salamanca (Spain)
Juan F. De Paz – University of Salamanca (Spain)
Sara Rodríguez – University of Salamanca (Spain)
Javier Bajo – University of Salamanca (Spain)
Alberto Saavedra – Tulecom Solutions (Spain)
Óscar García – Tulecom Solutions (Spain)
Ramón Hervás – University of Castilla-La Mancha (Spain)
Aitor Mata – University of Salamanca (Spain)
Rosa Cano – University of Salamanca (Spain)
Cristian Pinzón – University of Salamanca (Spain)

Contents

A Framework for the Reconfiguration of Ubicomp Systems*

Pau Giner, Carlos Cetina, Joan Fons, and Vicente Pelechano

Centro de Investigación en Métodos de Producción de Software
Universidad Politécnica de Valencia
Camino de Vera s/n, 46022, Spain
{pginer,ccetina,jjfons,pele}@dsic.upv.es

Summary. Ubiquitous Computing (Ubicomp) proposes a computation paradigm where services become invisible to the user and interaction is offered in a natural way. Considering the high heterogeneity of technologies and user requirements involved in Ubicomp systems, these systems become difficult to adjust to the specific user needs. The present work, introduces a framework for the reconfiguration of Ubicomp systems. This framework is based on the modeling of system variability and the definition of service policies that describe the user requirements about the different services. In order to validate the proposal, the presented ideas have been applied to PervML, an existing software development method for Ubicomp systems. As a result, the obtained systems are enhanced with reconfiguration capabilities.

1 Introduction

The Ubiquitous Computing (Ubicomp [1]) vision supposes a radical paradigm shift in the way users interact with systems. Computing resources become invisible to the user in order to allow a natural interaction with the system.

Ubicomp systems can offer their services in many different ways according to the available devices and the user requirements. It is unlikely that a single entity will be able to develop one-size-fits-all applications that match both, heterogeneous environments and heterogeneous user requirements [2]. In addition, configuration options should be presented to the user in an intuitive way [3]. Dealing with the configuration options of all the devices and sub-systems that compose the security system of a smart home is not feasible for common users (at least in a "natural and invisible" way).

In order to reduce configuration complexity in Ubicomp systems, following the separation of concerns principle, we propose to deal on the one hand with service composition heterogeneity (the different ways in which a service can be offered); and on the other hand, with the user requirements (which of the possible service compositions is preferred).

The main contribution of this work is a reconfiguration framework based on (1) the specification of system variability using Feature Modeling techniques (to

* This work has been developed with the support of MEC under the project SESAMO TIN2007-62894 and cofinanced by FEDER.

J.M. Corchado, D.I. Tapia, and J. Bravo (Eds.): UCAMI 2008, ASC 51, pp. 1–10, 2009.
springerlink.com

describe in a coarse-grained style service composition) and (2) the definition of simple policies that specify the conditions in which a service should be offered according to user requirements. When a predefined policy gets active, the system is reconfigured automatically to provide a service composition that fits the requirements stated in this policy.

The separation of concerns introduced in this work (decoupling service composition from user requirements) permits a clear definition of the roles involved in the development of Ubicomp systems and makes the system easy to configure for end users. In order to validate these ideas, we have applied them to PervML [4]. PervML is a Model Driven Engineering (MDE) method for the development of Ubicomp systems. By integrating the proposed framework in PervML, the obtained applications are provided with reconfiguration support.

The remainder of the paper is organized as follows: Section 2 gives an overview of the proposal. Section 3 presents the feature modeling technique used to capture system variability. Section 4 presents how service policies are specified and how they are related to system features to allow reconfiguration. Section 5 describes the reconfiguration framework presented in this work. Section 6 discuses related work and Section 7 concludes the paper.

2 Proposal Overview

The reconfiguration strategy proposed in the present work has an impact on both, development and run-time (see Fig. 1). The development phase defines the possible service compositions. At run-time, changes in user requirements are interpreted by the reconfiguration framework and the system is reconfigured accordingly.

PervML defines a strategy to **specify the services** and the code generation. PervML is a Domain Specific Language (DSL) for specifying pervasive systems in a technology independent fashion [4]. PervML applies the MDE principles; once the defined models are specified, code for the final system is generated

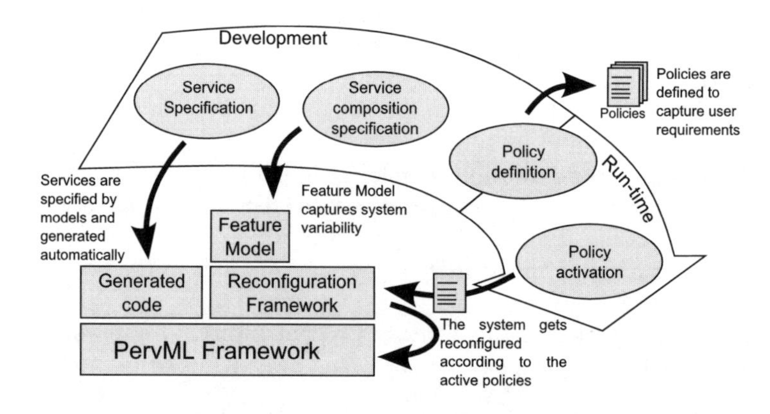

Fig. 1. Overview of the proposal

automatically by means of model transformations. Support for the edition of graphical models and the code generation is provided by PervGT tool [5].

Then, the variability for the **service composition** is captured and expressed by means of a Feature Model. This model indicates the alternatives for offering a certain service (e.g., notification to the user can be SMS or e-mail based). Only technical limitations are considered, subjective concepts (such as price, user preferences, etc.) are expressed by means of policies instead.

Policies are defined (either by developers or by skillful users) to indicate possible sets of properties that the system should fulfil (e.g., state that during a long journey the house should be empty). Developers can define a set of standard policies, but it is also possible for end-users to define their own.

Finally the user (or an automated system in behalf of her) **activates the desired policies**. For example, the user can activate the "long journey" policy when she is about to go on holidays. In this way the system is reconfigured to adapt the security system to the policy requirements (check that the house is empty, simulate presence to avoid robberies, etc.).

The Feature modelling technique, the definition of policies and the reconfiguration support offered by the framework are detailed in following sections.

3 Feature Modeling Technique

The feature modeling technique [6] (see Fig. 2) is widely used to describe a system configuration and its variants in terms of features. A feature is an increment in system functionality. The features are hierarchically linked in a tree-like structure through variability relationships and are optionally connected by cross-tree constraints. In detail, there are four relationships related to variability concepts in which we are focusing:

- **Optional.** A feature is optional when it can be selected or not whenever its parent feature is selected. Graphically it is represented with a small white circle on top of the feature.
- **Mandatory.** A feature is mandatory when it must be selected whenever its parent feature is selected. Graphically it is represented with a small black circle on top of the feature.
- **Or-relationship.** A set of child features have an or-relationship with their parent feature when one or more child features can be selected simultaneously. Graphically it is represented with a black triangle.
- **Alternative.** A set of child features have an alternative relationship with their parent feature when only one feature can be selected simultaneously. Graphically it is represented with a white triangle.

The different ways a service can be offered are stated explicitly in the Feature Model. In the example of Fig. 2, for the security service it is stated that presence detection can be achieved using the *Perimeter Detection* feature or the *In Home Detection* feature in a non-exclusive way. The notification for the Security system can either be performed by sending a textual *SMS*, an *e-mail* or a *video* stream.

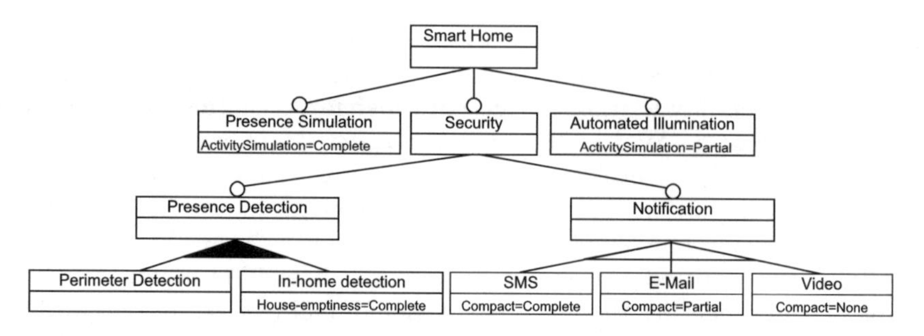

Fig. 2. Smart home feature model

In [7] feature modeling was proposed for developing dynamically-adaptive pervasive systems. Additionally, in the present work we have extended the feature modeling technique with some attributes. These attributes describe the capabilities and limitations of features in order to allow the application of policies (as it is detailed in next section). Each attribute defines the coverage degree (complete, partial or none) of a feature for a given property of the system. For instance, the *SMS* feature supports *compact interaction* in a *complete* manner (Fig. 2). If compact interaction were required (by some policy) the notification for the security service would be SMS-based.

4 Policy Description

A policy states some assertions about service requirements and capabilities. We define a policy as a set of assertions. Each assertion is formed by a pair (P, Q) where P is a property and Q is a qualifier for this property. Properties are domain-dependent concepts such as *activity simulation* (i.e., make the house look like it is habited) or *house emptiness* (i.e., ensure that nobody is inside the house). The qualifier Q indicates the enforcement degree for a property. We have considered four qualifiers: *required*, *preferred*, *discouraged* and *forbidden*.

For example, in order to configure the security system for a long journey, a policy is defined. In this policy *house emptiness* and *activity simulation* properties are required.

4.1 Driving System Reconfiguration

We propose the use of policies to guide system reconfiguration. The features that best support the policy assertions will be used to carry out the service. Fig. 3 illustrates the basic idea of using policies for choosing the most adequate alternative for the security service.

The *long journey* policy defines different assertions about three properties: the *activity simulation* and *house emptiness* properties are *required*, and the *compact* interaction is *preferred*. These assertions are used by the reconfiguration framework to determine how the security service is offered.

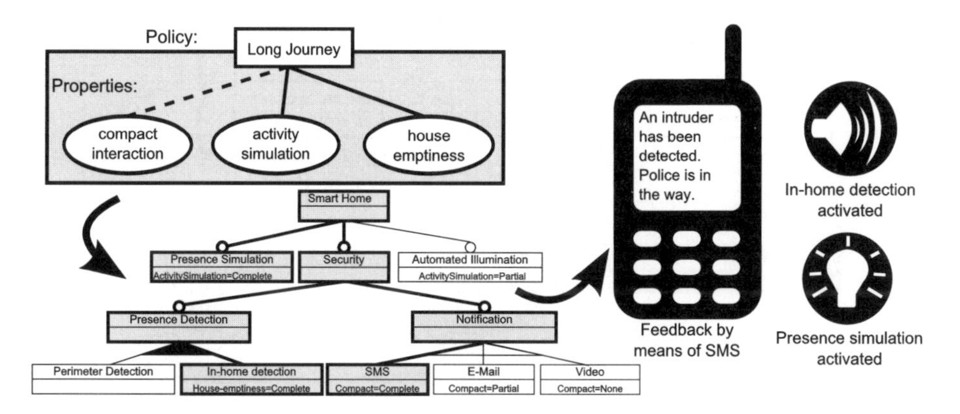

Fig. 3. The use of policies to configure a service

The reconfiguration framework chooses the features from the Feature Model (see Fig. 2) that better fulfill each property. The selected features for the example are *presence simulation* and *in-home detection* since they have a complete coverage of the required properties. For the notification, SMS has been selected over e-mail or video stream since (1) *compact interaction* is preferred according to the policy and (2) the SMS notification mechanism is the one with better coverage for this property in the Feature Model.

When the *long journey* policy is activated the security system will be offered in the way previously described. The transition from the previous configuration to the one that satisfies the active policy constitutes the reconfiguration process. This process could imply several actions such as the activation of the presence detection and simulation sub-systems and the substitution of the notification medium from a video-based one (if it was the active one) to the SMS-based notification. In this way, we ensure that the system will provide the security service in the most suited way. The steps that are followed for this reconfiguration and its run-time support are detailed below.

5 Reconfiguration Framework for Ubicomp Systems

In order to put our ideas into practice, we have extended the PervML implementation framework [8]. The role of PervML framework is to reduce the abstraction gap between PervML concepts and the target implementation technology. In this way, the code generated from models only constitutes a small part of the whole application.

The generated code extends the PervML framework to cover the specific needs of each development, while domain-independent concepts are kept in the framework. PervML framework is based on OSGi and follows a layered approach using the Model-View-Controller pattern.

We have extended this framework to allow different configurations for the services when different policies are applied. The components (in the form of

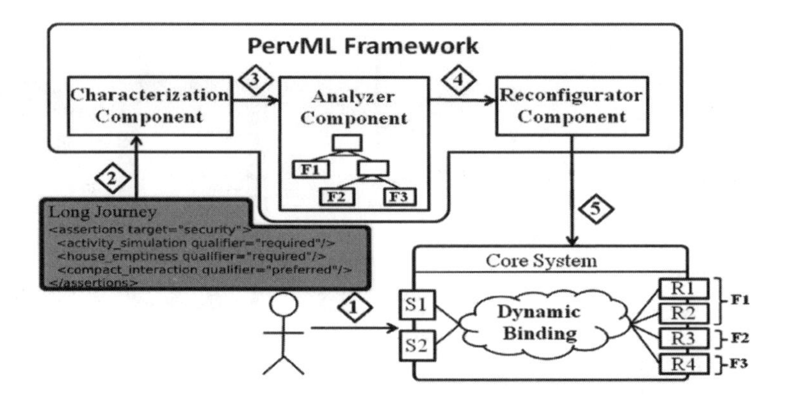

Fig. 4. Reconfiguration Strategy

OSGi bundles) incorporated to the framework for supporting the reconfiguration strategy are described below.

5.1 Reconfiguration Strategy

As stated in the previous section, policies provide criteria to dynamically reconfigure the system. At run-time, the framework must adapt the system to the active policies. It performs dynamic service-resource bindings taking care of the Feature Model constraints (e.g., not allowing multiple features simultaneously if they are defined as exclusive). Fig. 4 represents the steps of the reconfiguration strategy graphically.

1. Different policies are selected either explicitly by the user or by other means.
2. The Characterization component identifies and reports the assertions of the active policies to the Analyzer component as a list of properties.
3. The Analyzer component filters the feature model according to the policy constraints. The preferred features (representing system resources) are selected taking into account their attributes and their qualifiers.
4. The Reconfiguration component performs the bindings between services and resources.
5. Once all the bindings between services and resources are performed, the system services have been customized according to the active policies.

To support these steps, we have developed the following components using the Eclipse Equinox implementation of OSGi: Characterization Component, Analyzer Component and Reconfigurator Component.

Characterization Component

The active policies are provided to the system in order to indicate the set of assertions services should fulfil. Active policies (serialized in XML) are provided to the reconfiguration framework. This information is processed to identify the different assertions and express them by means of a list of required, preferred, discouraged and forbidden properties. This list aggregates the properties defined in the different active policies. These policies can be targeted to a specific feature (such as the security system) or the whole ubiquitous system.

When combining several policies some inconsistencies can appear (i.e., one property with different qualifiers). When this happens the qualifiers are prioritized. For example, the most restrictive (i.e., *required* over *preferred*) qualifiers are chosen. If the problem persists, it is reported to the user as an error and reconfiguration process is aborted.

Analyzer Component

The analyzer component actuates as a filter for the alternatives that are expressed in the Feature Model. Only the features that can fulfil the assertions specified in active policies are considered.

The Feature Model is maintained and queried at runtime. Object Constraint Language (OCL [9]) is used to query the Feature Model in order to obtain the most adequate feature given a list of properties. Features from the feature model are ranked according to the support they give to the properties in the active policies. The following OCL expression, given a feature ftr and a set of assertions ctx, obtains a prioritized collection of the selectable features according to the policy assertions:

```
ftr.child-> select(f|f.complete-> union(f.partial)
->containsAll(ctx.required))->reject(f|f.complete
->union(f.partial)-> containsAny(ctx.forbidden))
->sortedBy(f|ranking(ctx.preferred, ctx.discouraged, f))
```

First, features that support the required properties are selected; then, the ones that support forbidden properties are excluded. The remaining features are ordered according to their preferred and discouraged properties by means of the ranking function. The prioritized Feature model is used to decide the new configuration of the system. When there is an exclusive choice, the feature with a better ranking is selected (SMS for notification in the example).

After this process a concrete configuration for the system is obtained expressed by means of a set of selected features. This configuration is consistent with the Feature Model (possible compositions) and the active policies (user requirements).

Reconfigurator Component

Once the Analyzer component determines the suitable features, it is necessary to identify the resources that support these features and establish the resource-service bindings. In the case of PervML, the bundles representing services and resources are generated by the PervML tool from their specification.

Given a specific configuration (provided by the analyzer component), the reconfiguration component locates the references to the resources tagged with one of the feature names. Finally, the service-resource bindings are implemented using the OSGi Wire Class (an OSGi Wire is an implementation of the publish-subscribe pattern oriented to give support to dynamic systems).

The following Java method describes the actions to retrieve the reference of the selected resources:

```
public ServiceReference[] getServiceReference(String
filterString){
   Filter filter = context.createFilter(filterString);
   ServiceTracker tracker = new ServiceTracker(context,filter,null);
   tracker.open();
   ServiceReference [] selected = tracker.getServiceReferences();
        if (selected!=null) return selected;
        else return new ServiceReference [0];}
```

The above Java code uses the OSGi ServiceTracker to filter OSGi services (both PervML resources and PervML services are wrapped into OSGi services). The Service Tracker matches the services' properties against a filter parameter to narrow down the query results. The OSGi filter is based on the LDAP search filters string representation. For example, the filter to find the resources related with the In-Home-Detection feature is: *"(Feature= In-Home-Detection)"*.

Finally, the resources are bind with the service demanded by the user. The following Java method performs these bindings:

```
protected void createWires(WireManager wm, ArrayList
resourcesPIDs,
 String servicePID){
   wm.deleteWires(consumerPID); //Delete the previous binding
   //Establish the new binding:
   String wireID = wm.createPID(producerPIDs, consumerPID);
   Hashtable props = new Hashtable();
   props.put("PervML_WireID", wireID);
   wm.createWires(producerPIDs, consumerPID, props); }
```

Before the new bindings are created, the old ones must be deleted. New bindings are created using OSGi Wire objects. A Wire object connects a resource (producer) to a service (consumer). Producer and Consumer services communicate with each other via Wire messages. Both producers and consumers implement OSGi interfaces (Consumer and Producer) to send/receive Wire messages. Once these wires are created, the system services have been customized. The services messages are communicated to the resources that better fulfils user policies.

6 Related Work

Some proposals exist for the automatic reconfiguration of Ubicomp systems without user intervention. PCOM [10], Plaint [11] or iROS [12] are some examples. These approaches are focused on the invisibility concept of Ubicomp services. However we think that the user opinion cannot be completely overlooked. With our approach, although guided and limited by the design of the Feature Model, the user keeps her decision power by activating predefined system policies.

Weis in [2] qualifies the personalization of pervasive applications as a fundamental feature and presents a development process for customizable pervasive applications. Configuration is defined by users by means of a graphical programming language which allow for a great level of detail in personalization but increases its complexity compared to our policy system. This work intention is very close to ours, but we face configuration of Ubicomp systems from a coarse-grained service composition perspective based on feature modelling.

Rodden et al. in [3] stresses the relevance to give inhabitants the possibility to reconfigure the Ubicomp systems in the home. A jigsaw-based metaphor is used to present options to the user. They work is focused on the explicit composition of services, while in our approach thanks to the use of models, the composition is done in a declarative way (stating what properties a service must fulfil). In the present work we do not consider a concrete representation for the interaction with the user. The use of the jigsaw mechanism to indicate which policies can be combined constitutes an interesting interaction mechanism for our systems. In this way, the selection of policies can be made in a natural way.

7 Conclusions

In this paper, we have presented an approach to introduce reconfiguration capabilities in Ubicomp systems. The approach focuses on customizing Ubicomp services according to service policies. To achieve this goal, we have introduced a reconfiguration strategy based on Feature Models. The strategy involves a Policy Characterizer, an Analyzer and a Reconfiguration component. We have implemented all of these components as part of the PervML Framework using the Equinox implementation of OSGi.

Although this work extend PervML framework, the strategy is not coupled to this method and can be used independently. To prove this, some of the validation tests were developed using the reconfiguration framework as an implementation framework without following the MDE approach. However, by the integration of the framework in a MDE method, such as PervML, automatic code generation benefits are obtained.

The validation of the reconfiguration framework by real users (out of the lab), to evaluate the expressiveness of policies specifications and end-user satisfaction with the obtained reconfiguration, becomes the next step for this research.

References

1. Weiser, M.: The computer for the 21st century. Scientific American 265(3), 66–75 (1991)
2. Weis, T., Handte, M., Knoll, M., Becker, C.: Customizable pervasive applications. In: PERCOM 2006: Proceedings of the Fourth Annual IEEE International Conference on Pervasive Computing and Communications, Washington, DC, USA, pp. 239–244. IEEE Computer Society, Los Alamitos (2006)

3. Rodden, T., Crabtree, A., Hemmings, T., Koleva, B., Humble, J., Åkesson, K.P., Hansson, P.: Configuring the ubiquitous home. In: Proceedings of 6th International Conference on Designing Cooperative Systems, French Riviera (2004)
4. Muñoz, J., Pelechano, V.: Building a software factory for pervasive systems development. In: Pastor, Ó., Falcão e Cunha, J. (eds.) CAiSE 2005. LNCS, vol. 3520, pp. 342–356. Springer, Heidelberg (2005)
5. Cetina, C., Serral, E., Muñoz, J., Pelechano, V.: Tool support for model driven development of pervasive systems. In: MOMPES 2007, pp. 33–41 (2007) ISBN:0-7695-2769-8
6. Schobbens, P.Y., Heymans, P., Trigaux, J.C., Bontemps, Y.: Generic semantics of feature diagrams. Comput. Netw. 51(2), 456–479 (2007)
7. Cetina, C., Fons, J., Pelechano, V.: Applying software product lines to build autonomic pervasive systems. In: 102th International Software Product Lines Conference (SPLC) (2008) (pending publication)
8. Muñoz, J., Pelechano, V., Fons, J.: Model driven development of pervasive systems. In: I International Workshop on Model-Based Methodologies for Pervasive and Embedded Software (MOMPES), Hamilton, Canada (June 2004)
9. OMG: OCL 2.0 Specification, ptc/2005-06-06 (June 2005)
10. Handte, M., Becker, C., Rothermel, K.: Peer-based automatic configuration of pervasive applications. In: Proceedings of International Conference on Pervasive Services, 2005. ICPS 2005, vol. 1, pp. 249–260 (July 2005)
11. Arshad, N., Heimbigner, D., Wolf, A.: Deployment and dynamic reconfiguration planning for distributed software systems. In: Proceedings of 15th IEEE International Conference on Tools with Artificial Intelligence, 2003, pp. 39–46 (November 2003)
12. Ponnekanti, S.R., Johanson, B., Kiciman, E., Fox, A.: Portability, extensibility and robustness in iros, p. 11 (2003)

HI^3 Project: Software Architecture System for Elderly Care in a Retirement Home

A. Paz-Lopez, G. Varela, J. Monroy, S. Vazquez-Rodriguez, and R.J. Duro

Grupo Integrado de Ingeniería
Universidade da Coruña
c/Mendizábal s/n, 15403, Ferrol, Spain
www.gii.udc.es

Summary. This paper focuses on the application of ambient intelligence to elderly care and, in particular, elderly healthcare in the context of a retirement home, with the main goal of improving the residents quality of life and additionally make the work of the healthcare staff lighter. The objective of the paper is to describe a system that allows for the implementation of services and applications in an ambient intelligence and pervasive computing context with the above mentioned aim. For this purpose HI^3 , a multi-layer intelligent agent based software architecture developed by our group, is used to implement all the tasks and processes. A scenario is defined to build a particular application of the architecture for this settings in order to illustrate the facilities provided by the proposed solution.

Keywords: ambient intelligence, pervasive computing, elderly care, multi-agent systems.

1 Introduction

Healthcare is one of the areas where the current development of pervasive computing and telecommunications technologies have made it possible to improve tasks, processes and performances and to address new goals not yet considered [1]. As an evolution of traditional desktop telemedicine healthcare systems, wireless communication solutions, mobile computing and wearable medical sensors have led to m-health technologies [2]. Wireless technologies allow to reduce or to eliminate some of the problems in healthcare services [3]. In fact, thanks to their affordability, portability, reusability, flexibility and universal access anytime from anywhere [4], they lead to the reduction of medical errors due to the availability of information [5], and, consequently, help to reduce the overall cost of healthcare services [6].

In terms of the segmentation of the possible objective population, nowadays, the number of senior citizens is rising in developed countries [9]. They are the main community users of healthcare services and they are obvious beneficiaries of the use of m-health technologies. This fact has been recognized by different researchers and several research groups are directing their efforts towards them. An example of this is [10], where an assisted living facility and the implementation of pervasive

J.M. Corchado, D.I. Tapia, and J. Bravo (Eds.): UCAMI 2008, ASC 51, pp. 11–20, 2009.
springerlink.com

computing technologies for the healthcare of residents and the assistance of the professional staff are presented.

The use of wireless communication networks in conjunction with pervasive and mobile computing have allowed developing personal area networks where minimally invasive wireless sensors sample medical parameters and send them to remote health-monitoring systems [7]. However, the information being sought is not always directly related to a given sensed variable but, rather, to the summation of multiple data sources and evidence obtained from their time related evolution and their context. This involves taking into account multiple sources of information that are not traditionally used in healthcare monitoring, such as the position of the individual, the trajectories he/she has followed in time and how they relate to their habits, etc. Some work has been carried out in this line, such as in [8], where the authors present a system where data is extracted from the user's context information, consisting of pulse rate, galvanic skin reflex, skin temperature and a three-axis acceleration sensor, supplied by a wearable healthcare support system. However, the systems obtained have usually been implemented in an ad hoc manner with very little in terms of standardization and, in particular, of methodological approaches or architecture based approaches for the creation and adaptation of the systems to particular problems in a modular and scalable way.

To address these problems, in the last few years we have been working on the development and application to different environments of a multi-layer intelligent agent based software architecture called HI^3 [11]. The objective of this architecture is to provide the methodology, structure and base services to easily produce modular scalable systems in ambient intelligence settings that can operate without downgrading their performance as the system grows. The development of middleware for ubiquitous and ambient intelligence environments is a hot topic, as can be seen by the large number of research projects working in the area. Prominent examples are the iROS system [12] a meta-operating-system for applications on interactive workspaces. The Gaia project, a middleware infrastructure to support the development of applications for active spaces, and more recently we find developments like the Amigo Project [14], that are using multiagent systems to achieve adaptivity and dymanic composition of services, and the CHIL Project [15], with its developments on an architecture for ambient intelligence systems. Nevertheless, these projects are focused on office or home environments, and as we stated before, the projects carried out in the area of healthcare are generally ad-hoc and developed by private enterprises. Another usual weak point associated with developments like Gaia and others lies on its lack of hardware abstraction ability, that is one of the fundamentals of our proposal.

In the context of this paper, our interest is focused on the healthcare of elderly people in a retirement home. A system was created that implements services and applications in an ambient intelligence, pervasive computing, wireless communications and social context with the aim of improving the residents' quality of life and making the work of staff lighter. For this purpose we created services and applications, such as people tracking, medical and three axis acceleration metering, speech recognition, natural multimodal interfaces, facility management and maintenance and staff assistance, in the context of the HI^3 architecture in order to construct a residential monitoring system. Here, we will briefly describe the whole system and concentrate on one of the subsystems in order to provide implementation details about the abilities provided by HI^3 .

The paper is organized as follows: The context in which the work is carried out is described first. After this the HI^3 architecture and its components are presented and some design and implementation details will be discussed. In the next section, we explain how applications are implemented and how their different components interact with each other. Then we will use a particular application based on HI^3 to describe the implementation details. Finally some conclusions will be presented.

2 Description of the HI^3 Architecture

In order to provide support to applications and technologies in the framework of ambient intelligence and pervasive computing, we have defined an architecture that complies with the main objective of developing a general architecture with the capacity of transparently integrating multiple technologies.

Figure 1 displays the conceptual structure of the proposed solution. It is a layer based design that enables the division of system elements into levels, reducing the coupling between modules, facilitating abstraction and facilitating the distribution of responsibilities. Communication between layers is vertical so that a given layer can only communicate with the one right on top or below it. Within a layer, horizontal communication may also take place to allow its elements to cooperate to achieve greater functionalities. These layers are:

- Device Access Layer: It is the layer that contains those elements that are in charge of providing access to the physical devices. It is divided into two sublayers [11].
  - Device Drivers: A conceptual layer that contains the device drivers (both, proprietary and open) of the different devices connected to the system.
  - Device access API: It defines an interface so as to homogeneously access devices and abstracts the rest of the elements from the particular implementations and characteristics of the hardware devices. The elements of this layer play the role of adapters.
- Sensing and actuation layer: There are two clearly different types of components in this layer: sensing elements and actuation elements. Sensing elements are in charge of accessing the information provided by the "sensor" type devices (through the device access API). In the same way, the actuation elements must provide, again through the device access API, the necessary functionality to command the actuation of the "actuator" type devices.
- Service layer: A service is defined as an element designed to solve a particular high level task and which, by itself, doesn't provide a complete solution for a system use case. Service composition is allowed, making possible for one

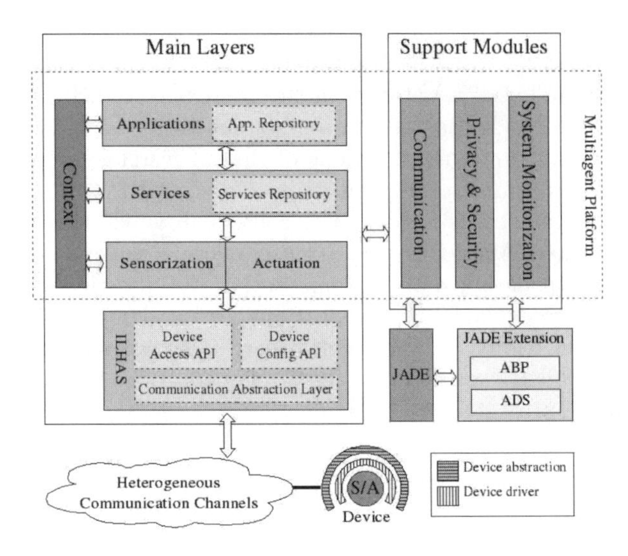

Fig. 1. HI^3 architecture block diagram

service to use other services to carry out its tasks. The elements of this layer make use of sensing/actuation layer elements to access the devices. Services are managed through a module that provides a service repository. This repository enables the discovery of services needed by others to perform their task. Each service is defined following a template specified by the architecture.

- Application layer: This is the highest level layer and the one that hosts the elements representing and implementing components that solve particular functionalities that a user expects from the system. These components make use of services registered in the system to carry out its tasks. Following the proposed layer based philosophy applications will not access elements of the sensing and actuation layer directly. This task is delegated to the service layer.

Furthermore, there exists a common component for the three higher level layers, the Context, its objective is the exchange and management of information generated throughout the system, and information necessary to characterize one entity situation, being an entity a person, an object or a place that is relevant to the interaction between a user and an application. One of its main goals is to represent the current state of the environment.

Finally, this architecture defines a set of support modules. These encompass those components that provide basic functionalities for the correct operation of the whole system. Among them we must point out the modules devoted to handling all the aspects related with communications, security and privacy or the continuous monitoring of the system itself.

In order to implement the previously presented conceptual model, we have chosen a multiagent technology based approach, supported by the JADE agent framework. Many of the characteristics of this paradigm fit perfectly with the main objectives for our architecture, highlighting autonomous operation, distribution and collaboration on task resolution, proactivity, intrinsic communication capabilities, mobility and adaptability.

The main elements of the high level layers (applications, services, sensors and actuators) are implemented as software agents that collaborate for task resolution. In addition, these agents use ontologies that provide common semantics for agent interaction and allow the Agent Discovery Service (ADS) to perform functionality based searches over the system agents.

Currently we have developed the basic elements of the architecture that provide us with a basic platform to implement and deploy applications and services in the ambient intelligence area. In the next sections we will describe some of our efforts aimed to the conceptualization and development of services and applications on a particular scenario, elderly care.

3 Application to an Elderly Care Scenario

As mentioned above, we have enhanced the HI³ architecture with tasks and processes in the context of a retirement home. The objective is to design and implement new functionalities that permit users, software and environment to interact with each other. Two kinds of users may be considered: on one hand we find elderly people as the major benefiting users. In some cases, their motion, sensing or mental faculties are diminished; therefore, the system should take the initiative to interact with them without demanding active participation. When this is not the case, the system should communicate with the users in a natural, intuitive and multimodal manner. On the other hand, the staff of the retirement home, that is, medical, auxiliary, maintenance and management staff, is also a system user category. Their needs are different from that of the elderly and their knowledge and use of technologies may be different too. They probably seek to have tasks made easier and to access information and resources from anywhere anytime.

We have been working in collaboration with professionals to identify the main problems associated to elderly people assistance tasks. As a result of this analysis we have obtained a set of key functionalities that define an environment in which the elderly and the staff are assisted by the system in a natural and non-intrusive manner. We conceive a system that identifies and localizes people in a multimodal and passive way, monitor people activities, assists the staff on the prevention and management of incidences and adapts its operation to its inhabitants particularities.

Many of these functionalities can be achieved by introducing a reduced set of technological solutions, summarized in the following categories:

- People location and tracking: It consists of a set of portable wireless units that continuously communicate with a network of multiple static devices. This type of service provides information that can be used, for instance, when faced with a health emergency.

- Personal context processing: Depending on their health conditions some residents should benefit from the use of a non-intrusive personal area network equipped with medical sensors and three axis accelerometers. This wearable network should also include a wireless device for location and communication. As an example, medical staff may monitor medical parameter evolution just as an estimation of the physical activity for certain periods of time. Context information should also be processed to detect abnormal events, such as falls or people getting disoriented.
- Natural, intelligent and multimodal interfaces: Since some elderly are affected by motion, sensorial or mental disabilities, the environment should take an active role during communications. Natural communication among people is multimodal. We unconsciously process information coming from different kinds of sources and this has to be taken into account in order to make an interface natural. Obviously, in human communications sound is often present and this makes it necessary to have audio command recognition implemented as part of a natural multimodal interface.
- Staff assistance: All the functionalities that make the staffs work easier should lead to improvements in the healthcare services provided and, therefore, to a rise in the overall quality of life of senior citizens. For instance, the system should manage digital medical records of residents, staff scheduling or remote monitoring and assistance.

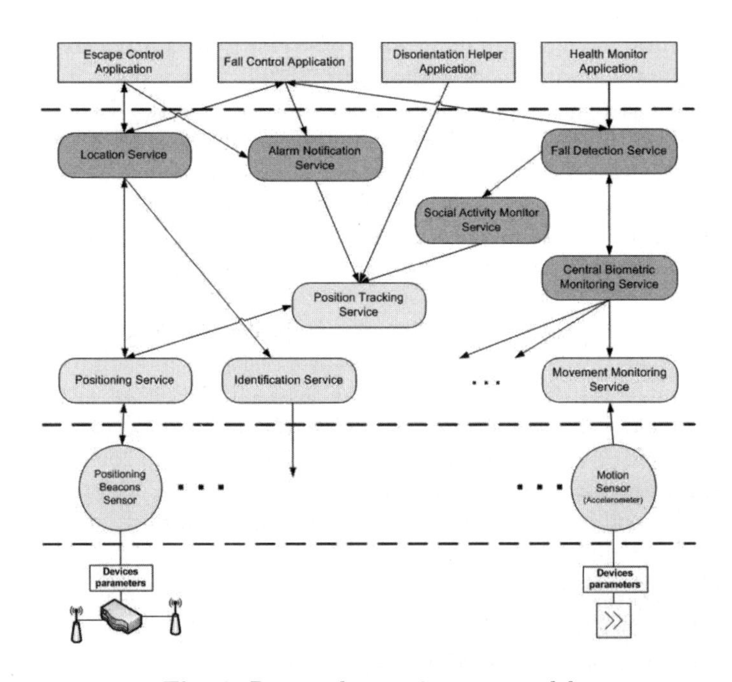

Fig. 2. Proposed scenario agent model

A set of services and applications that following the HI^3 architecture specifications provide the functionalities defined in the previously described scenario have been implemented. Figure 2 shows a component and interaction diagram, illustrating how these components are structured in layers according to the HI^3 architecture and how they cooperate to solve tasks.

As can be seen in figure 2, the system is made up of several applications such as a Fall Control Application to support the automatic detection of accidents by using motion, location and identification information. For this purpose, the application relies on some reusable services like the Localization Service or the Fall Detection Service. These services are an example of composed services that need other low level reusable services to complete their tasks. These low level services delegate the access to the hardware devices on elements from the sensing/actuation layer through the Device Abstraction Layer (ILHAS) [11].

4 Particular Examples

Based on the scenario previously described and out of all the services and applications that were implemented, for the purpose of illustrating the operation of the system, we will concentrate on one of the subsystems. We use this example to test how well the architecture features and guidelines allow an easier development of ambient intelligence systems.

The selected subsystem was the Fall Control Application, because it provides a really interesting functionality to the system, allowing it to monitor possible accidents (a really common incidence in this type of environments) and provide information about them to the staff in real time.

The component structure developed for the example can be seen in figure 3. It follows the structure presented in the previous section, but simplified to show only the components involved in the example proposed.

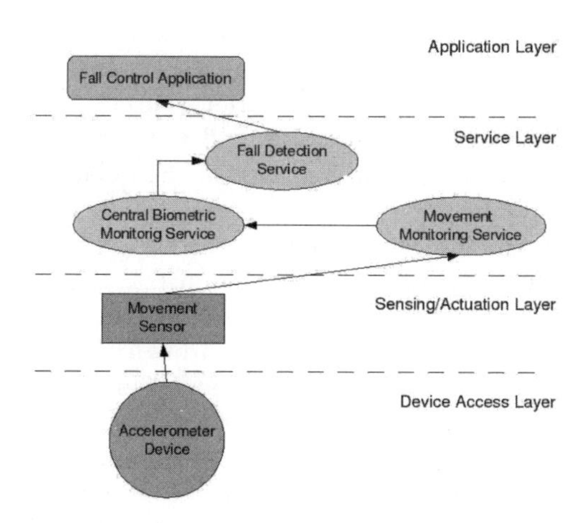

Fig. 3. Fall Control Application interactions

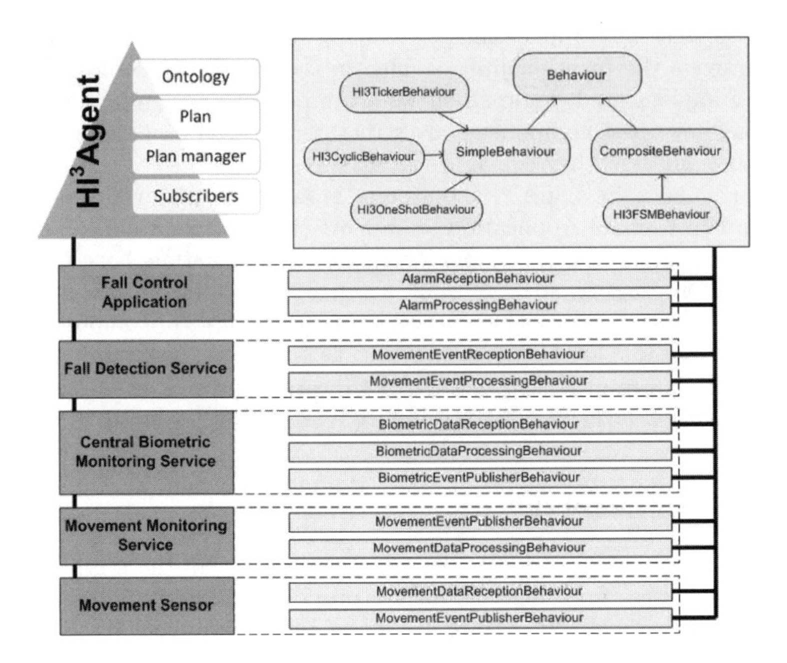

Fig. 4. Fall Control Application low level software structure

The example operation is as follows. The Device Access Layer software component of the accelerometer device is configured to publish events [11] when its data value is changed. Consequently, it is sending events to its subscribed components, the Motion Sensor, with information about the movements of a person. The sensor element sends that information to its subscribers, the Movement Monitoring Service, that will convert accelerometer data into motion information in order to track movements of all the users (taking into account contextual information about them). The movement service will notify the Central Biometric Monitoring Service about motion patterns. This biometric service will act as a central point for biometric information retrieval, useful to share this information with multiple services, in this case the Fall Detection Service, that will be searching for fall patterns in the motion data of the monitorized persons, so as to notify the Fall Control Application.

Figure 4 illustrates the low level software structure adopted for the example components. All of the components were developed using the HI^3 architecture paradigm and abilities, so they are all software agents, except for the Accelerometer Device, that is presented to the system through the Device Access Layer [11]. All the agent based components were developed as HI^3 Agent extensions, so they can manage ontologies (that specify the functionality provided by the component) to configure the execution plan of the agent (wich is loaded at runtime from a configuration file, and specifies the behaviors that make up the agent and its transitions) as well as abilities to support the dynamic subscription of other components.

All the agent's behaviors were also implemented by making use of the features provided by the HI^3 architecture, so they are all extensions of the base behaviors defined within the architecture. For example, the AlarmReceptionBehaviour of the Fall Control Application agent is an HI3CyclicBehaviour, that continuously checks for the reception of fall alarms, while the AlarmProcessingBehaviour is an HI3OneShotBehaviour, that, when fired, is executed only once.

As can be seen, the guidelines and facilities offered by the architecture provide us with a good starting point to develop ambient intelligence systems. On one hand, the layered architecture helps us on the distribution of responsibilities and at the time the multiagent based approach makes the creation of highly distributed and scalable systems easy. On the other hand, the HI^3 architecture provide solutions to basic necessities, like a dynamic behaviour planning system and a communications paradigm that uses the functionality presented by the components to search for and connect them. This, together with our hardware abstraction layer, clearly reduces development time and effort of ambient intelligence projects, in which distributed operation and communications, as well as heterogeneous hardware control are key aspects.

5 Conclusions

In this paper, we have implemented a service and software application system to provide care to elderly living in a retirement home in ambient intelligence, pervasive computing, wireless communication and social context. All the developments presented are consistent with the HI^3 architecture, developed by the authors. This model leads to a multiagent software technology based approach, supported by the JADE agent framework. Many of the characteristics of this paradigm fit perfectly with the main objectives for our architecture, highlighting autonomous operation, distribution and collaboration on task resolution, proactivity, intrinsic communication capabilities, mobility and adaptability.

Finally, we define an example to test how well the architecture capabilities and guidelines permit the development of ambient intelligence systems.

Acknowledgement. This work was partially funded by the Ministerio de Educación y Ciencia of Spain through project DEP2006-56158-C03-02 and Xunta de Galicia DXID under project PGIDIT06TIC025E.

References

1. Lin, J.C.: Applying telecommunications technology to health-care delivery. IEEE Eng. Med. Biol., 28–31 (July-August 1999)
2. M-Health: beyond seamless mobility and global wireless healthcare connectivity. IEEE Transactions on IT in Biomedicine 8(4) (December 2004)
3. Varshney, U.: Pervasive healthcare and wireless health monitoring. Mobile Networks and Applications 12(2-3), 113–127 (2007)

4. Raatikainen, K., Christensen, H., Nakajima, T.: Application Requirements for Middleware for Mobile and Pervasive Systems. ACM Mobile Computing and Communications Review (MC2R) 6(4), 16–24 (2002)
5. Estimating hospital deaths due to medical errors. JAMA 286(4) (2001)
6. Pattichis, C.S., Kyriacou, E., Voskarides, S., Pattichis, M.S., et al.: Wireless telemedicine systems: an overview. IEEE Antennas Propag. Mag. 44(2) (2002)
7. Boric-Lubecke, O., Lubecke, V.M.: Wireless house calls: using communications technology for health care and monitoring. IEEE Microw. Mag., 43–48 (2002)
8. Suzuki, T., Doi, M.: LifeMinder: an evidence-based wearable healthcare assistant. In: Proc. ACM CHI Conference (2001)
9. US Administration on Aging Report on Demographic Changes, `http://www.aoa.dhhs.gov/aoa/stats/aging21/demography.html`
10. Stanford, V.: Using pervasive computing to deliver elder care. IEEE Pervasive Computing Magazine, 10–13 (2002)
11. Varela, G., Paz-López, A., Vázquez-Rodríguez, S., Duro, R.J.: HI3 Project: Design and Implementation of the Lower Level Layers. In: IEEE International Conference on VECIMS 2007, p. 6 (2007)
12. Johanson, B., Fox, A., Winograd, T.: The Interactive Workspaces project: experiences with ubiquitous computing rooms. Pervasive Computing 1(2), 67–74 (2002)
13. Román, M., Hess, C.K., et al.: Gaia: A Middleware Infrastructure to Enable Active Spaces. In: IEEE Pervasive Computing, October-December, 2002, pp. 74–83 (2002)
14. Vallée, M., et al.: Dynamic Service Composition in Ambient Intelligence Environments: a Multi-Agent Approach. In: Proceedings of the First European Young Researcher Workshop on Service-Oriented Computing (2005)
15. Soldatos, J., Dimakis, N., et al.: A breadboard architecture for pervasive context-aware services in smart spaces: middleware components and prototype applications. Personal Ubiquitous Comput. 11(3), 193–212 (2007)

An Architecture for Secure Ambient Intelligence Environments

Daniel Serrano[1], Antonio Maña[2], Pedro Soria-Rodríguez[1], Ana Piñuela[1], and Athanasios-Dimitrios Sotirious[3]

[1] ATOS Origin Research and Innovation
{serrano,amg}@lcc.uma.es,
{pedro.soria,ana.pinuela}@atosorigin.com
[2] University of Málaga
[3] Athens Technology Center
sotiriou@atc.gr

Abstract. The vision of Ambient Intelligence (AmI) depict scenarios where people are surrounded by intelligent and intuitive interfaces embedded in everyday objects, integrated in an environment recognising and responding transparently to the presence of individuals.

The realisation of this vision involves challenges for both software and hardware elements. Nowadays, advanced devices are available to society, with complex microprocessors, realistic screens, long battery life, and other interesting features that can make them useful for the realisation of AmI. However, the software embedded in such devices is a step behind those hardware architectures. Furthermore, in most cases the software that must be developed for these devices has stringent security requirements that are very difficult to guarantee using the current development approaches and the traditional static software architectures.In this paper we present a new model for AmI applications that covers both the development-time aspects (including the development of security solutions and also of secure applications making use of those solutions) and the provision of security at run-time (by means of an infrastructure for the dynamic selection, use and monitoring of security solutions). Our approach is based on the development of certified security and dependability solutions, called *S&D Solutions* that have a precise description (called *S&D Pattern*) which allows us to automate their management. It is important to note that the model is very appropriate for many other highly distributed, dynamic and heterogeneous scenarios.

Keywords: Security patterns, ambient intelligence ecosystem, run-time security provision, run-time monitoring, secure development process.

1 Introduction

Ambient Intelligence (AmI) environments are an emerging paradigm that companies are beginning to pay attention to. In AmI scenarios people are surrounded by intelligent and intuitive interfaces embedded in everyday objects, integrated in an environment recognizing and responding transparently to the presence of individuals. This vision builds on three major concepts: ubiquitous computing, ubiquitous communication and intelligent user interfaces. Defined by the EC Information Society Technologies Advisory Group

J.M. Corchado, D.I. Tapia, and J. Bravo (Eds.): UCAMI 2008, ASC 51, pp. 21–29, 2009.
springerlink.com

as a vision of the Information Society [1], Ambient Intelligence places emphasis on greater user-friendliness, more efficient services support, user-empowerment, and support for human interactions.

An AmI environment is realized with both software and hardware. Nowadays, advanced devices are available to society, with complex microprocessors, realistic screens, long battery life, and other interesting features. However, the software embedded in such devices is a step behind those hardware architectures. In fact, software systems are becoming a critical part of these scenarios, and errors in devices are often software related bugs or malfunctions. The main challenge for this kind of software is that AmI presents a shift from closed applications, fixed systems and predefined architectures to scenarios where components from different sources are dynamically assembled to create open architectures that will be sensitive, adaptive, context-aware and responsive to users' needs and habits. Furthermore, in most cases the software developed for these devices has stringent security requirements. Due to the appealing futuristic nature of other AmI challenges, security aspects of AmI applications development have been largely overlooked until now. The common practice is limited to the use of traditional security systems adapted for Ambient Intelligence scenarios.

The AmI applications introduce new risks due to the wide range of possible attacks. Because of this, the application of traditional security mechanisms is not an appropriate strategy. For instance, it is important to take into account that in AmI scenarios a device must communicate with unknown, and probably untrusted, devices. This paper presents an infrastructure supporting the run-time provision and monitorization of security and dependability (S&D) solutions. This infrastructure has been developed as part of the Serenity project [2]. Serenity covers both the development and the use of these security solutions. But the central aspect is the provision of security solutions at run-time for Ambient Intelligence applications. This paper is structured in 5 sections, this introduction being the first. Section 2 presents the related works, where we present a review of technologies and systems closer to our proposal. Section 3 introduces some concepts, developed as part of Serenity project, that are important in order to understand the architecture of the infrastructure presented. Section 4 presents the Serenity Run-time Framework. Finally, Section 5 presents the conclusions and outlines some unresolved issues that are on the focus of future work.

2 Related Works

The closest approaches to the one presented in this paper are those devoted to provide adaptable software components at run-time, where the most representative are Component-based and Frameworks approaches, due to the importance and the impact of them in current scientific state of the art. The well-known Component-Based Software Development (CBSD) approach has been improved from [3] to [4] and now it is a mature technology. There are extensive works on component-based systems security [5 - 6].

Regarding the deployment of security solutions by means of components, in [6] the authors propose a security characterization structure of software components and their composition. Unfortunately, the CBSD produces complete systems that are closed at run-time. So the aforementioned (and others) approaches can be applied at development time, but they are useless at run-time.

Regarding to CBSD systems evolution at run-time, Gupta et al. [7] present a formal approach to modeling run-time changes at the statement, and procedure, level for a simple theoretical imperative programming language. Their technique uses control program control points where a change can occurs successfully. In [9], authors present an architecture-based approach to run-time software evolution focused on the role of software connectors. The authors of [9] present an approach based on self-organising software architecture where components automatically configure their interaction in a way that is compatible with an overall architectural specification. Finally, [10] presents a novel approach to recovering software architecture at run-time via manipulations on the recovered software architecture. Current architecture evolution proposals fail into take into consideration security aspects of software under evolution.

3 Serenity, Making Security Expertise Available

The objective of the Serenity project is to provide a framework for the automated treatment of security and dependability issues in AmI scenarios. To achieve this goal two key aspects have been identified by the members of the Serenity consortium that are the main focus of the Serenity project: capturing the specific expertise of the security engineers in order to make it available for automated processing; and providing means to perform run-time monitoring of the functioning of the security and dependability mechanisms. These two cornerstones have been deployed by means of:

1. A set of modelling artefacts used in order to capture security expertise. We use the term S&DSolution to refer to an isolated component that provides a security and/or dependability service to an application. These artefacts represent S&DSolutions at different levels of abstraction. The use of different levels of abstraction responds to the needs of different phases of the software development process.
2. An infrastructure for the development and the validation of S&DSolutions by means of these Serenity artefacts. This infrastructure includes concepts, processes and tools used by security experts for the creation of new S&DSolutions ready for automatic processing.
3. An infrastructure for secure application development. Under the name of Serenity Development-time Framework there is an infrastructure that supports the development of secure applications. These secure applications, called Serenity-aware applications, include references to S&DSolutions that will be provided at run-time.
4. A Run-time framework, called Serenity Run-time Framework (SRF). The SRF provides support to applications at run-time, by managing S&DSolutions and monitoring the systems' context. The SRF is presented in Section 4.

This paper presents the SRF. In section 4, we show how Serenity-aware applications are supported at run-time by completing its architecture with security services. In order to facilitate the understanding of the SRF and to make it self-contained, the following introduces how it is possible to capture security expertise by means of the Serenity artefacts.

In order to provide a logical way to represent S&DSolutions the Serenity project provides five main artifacts: *S&DClasses*, *S&DPatterns*, *IntegrationSchemes*, *S&DImplementations* and *ExecutableComponents*. These artifacts, depicted in figure 1, represent S&DSolutions using semantic descriptions at different levels of abstraction. The main reason for using different artifacts, each one for covering an abstraction level, is that doing this it is possible to cover the complete life cycles of secure applications, and in particular development and run-time phases.

- To start, S&DClasses represent abstractions of a security service that can be provided by different S&DSolutions characterized for providing the same S&D Properties and complying with a common interface.

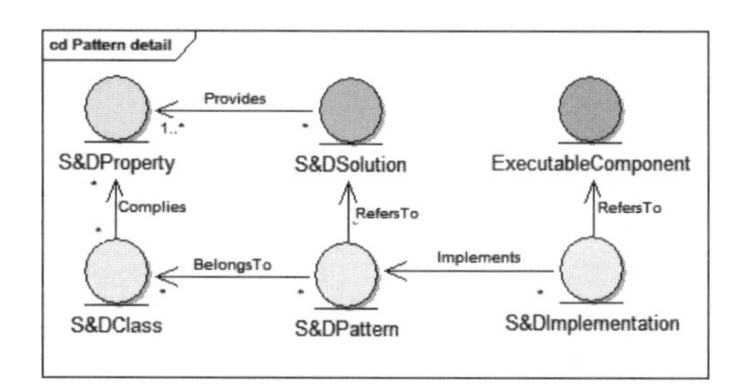

Fig. 1. Simplified model showing relations between Serenity artefacts

- S&DPatterns are detailed descriptions of abstract S&DSolutions. As presented in the conceptual model, each S&DPattern belongs at least to one S&DClass. The descriptions they represent contain all the information necessary for the selection, instantiation, adaptation, and dynamic application of the security solution represented in the S&DPattern. Since each S&DPattern may have a different interface, they contain a specification that allows mapping the abstract calls defined in the S&DClass into the specific calls required by the S&DPattern. Among others, S&DPatterns include two main elements: the representation of the solution itself corresponding to the service and behavioural description, and the pattern semantics, which are related to the semantics of the security properties provided and the restrictions imposed by the solution. It is important to take into account that the S&DPatterns we are using in the Serenity Project differ from the current concept of patterns in software engineering. S&DPatterns are components containing all security solution represented related information.
- IntegrationSchemes are an especial type of S&DPattern that represent S&D Solutions that are built by combining several S&DPatterns. While S&DPatterns are independent or atomic descriptions of S&D Solutions, IntegrationSchemes describe solutions for complex requirements achieved by the combination of smaller S&D Solutions.
- S&DImplementations represent the components that realize the S&D Solutions. S&DImplementations are not real implementations but their representation/ description.

- Finally, an ExecutableComponent is the actual implementation (the piece of code) of an S&DImplementation.

S&DClasses, S&DPatterns and S&DImplementations are development-time oriented artefacts, and ExecutableComponents are especially suitable for run-time. Depending on the artefact level of abstraction used by an application developer, at run-time the SRF is more flexible when selecting S&D Solutions. The main purpose of introducing this hierarchy is to facilitate the dynamic substitution of the S&D Solutions at run-time, while facilitating the development process. For instance, at run-time all S&DPatterns (and their respective S&DImplementations, see figure 1) belonging to an S&DClass will be selectable by the framework in response to an S&DClass-based request by the application. Interested readers can find a complete description of all these concepts in [11].

4 Serenity Run-Time Framework, Providing Security Solutions on the Fly

At run-time the SRF supports Serenity-aware applications. The SRF is implemented as a service running in a device, on top of the operating system, and listening to applications requests. Applications send requests in order to fulfil their security requirements. These requests, once processed by the SRF, are translated into S&D Solutions that may be used by the application. In order to do that, the SRF performs a selection process resulting in the instantiation of security solutions by means of ExecutableComponents. ExecutableComponents offer the S&DPatterns or S&DClasses interface to application, depending on the S&D artefact requested.

The architecture of the SRF has been designed taking into account the wide variety of target devices where the system will run. The main elements of the architecture have been split into separate components. This separation makes possible the implementation of each component by a separate way. Doing this, it is possible to implement each component for a specific platform, in the cases where this is necessary. The main architecture is shown in figure 2. External components are also displayed in order to provide a full view of the Serenity concept during run-time.

From the outside point of view, every instance of the SRF, as a system, provides interfaces in order to allow interaction with other systems, or other SRFs instances. The SRF provides two main interfaces. Begin with the negotiation interface. It is used in order to establish the configuration of the interacting SRFs. This interaction makes sense when two applications supported by different SRFs need to agree on the use of the same S&D Solution. Secondly, SRFs offer a monitoring interface. External systems interacting with an instance of the SRF are able to monitor that the behaviour of the framework is correct. These interfaces provide support for the dynamic supervision and adaptation of the system's security to the transformations in ever-changing AmI ecosystems.

Outside of the SRF, but very related to it, we find Serenity-aware applications, ExecutableComponents and Monitoring Services. Every time an application needs a security service (provided by an S&D Solution) it sends an SDrequest to the SRF. The

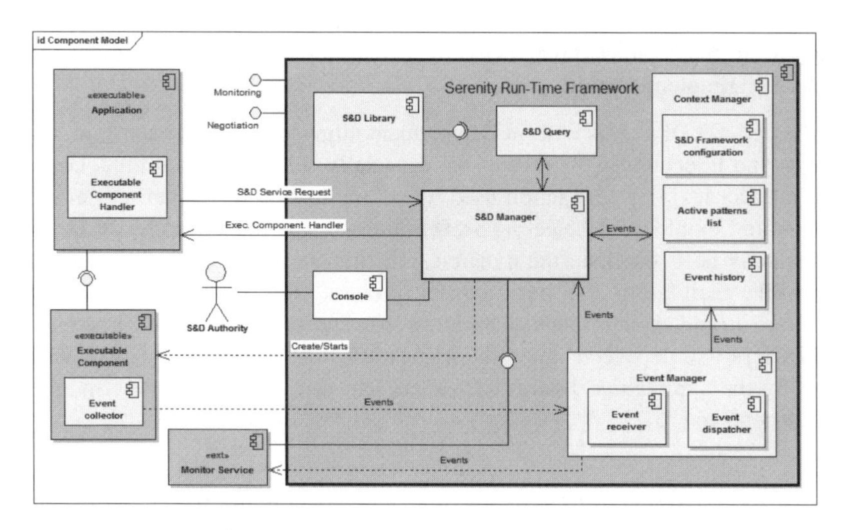

Fig. 2. Serenity Run-time Framework architecture

SRF responds with a ready to use ExecutableComponent, if and only if there is an S&D Solution meeting the SDrequest. Monitoring services are external entities having the responsibility of performing dynamic analysis of the properly operation of S&D Solutions. They count on monitoring rules associated to each S&D Solution. Monitoring services operation is based on event calculus [12], the events are collected from events collectors implemented as part of the ExecutableComponents. The SRF reacts performing recovery task taking into account the information provided by monitoring services.

The S&D Manager is the core of the system. It controls the rest of the components of the system. It processes SDrequests from Serenity-aware applications returning back ExecutableComponents. It is responsible for the activation and deactivation of the ExecutableComponents; it also takes the necessary recovery actions when it is informed by the monitoring services of a monitoring rule violation. The S&D Manager component manages monitoring services by sending the monitoring rules to the Monitoring Service.

The S&D Library is a local S&D Solution repository, it stores the S&DClasses, S&DPatterns and S&DImplementations that are specific to the platform and that may be used in the device. The S&D Query component is the responsible for consulting the S&D Library database and retrieving the artefacts fulfilling the requests.

The Context Manager Component records the data related to the context (S&D Framework configuration, active patterns, events history). The context is used by the S&D Manager to select the most appropriate S&D Solution for a given scenario. The Console is the main interface through which the S&D Manager contacts with the SRF administrator and vice versa. Different implementations of this console provide ways of defining, viewing and changing the S&D configuration through a user-friendly interface.

The Monitoring Services are in charge of analysing events and mapping them on monitoring rules in order to identify possible violations and inform the S&D Manager about them. The Event Manager Component is in charge of receiving the events from the Event Collectors and sending them to the Monitoring Service to be analysed and to the Context Manager to be stored. Finally, each ExecutableComponent implements Event Collectors for all the events described in the corresponding S&DPattern monitoring rules.

The rest of this section describes how the SRF selects and activate security solutions at run-time.

4.1 S&D Solution Selection

The S&D Solution Selection process describes a scenario where a running Serenity-aware application needs the services provided by an S&D Solution. In this activity diagram shown in Figure 3, we present a swim line for each component of the SRF. There are two more swim lines, one for the Serenity-aware application and another for the S&D Solution.

Looking from an abstract point of view, there are three independent systems running in this scenario: the SRF, the Serenity-aware application, and the Executable-Component. The application triggers the process by sending an SDrequest to the SRF. For each SDrequest the S&D Manager builds a search query that is sent to the S&D Library. The S&D Query component translates the SRF standard query to the specific S&D Library query language. Finally, the S&D Library provides a set of S&D Solutions that could be applicable under specific conditions. In the case that there is no S&D Solution available the SDrequest is denied. This scenario is very improbable, because the S&D Library usually is updated when a new Serenity-aware application is installed in the device.

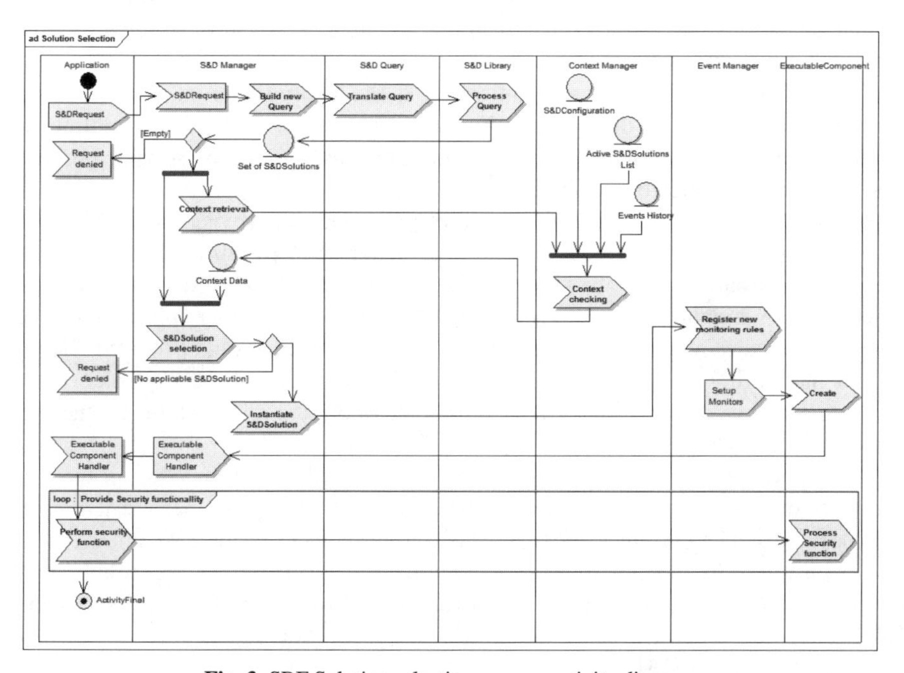

Fig. 3. SRF Solution selection process activity diagram

In order to decide the S&D Solution to be selected, the S&D Manager needs to know devices' contextual conditions. The Context Manager uses three sources for evaluating the context conditions: the S&DConfiguration, the Active S&DPatterns List and the Events History. The S&D Manager uses the set of applicable S&D Solutions and the Context Data for performing the S&D Solution selection. Resulting of this process, the SRF instantiates an S&D Solution. At this point, the S&D Manager sends the S&D Solution monitoring rules to the Event Manager. The Event Manager component sends the monitoring rules to the corresponding monitoring service.

The rest of the process is devoted to the deployment of the S&D Solution. If the ExecutableComponent starts properly the S&D Manager sends its handler to the Serenity-aware application. At this point, the Serenity-aware application uses the handler in order to perform as many calls to the ExecutableComponents as it needs.

5 Conclusions and Future Works

In this paper we have presented a new framework for addressing the security issues in the development of distributed applications and systems for very dynamic and heterogeneous scenarios such as Ambient Intelligence. The model is based on the concept of Security and Dependability Pattern.

On the one hand, the results of our work introduce a new model for the development of applications for these scenarios, supported by the use of a library of security and dependability patterns. We have called this part of the framework *Serenity Development-time Framework (SDF)*. This library is not a simple set of 'best practices' or recommendations like those proposed in the literature as 'security patterns repositories', but a precise, well-defined and automated processing-enabled repository of security mechanisms. The most relevant feature is the incorporation of rich and precise semantic descriptions of the patterns. The extensive use of semantic descriptions facilitates the use of automated reasoning mechanisms capable of solving problems such as pattern composition and adaptation.

On the other hand, in this paper we have presented the provision of security and dependability solutions (represented as S&D patterns) at run-time. This part of the framework is called *Serenity Run-time Framework (SRF)*. The SRF completes the open software architectures designed using the SDF by supporting applications when they need security solutions at run-time. It is important to highlight that the SRF includes monitoring mechanisms that guarantee that the selected security solutions are running properly. In this paper we have presented both the architecture and the main process of the SRF.

Currently, we count on two prototypes, one of the SDF and the other of the SRF. Both prototypes are operational, and particularly the SRF one has been tested in several scenarios. The current work focuses on the refinement of the SRF prototype and the extension of its functionalities (at the moment not all functions are included). We are also involved in the definition of a development process model that covers secure applications development using the defined artefacts. This latter work is part of an ongoing PhD Thesis [13].

Acknowledgements. Work partially supported by E.U. through projects SERENITY (IST-027587) and OKKAM (IST- 215032).

References

1. Shaping Europe's Future Through Ict the Information Society Technologies Advisory Group (ISTAG) (2006)
2. Serenity Project. Funded by European Commission. Directorate General Information Society & Media. Unit D4 - ICT for Trust and Security, under grant IST-027587, http://www.serenity-project.org
3. Mcilroy, D.: Mass-Produced Software Components. In: Proceedings of the 1st International Conference on Software Engineering, Garmisch Pattenkirchen, Germany, pp. 88–98 (1968)
4. Senthil, R., Kushwaha, D.S., Misra, A.K.: An improved component model for component based software engineering. SIGSOFT Softw. Eng. Notes (2007)
5. Jaeger, T., Liedtke, J., Panteleenko, V., Park, Y., Islam, N.: Security architecture for component-based operating systems EW 8. In: Proceedings of the 8th ACM SIGOPS European workshop on Support for composing distributed applications, pp. 222–228. ACM, New York (1998)
6. Khan, K.M., Han, J., Zheng, Y.: Security Characterization of Software Components and Their Composition tools, p. 240. IEEE Computer Society, Los Alamitos (2000)
7. Gupta, D., Jalote, P., Barua, G.: A formal framework for on-line software version change. IEEE Trans. Softw. Eng. 22(2), 120–131 (1996)
8. Oreizy, P., Medvidovic, N., Taylor, R.N.: Architecture-based runtime software evolution. In: ICSE 1998: Proceedings of the 20th international conference on Software engineering, Washington, DC, USA, pp. 177–186. IEEE Computer Society, Los Alamitos (1998)
9. Georgiadis, I., Magee, J., Kramer, J.: Self-organising software architectures for distributed systems. In: WOSS 2002: Proceedings of the first workshop on Self-healing systems, pp. 33–38. ACM, New York (2002)
10. Huang, G., Mei, H., Yang, F.-Q.: Runtime recovery and manipulation of software architecture of component-based systems. Automated Software Engg. 13(2), 257–281 (2006)
11. Maña, A., Muñoz, A., Sanchez-Cid, F., Serrano, D., Androutsopoulos, K., Spanoudakis, G., Compagna, L.: Patterns and integration schemes languages. Serenity Public Report A5.D2.1 (2006)
12. Kloukinas, C., Spanoudakis, G.: A Pattern-Driven Framework for Monitoring Security and Dependability. In: Lambrinoudakis, C., Pernul, G., Tjoa, A.M. (eds.) TrustBus 2007. LNCS, vol. 4657, pp. 210–218. Springer, Heidelberg (2007)
13. System development based on MDA. Technical Report ITI-07-03, Computer Science Departament (University of Malaga) (2007)

An Architecture for Ambient Intelligent Environments

Carlos Ramos

GECAD – Knowledge Engineering and Decision Support Group
Institute of Engineering – Polytechnic of Porto, Portugal
csr@dei.isep.ipp.pt

Abstract. This paper presents the proposal of an architecture for developing systems that interact with Ambient Intelligence (AmI) environments. This architecture has been proposed as a consequence of a methodology for the inclusion of Artificial Intelligence in AmI environments (ISyRAmI - Intelligent Systems Research for Ambient Intelligence). The ISyRAmI architecture considers several modules. The first is related with the acquisition of data, information and even knowledge. This data/information/knowledge deals with our AmI environment and can be acquired in different ways (from raw sensors, from the web, from experts). The second module is related with the storage, conversion, and handling of the data/information/knowledge. It is understood that incorrectness, incompleteness, and uncertainty are present in the data/information/knowledge. The third module is related with the intelligent operation on the data/information/knowledge of our AmI environment. Here we include knowledge discovery systems, expert systems, planning, multi-agent systems, simulation, optimization, etc. The last module is related with the actuation in the AmI environment, by means of automation, robots, intelligent agents and users.

Keywords: Ambient Intelligence, Artificial Intelligence, Ubiquitous Computing, Context Awareness.

1 Introduction

The European Commission's IST Advisory Group (ISTAG) has introduced the concept of Ambient Intelligence (AmI) in the period 2001-2003 [1,2,3]. Rapidly Ambient Intelligence received some attention from Artifiical Intelligence (AI) community. In 2003, Nigel Shadbolt has written the editorial of IEEE Intelligent Systems magazine dedicated to the Ambient Intelligence topic [4]. More recently, we may refer the Ambient Intelligence Workshops at ECAI'2006 [5] and ECAI'2008 (European Conference on Artificial Intelligence) and IJCAI'2007 [6] (International Joint Conference on Artificial Intelligence) and the Special Issue on Ambient Intelligence published in the IEEE Intelligent Systems magazine [7].

Ambient Intelligence (AmI) deals with a new world where computing devices are spread everywhere, allowing the human being to interact in physical world environments in an intelligent and unobtrusive way. These environments should be aware of the needs of people, customizing requirements and forecasting behaviors. AmI environments may be so diverse, such as homes, offices, meeting rooms, schools, hospitals, control centres, transports, touristic attractions, stores, sport installations, music

J.M. Corchado, D.I. Tapia, and J. Bravo (Eds.): UCAMI 2008, ASC 51, pp. 30–38, 2009.
springerlink.com

devices, etc. In the aims of Artificial Intelligence, research envisages to include more intelligence in the AmI environments, allowing a better support to the human being and the access to the essential knowledge to make better decisions when interacting with these environments [8].

However, today some systems use Ambient Intelligence like a buzzword, and most of times a limited amount of intelligence is involved. Some researchers are building AmI systems without AI, putting the effort just in operational technologies (sensors, actuators, and communications). However, soon or later, the low level of intelligence will be a clear drawback for these systems. The acceptability of AmI will result from a balanced combination from AI and operational technologies.

The main objective of ISyRAmI (Intelligent Systems Research for Ambient Intelligence) is to develop a methodology and architecture for the development of Ambient Intelligence systems considering Artificial Intelligence methods and techniques, and integrating them with the operational part of the systems. It is important to notice that the ISyRAmI architecture is centered in the concepts and is oriented for the intelligent behaviour, in spite of being oriented for technologies. In this way, it makes no sense to compare ISyRAmI with other technology-oriented approaches previously proposed for Ambient Intelligence.

2 The ISyRAmI Methodology

Artificial Intelligence methods and techniques can be used for different purposes in AmI environments and scenarios. Several AI technologies are important to achieve an intelligent behaviour in AmI systems. Areas like Machine Learning, Computational Intelligence, Planning, Natural Language, Knowledge Representation, Computer Vision, Intelligent Robotics, Incomplete and Uncertain Reasoning, Multi-Agent Systems, and Affective Computing are important in the AmI challenge [9].

The ISyRAmI methodology involves the identification of the roles in which Artificial Intelligence methods and techniques are important. Seven important roles in which these methods and techniques are useful have been identified in [8]. The identified roles are:

- helping in the interpretation of the environment situation;
- representing the information and knowledge associated with the environment;
- modelling, simulation, and representation of the entities involved in the environment;
- planning about the decisions or actions to take;
- learning about the environment and associated aspects;
- interaction with the human being;
- action on the environment.

A better description of these roles is found in [8] and will not be described in this paper. The objective of this paper is to describe the ISyRAmI architecture, and this will be done in the next section.

3 The ISyRAmI Architecture

An updated version of the Artificial Intelligence view on Ambient Intelligence is found in figure 1, adapted from [8].

In this new view the data/information/knowledge for the AmI system arrives not only by the automatic sensing, but also from other sources (web, social contacts, media, etc). Besides, the automatic action is not directed just to the AmI environment, but also for the other actuators of the AmI environment (other persons or autonomous devices) and to the other sources (for example for updating an information). The external events are separated in Other Actuators (e.g. other persons or autonomous devices in the same environment) and Unexpected Events (e.g. a storm). The interaction between the human being supported by our AmI system and the other actuators is now more clear.

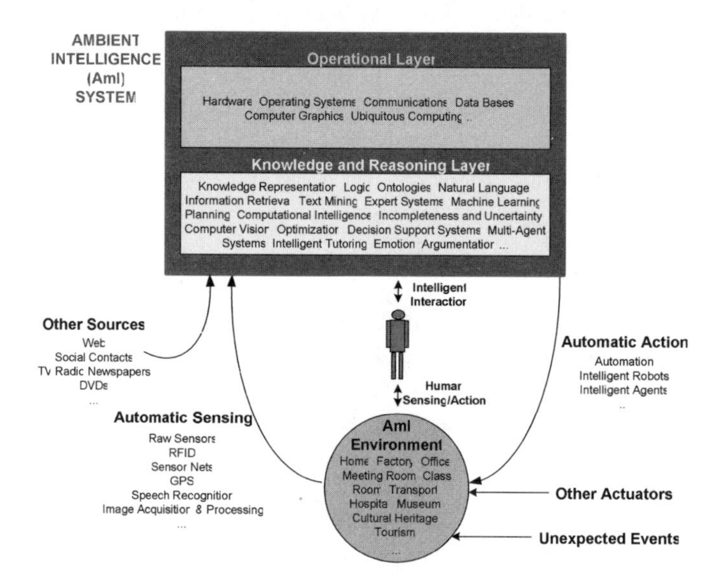

Fig. 1. The updated Ambient Intelligence view from Artificial Intelligence perspective, adapted from [8]

Figure 2 illustrates how the ISyRAmI architecture will be included in the overall system and their internal modules.

Our AmI system is able to obtain data, information and knowledge from two distinct sources:

- from the AmI environment, using a set of automatic sensing devices (raw sensors, cameras, microphones, GPS, RFID, sensor nets);
- from other sources, here we include the web, social contacts, general persons, experts, radio, TV, newspapers, books, DVD, etc.

Our AmI system may act in four different ways:

- operating directly on the AmI environment, using automation devices, robots, and intelligent agents;

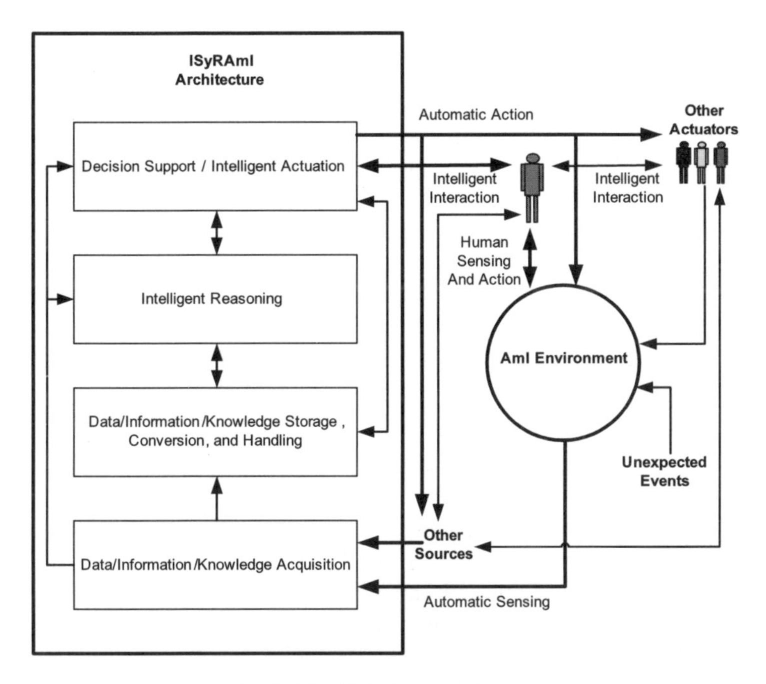

Fig. 2. The ISyRAmI architecture

- interacting with human beings by means of a decision support system, and the actuation on the AmI environment is done by this/these person(s);
- interacting directly with the other external actuators with the ability to act on the AmI environment (e.g. asking one person or robot in order to do some specific task in the environment);
- updating the other sources of data/information/knowledge.

We also assume that the human beings supported by the AmI system have access to the other sources and can interact themselves with the other external actuators.

It is important to notice that our architecture accepts a wide range of data/information/knowledge sources, from sensors to persons, passing by all media sources. To illustrate this we will consider a tourist that will visit a city during some days. Our Tourism points of interest can be seen as our AmI environment. For instance, the weather conditions will be important in order to decide if this tourist will start with a visit to an open air attraction or to an attraction into a building, e.g. a museum. We can obtain these weather conditions directly from sensors (temperature, humidity, etc), or by some media report, like a newspaper or a web-based weather forecasting system, or by means of a friend living in this city. Notice that the granularity and quality are different according to the sources. A temperature sensor is able to inform that we have 34.6 ºC, on the other hand the tourist friend is able to say that it is too hot, but that usually it is cooler in the early morning.

Another interesting aspect is that our AmI system is able to act directly on the AmI environment, or to interact with someone that will act on the environment. This means that our AmI system is able to operate like an Intelligent Agent or like a Decision

Support System, or like both since some tasks can be done autonomously and other may require the intervention of human beings. The human beings, supported by the AmI system, have the responsibility to select which tasks can be done autonomously and which tasks need to pass by the human interaction. If the human beings agree with the suggestions of the decision support system in some specific tasks, there is a trend that in the future these tasks will be performed autonomously (by means of an intelligent agent or another autonomous device).

The AmI system is also according to the social computing paradigm, since the interaction with the external actuators of the AmI environment or with other persons acting like external sources is allowed.

Several Artificial Intelligence technologies are important to achieve an intelligent behaviour in AmI systems. Areas like Machine Learning, Computational Intelligence, Planning, Natural Language, Knowledge Representation, Computer Vision, Intelligent Robotics, Incomplete and Uncertain Reasoning, Multi-Agent Systems, and Affective Computing are important in the AmI challenge. The ISyRAmI architecture was developed to define the place of these technologies in the AmI effort.

The ISyRAmI architecture considers several modules. The first is related with the acquisition of data, information and even knowledge. This data/information/knowledge deals with our AmI environment and can be acquired in different ways (from raw sensors, from the web, from experts). The second module is related with the storage, conversion and handling of the data/information/knowledge. It is understood that incorrecteness, incompleteness, and uncertainty are present in the data/information/knowledge. The third module is related with the intelligent reasoning on the data/information/knowledge of our AmI environment. Here we include knowledge discovery systems, planning, multi-agent systems, simulation, optimization, etc. The last module is related with the actuation in the AmI environment, by means of automation, robots, intelligent agents and users, using decision support systems in this last case.

The first module of the ISyRAmI architecture can be seen as a module with the ability to acquire and join data, information and knowledge. This module has not the responsibility to decide if some part of the data/information/knowledge is correct or complete or affected by uncertainty. Notice that in real life human beings are also affected by these data/information/knowledge problems, and this dos not avoid human beings to reason and act in the real world.

Data/Information/Knowledge treatment and fusion is the responsibility of the second module of the architecture. Using the same example of the tourist and the weather in a city, this module has the responsibility to integrate different visions about the weather, like the following: now the thermometer indicates 20.2 ºC and no rain; the weather forecast from the web refers high temperatures and the possibility of rain; our friend is saying that temperature is moderate and that it is raining.

The Intelligent Reasoning module has the responsibility to perform high-level reasoning processes like Planning, Image and Language Semantics Understanding, Machine Learning, Multi-Agent based Modeling and Simulation, Optimization, Knowledge-based inference, etc. The components of this module will be useful for the top module of Decision Support / Intelligent Actuation. This module uses the data/information/knowledge of the two modules previously described, being able to generate more knowledge for the module immediately bellow (e.g. by means of a Machine Learning algorithm or using an Expert System).

The top module, Decision Support / Intelligent Actuation, is responsible for the direct or indirect actuation on the AmI environment. We claim that some tasks can be assigned to the user, while others can be assigned to the AmI system.

In the first case, decision support is important. The AmI system suggests actuations and should be able to explain the proposed actuation to the user. It should be able to accept the user changes in the solution, having a kind of "what-if" performance in order to respond to the user showing the implications of the user proposed changes. Another desired characteristic is the ability to learn from the user observation. If the user suggests changes to the proposed solutions it is possible that the user is not well modeled by the AmI system, namely if this kind of changes repeats over time. It is not expected that the user considers the AmI system as an intelligent part if this system is always suggesting solutions that the user does not like. Let us illustrate this with the following example:

> The tourist has an AmI system, installed in a PDA with GPS and other technologies, this system is suggesting points of interest to be visited in a city. The AmI system suggests a church, however the tourist does not visit that church since the AmI system has already suggested other 3 churches that he/she has visited. The tourist continues for the next point of interest, however the AmI system is always suggesting new churches to be visited, and the tourist is always denying these suggestions. The suggestions of the AmI system are based in the knowledge that the tourist likes to visit churches, but this knowledge must be updated informing that he/she likes to visit churches, but till a limited number.

Notice that if our system is able to learn from user observation in a specific task then the possibility of the user to pass this task to the automatic state increases significantly.

In some other tasks our AmI system may have the permissions to execute them automatically, by means of intelligent automata (robot, agent). Let us see another example:

> Our AmI system may have the ability to turn off the lights of the living room when the user turn off the TV and go to the bedroom. But again here, it is desired that the system should have the ability to learn from user observation. For example, if the user has started a new routine to turn off the TV, walking to the bedroom to take a book to be read at the living room, our AmI system must be able to understand this new routine, after its occurrence in a certain number of times.

The Decision Support / Intelligent Actuation module has also the responsibility to interact with the other actuators of the AmI environment (other persons or even automatic systems). Another functionality of this module is the ability to update the other sources of data/information/knowledge. For example, if the AmI system turn off the lights then this information can be explicitly given, since it can be important for the other actuators to know that this was done automatically by the AmI system and not by the user.

4 Some Examples of AmI Environments

The first example is related with Free flight sports, like gliding, hangliding and especially paragliding. Up to now free flight pilots received few support from flight equipments, basically numeric GPS information. This drawback led to the development of FlyMaster project, creating a product that received innovation prizes and originated the FlyMaster Avionics SA spin-off [10]. The main idea in the FlyMaster project was to create an on-board Ambient Intelligent environment composed by low-cost equipments able to assist pilots during the flight, giving them on-line decision support. A significant amount of intelligence was integrated in FlyMaster products, namely to allow: the decision about when a pilot should move to the start line; the pilot support when climbing in the rising air columns (thermal lift); the trajectory plan; the glide optimization; and the context awareness to allow the identification of the information to display since in this kind of sports both hands are necessary to control the gliding device. When matching FlyMaster with the ISyRAmI architecture we can observe the automatic sensing (temperature, pressure, velocity, position by GPS) that will allow the acquisition of data that is analyzed in order to infer knowledge (e.g. the identification of the proximity of a thermal lift). The intelligent reasoning part is used for trajectory planning and for the decision support to the user (pilot) suggesting him/her how to enter a thermal lift or when to go to the start line or the better trajectory to pass all goal areas (cylinders around goal points) of a competition. Thermal lifts can be detected from the acquired data or searched from thermal maps on the web (other sources). In this last case the information is not certain, since they refer to possible positions of thermal lifts, but they vary over time.

The second example is SPARSE, a project initially created for helping Power Systems Control Centre Operators in the diagnosis of incidents and power restoration [11]. It is a good example of context awareness since the developed system is aware of the on-going situation, acting in different ways according the normal or critical situation of the power system. This system is evolving for an Ambient Intelligence framework applied to Control Centres, to help the operators in the diagnosis and restoration of the Power System, but also to help in the training of new operators by means of an Intelligent Tutoring approach. When matching SPARSE with the ISyRAmI architecture we can observe the automatic sensing (alarm messages arriving by an industrial SCADA system) that will allow the acquisition of information about the different components of the power system network. Other sources are used, like experts in power systems diagnosis and restoration, for the knowledge acquisition. It is important to observe that incomplete information and uncertainty problems appear in this kind of problems, since messages may not arrive. New knowledge is inferred by the knowledge-based system. There is the possibility to perform the training of operators using an Intelligent Tutoring approach, based on the knowledge previously acquired. In this case, the modeling of the trainee learning process is very important, since our system needs to know which concepts are well assimilated and which are not, and the next problem to appear depends on that concept assimilation. The AmI system could be prepared to do some automatic actuations, however, the Power System Network company prefers to see this AmI system more like a decision support tool.

The third and last example is related with Decision Making, one of the most important activities of the human being. Nowadays decisions imply to consider many different points of view, so decisions are commonly taken by formal or informal groups of persons. Groups exchange ideas or engage in a process of argumentation and counter-argumentation, negotiate, cooperate, collaborate or even discuss techniques and/or methodologies for problem solving. Group Decision Making is a social activity in which the discussion and results consider a combination of rational and emotional aspects. ArgEmotionAgents is a project in the area of the application of Ambient Intelligence in the group argumentation and decision support considering emotional aspects and running in the Laboratory of Ambient Intelligence for Decision Support (LAID), a kind of Intelligent Decision Room. This work has also a part involving ubiquity support [12]. This is a very interesting AmI environment for the ISyRAmI architecture. ArgEmotionAgents can be seen as an AmI system to support a meeting participant. However, the meeting involves other participants that are other actuators in ISyRAmI architecture. ArgEmotionAgents model these participants in the Intelligent Reasoning module by means of a Multi-Agent approach. Thus we can simulate the group of participants in order to feel the meeting trends. Emotional aspects are considered in the participants' modeling. The output is a set of suggestions of arguments to exchange with other participants of the meeting, trying to find support for the proposals that the supported participant prefers, or, at least, to avoid the alternatives that he/she does not like to see approved.

5 Conclusions and Further Work

This paper proposed a new architecture for Ambient Intelligent Systems. This architecture is oriented for the intelligent behaviour of AmI environments and is divided in four modules described in the paper. The paper also represents the external parts of the system, namely the other actuators on the AmI environment, the other sources of data, information, and knowledge, the automatic sensing from the AmI environment, the automatic actuation on the AmI environment, the interaction with the user, and the unexpected events that can change the AmI environment state. Three different application domains were presented. These domains were the inspiration for the development of the ISyRAmI methodology and architecture. However, since they correspond to work developed before the ISyRAmI architecture development, they do not follow it explicitly. We expect to adapt these examples to the proposed architecture and to develop new AmI environments from the scratch, e.g. in tourism domain, using the ISyRAmI architecture. We also claim that it will be easier to develop new AmI systems by the reuse of some ISyRAmI systems previously developed. For example, an AmI system for the area of tourism may benefit from another AmI system developed for the area of city public transportation.

References

1. ISTAG, Scenarios for Ambient Intelligence in 2010, European Commission Report (2001)
2. ISTAG, Strategic Orientations & Priorities for IST in FP6, European Commission Report (2002)
3. ISTAG, Ambient Intelligence: from vision to reality, European Commission Report (2003)
4. Shadbolt, N.: Ambient Intelligence. Editorial of the IEEE Intelligent Systems 18(4), 2–3 (2003)
5. Augusto, J.C., Shapiro, D.: Proceedings of the 1st Workshop on Artificial Intelligence Techniques for Ambient Intelligence (AITAmI 2006), co-located event of ECAI 2006, Riva del Garda, Italy (2006)

6. Augusto, J.C., Shapiro, D.: Proceedings of the 2nd Workshop on Artificial Intelligence Techniques for Ambient Intelligence (AITAmI 2007), co-located event of IJCAI 2007, Hyderabad, India (2007)
7. Ramos, C., Augusto, J.C., Shapiro, D.: Special Issue on Ambient Intelligence. IEEE Intelligent Systems 23(2), 15–57 (2008)
8. Ramos, C., Augusto, J.C., Shapiro, D.: Ambient Intelligence: the next step for Artificial Intelligence. IEEE Intelligent Systems -Special Issue on Ambient Intelligence, Guest Editors' Introduction 23(2), 15–18 (2008)
9. Ramos, C.: Ambient Intelligence Environments. In: Rabuñal, J., Dorado, J., Sierra, A. (eds.) Encyclopedia of Artificial Intelligence. Information Science Reference (2008) ISBN 978-1-59904-849-9
10. Gomes, N., Ramos, C., Pereira, C., Nunes, F.: Ubiquitous Ambient Intelligence in Flight Decision Assistance System. In: Neves, J., Santos, M.F., Machado, J.M. (eds.) EPIA 2007. LNCS (LNAI), vol. 4874, pp. 296–308. Springer, Heidelberg (2007)
11. Vale, Z., Moura, A., Fernandes, M., Marques, A., Rosado, A., Ramos, C.: SPARSE: An Intelligent Alarm Processor and Operator Assistant. IEEE Expert- Special Track on AI Applications in the Electric Power Industry 12(3), 86–93 (1997)
12. Marreiros, G., Santos, R., Novais, P., Machado, J., Ramos, C., Neves, J., Bulas-Cruz, J.: Argumentation-based Decision Making in Ambient Intelligence Environments. In: Neves, J., Santos, M.F., Machado, J.M. (eds.) EPIA 2007. LNCS (LNAI), vol. 4874, pp. 309–322. Springer, Heidelberg (2007)

A Hardware Based Infrastructure for Agent Protection

Antonio Muñoz and Antonio Maña

University of Malaga
{amunoz,amg}@lcc.uma.es

Summary. Mobile agents are software entities consisting of code and data that can migrate autonomously from host to host executing their code. Despite its benefits, security issues strongly restrict the use of code mobility. The protection of mobile agents against the attacks of malicious hosts is considered the most difficult security problem to solve in mobile agents systems. The centre of our work is a new agent migration protocol that takes advantage of TPM technology. The protocol has been validated using AVISPA model checking toolsuite. In order to facilitate its adoption, we have developed a software library to access TPM functionality from agents and to support their secure migration. This paper presents hardware-based system to protect agent systems. Concretely our work is based on trusted platform module (TPM) protocol, which is the basis to build the solution. In order to build our solution on a robust basis, we validate this protocol by means of a model checking tool called AVISPA. Then as final result we provide a library to access to TPM (Trusted Platform Module) functionality from software agents. Along this paper we detail more relevant aspects of this library both in the development stage of it and while we use it to develop a system based agent.

1 Introduction

A well accepted definition of software agent is found in [10]: "An agent is an encapsulated computer system that is situated in some environment and that is capable of flexible, autonomous action in that environment in order to meet its design objectives". More important characteristics of an agent are that are highly agree upon to include: Autonomy (behaves independent according to the state it encapsulates), proactiveity (able to initiate without external order), reactivity (react in timely fashion to direct environment stimulus), situatedness (ability to settle in an environment that might contain other agents), directness (agents live in a society of other collaborative or competitive groups of agents), and the social ability to interact with other agents, possibly motivated by collaborative problem solving.

Then an agent is supposed to live in a society of agents; multi-agent systems (MAS). A MAS is known as a system composed of several agents collectively capable of reaching goals that are difficult to achieve by an individual agent of monolithic system. Also a MAS represents a natural way of decentralization, where there are autonomous agents working as peers, or in teams, with their own behaviour and control. Mobile agents are a specific form of mobile code and software agents paradigms. However, in contrast to the Remote evaluation and Code on demand paradigms, mobile agents are active in that they may choose to migrate between computers at any time during their execution. This makes them a powerful tool for implementing distributed applications in a computer network. Development process required some considerations, such as a JADE [2]

J.M. Corchado, D.I. Tapia, and J. Bravo (Eds.): UCAMI 2008, ASC 51, pp. 39–47, 2009.
springerlink.com

system is composed for a platform that contains a main container in which agents are deployed.

Additional containers can be added to this platform, some of them can be remote containers, and different platform can interacts among them, allowing the migration of the agents among them. Them containers play agencies roles in JADE on which agents will be deployed. For this purpose we based our work on JADE, the main reason to choose this platform was its interplatform migration mechanism. Taking into account JADE structure, we conclude that two different kinds of migration exists, migration among containers from different platforms and migration from containers from the same platform. In the case that the migration is from containers from different platforms, the agent migrates from a container from source agency to the destination agency main container. In such a case that destination agency is not a JADE built-on platform architecture can be different, depending on the platform. In the other case, the agent migrates from a container to another one but in the same platform. Both migration processes imply some security concerns. Platform migration is not secure because main container from source platform can be untrusted. Also containers migration has the same problem, it is, if destination container is not trusted then the migration is not secure. Secure migration library solves arisen problems.

Agent migration consists on a mechanism to continue the execution of an agent on another location [1]. This process includes the transport of agent code, execution state and data of the agent. In an agent system, the migration is initiated on behalf of the agent and not by the system. The main motivation for this migration is to move the computation to a data server or a communication partner in order to reduce network load by accessing a data server a communication partner by local communication. Then migration is performed from a source agency where agent is running to a destination agency, which is the next stage in agent execution. This migration can be performed by two different ways. First way is by moving, that is, the agent moves from the source agency to the destination one. And the second way is by cloning. In this case, the agent is copied to the destination agency. From this moment on, the two copies of the agent coexist executing in different places. In the remainder of this paper, and at least stated explicitly we will use the term migration to refer to both agent cloning and moving. Software agents are a promising computing paradigm.It has been shown [13] that scientific community has devoted important efforts to this field. Indeed, many applications exist based on this technology. However all this efforts are not useful in practice due to the lack of a secure robust basis to build applications. In [13] two different solutions are presented to provide a security framework for agents building. The first of these is a software solution based on the protected computing approach [3]. The second one consists on a hardware based solution, that takes advantage of the TPM technology. In this paper we use that theoretical and abstract solution to build a solution for the secure migration of software agents.

2 State of the Art

The hardware-based protection infrastructure described in the present paper takes advantage of the recent advances in trusted hardware. The basic idea behind the concept of Trusted Computing is the creation of a chain of trust between all elements in the computing system, starting from the most basic ones. Consequently, platform boot processes are modified to allow the TPM to measure each of the components in the system and securely store the results of the measurements in Platform Configuration Registers (PCR) within the TPM. This mechanism is used to extend the root of trust to the different elements in the computing platform. Therefore, the chain of trust starts with the aforementioned TPM, which analyses whether the BIOS of the computer is to be trusted and, in that case, passes control to it. This process is repeated for the master boot record, the OS loader, the OS, the hardware devices and finally the applications. In a Trusted Computing scenario a trusted application runs exclusively on top of trusted and pre-approved supporting software and hardware. Additionally the TC technology provides mechanisms for the measurement (obtaining a cryptographic hash) of the configuration of remote platforms. If this configuration is altered or modified, a new hash value must be generated and sent to the requester in a certificate. These certificates attest the current state of the remote platform.

Several mechanisms for secure execution of agents have been proposed in the literature with the objective of securing the execution of agents. Most of these mechanisms are designed to provide some type of protection or some specific security property. In this section we will focus on solutions that are specifically tailored or especially well-suited for agent scenarios. More extensive reviews of the state of the art in general issues of software protection can be found in [3] [4]. Some protection mechanisms are oriented to the protection of the host against malicious agents. Among these, SandBoxing [5], proof-carrying code [6], and a variant of this technique, called proof-referencing code [7]. One of the most important problems of these techniques is the difficulty of identifying which operations (or sequences of them) can be permitted without compromising the local security policy. Other mechanisms are oriented toward protecting agents against malicious servers. Among them the concept of sanctuaries [8] was proposed. Several techniques can be applied to an agent in order to verify self-integrity and avoid that the code or the data of the agent is inadvertently manipulated. Anti-tamper techniques, such as encryption, checksumming, anti-debugging, anti-emulation and some others [9] [11] share the same goal, but they are also oriented toward the prevention of the analysis of the function that the agent implements. Additionally, some protection schemes are based on self-modifying code, and code obfuscation [12]. Finally there are techniques that create a two-way protection. Some of these are based on the aforementioned protected computing approach [3].

3 Accessing TPM from Agents

Previously we argued that is essential to integrate strong security mechanisms in agent systems world to achieve a reasonable security level to support real world applications. For this reason we propose a hardware-based mechanism to provide security to agent systems. The TPM provides mechanisms, such as cryptographic algorithms, secure key storage and remote attestation that provides important tools to achieve a high level of security,

Unfortunately, the TPM technology is very complex and so are the procedures and action sequences to use it. The access to the device and the use of TPM functionality is not an easy task, especially for average software developers. Therefore, we have developed a library that provides access to the TPM from software agents. The main advantage for this approach is that developers of agent systems do not need to become security experts, and can access the security mechanisms without taking care of low level details of the TPM technology. Hardware-based solutions can be built on the basis of different devices. We have selected the TPM for our solution because it provides all necessary features. In particular it provides, cryptography capabilities, secure storage, intimate relation to the platform hardware, remote attestation, etc. Remote attestation is perhaps the most important feature; it allows the computer to recognize any unauthorized/authorized changes to software. If certain requirements are met then the computer allows the system to continue its operations. This same technology that TPM utilizes can also be applied in agent-bases systems to scan machines for changes to their environments made by any entity before the system boots up and accesses the network.

Additional reasons are the support of a wide range of industries and the availability of computers equipped with TPM In summary, this element is the cornerstone of our approach and the security of our system is focused on it. The use of TPM in these systems is as follows. Each agency takes measures of system parameters while booting to determine its security, such as BIOS, keys modules from Operating System, active processes and services in the system. Through these parameters an estimation of the secure state of the agency can be done. Values taken are stored in a secure way inside the trusted device, preventing unauthorized access and modification. Agencies have the ability to report previously stored configuration values to other agencies in order to prove their security.

Our system takes advantage of the remote attestation procedure provided by the TPM in order to verify the security of the agencies. In this way agents can verify the security of the destination agency before migration.

4 Secure Migration Library for Agents

This Section introduces the Secure Migration Library for Agents. It aims to serve as a platform for the agent migration on top of the JADE platform. In order to provide a secure environment this library makes use of a TPM. In this way, the migrations are supported by a trusted platform.

The SMLA is based on the one hand on the JADE platform and on the other hand on the TPM4Java library. TPM4Java provides access to the TPM functions from the Java programming language. However, as it is, it is not possible to use TPM4Java in agents. For this reason, we have extended TPM4Java, as described in next sections.

Among the central objectives for the development of this library we highlight the following; (i) to provide a secure environment in which agents can be securely executed and migrated. (ii) To facilitate the integration of TPM technology with JADE. (iii) To provide mechanisms that are simple to use from the programmers point of view. (iv) To provide a library that is easy to adapt to be based on other security devices such as smartcards. (v) To provide a library with a generic and flexible interface. Doing this allows users to use the library to solve a wide range of problems. (vi) To provide a library that can easily be extended in the future.

The most important function provided by the library is the secure migration mechanism. This secure mechanism is based on the remote attestation of the destination agency, before migration actually takes place, in order to guarantee that agents are always executed in a secure environment. Additionally we provide a secure environment for agents to run, in such a way that malicious agents are not able to modify the host agency.

The library was designed according to the following set of requirements. On the one hand, concerning functional requirements we considered that our main objective was to provide a mechanism for secure agent migration in JADE platform. The goal was to make possible for agents to extend their trusted computing base to the next destination agency before migrating to it. It is important to mention that each agency must provide local functionality to perform the secure migration mechanism, which requires the presence of the TPM for remote attestation. Similarly each agency must provide the functionality to allow to other agencies to take remote integrity measures to determine whether its configuration is secure. In order to support the previous processes, the library includes a protocol that allows two agencies to exchange information about their configurations.

This exchange establishes the bases of the mutual trust between agencies. Concerning the non-functional requirements, the library must seamlessly integrate in the JADE platform, in such a way that its use does not imply modifications in the platform. Additionally, the operation of the library must be transparent to the user. In order to facilitate the adaptation of the library to new devices we decided to create an additional abstraction layer that provides generic services. In this way the generic services remain unchanged. One final requirement has to support both types of migration (moving and cloning).

4.1 Secure Migration Protocol

Following we details all the protocols that allowed that our library provides the security to the migration. Firstly we analyse the different attestation protocols as well as secure migration protocols, studying their benefits, so that we can build our own secure migration protocol. From this point we use the concept of migration both to agent cloning and agent moving, because from protocol point of view there is no difference between them. A first approach of a secure migration protocol is in [3]. This protocol provides some important ideas to take into account during the design process of the final protocol, we highlight that the agent trusts in its platform to check the migration security. As well as the necessity of the use of TPM to obtain and report configuration data by a secure way.

Other protocol that provides interesting ideas to take into account when we develop a secure migration system is in [9]. More relevant ideas provided are, a protocol shows how an agent from the agency requests to TPM the signed values from PCRs. As well as the protocol shows how the agent obtain platform credentials. These credentials together with PCRs signed values allow to determine if destination configuration is secure. Finally we analyse the protocol from [9]. More relevant ideas from this protocol are, the use of an AIK to sign the PCR values, the use of a CA that validates the AIK and the use of configurations to compare received results from remote agency. We designed a new protocol based on the study of these three protocols, in such a way that we took advantage of the appeals provided by each of them. Our protocol has some characteristics. The agency provides to the agent the capacity to migrates by a secure way. Also the agency uses a TPM that provides configuration values stored in PCRs.

TPM signs PCRs values using a specific AIK for destination agency, in such a way that data receiver knows securely TPM identity which signed. A Certification Authority generates needed credentials to correctly verify the AIK identity. Together with signed PCRs values the agency provides AIK credentials in such a way that the requester can correctly verify that data comes from agency TPM. Following we define the 18 steps protocol, used to perform secure migration.

Algorithm 1. Secure migration protocol

1: Agent Ag requests to his agency the migration to Agency B.
2: Agency A (source agency) sends to agency B (destination agency) a request for attestation.
3: Agency B accept the request for attestation and send a nonce (this value is composed by random bits used to avoid repetition attacks) and indexes of PCRs values that needs.
4: Agency A request to its TPM PCR values requested by agency B together with the nonce all signed.
5: TPM returns requested data.
6: Agency A obtain AIK credentials from its credentials repository.
7: Agency A requests agency B for PCRs values requested and nonce all signed. Then it sends AIK credentials, which contains the public key corresponding with the private key used to sign data. Additionally, it sends a nonce and the indexes of PCRs that wants to know.
8: Agency B validates the authenticity of received key verifying the credentials by means of the CA public key which was used to generate those credentials.
9: Agency B verifies the PCRs values signature and the nonce received using the AIK public key.
10: Agency B verifies that PCRs values received belongs to the set of accepted values and them the agent configuration is valid.
11: Agency B requests to its TPM the PCR values requested by the agency A together with the nonce signed.
12: TPM returns requested data.
13: Agency B obtains AIK credentials from its credentials repository.
14: Agency B sends to agency A PCR values requested and the nonce signed. Also it sends AIK credentials, which contains the public key corresponding to the private key used to encrypt the data.
15: Agency A validates the authenticity of received key verifying the credentials by means of CA public key that generated those credentials.
16: Agency A verifies the PCR values signature and the nonce received using the AIK public key.
17: Agency A verifies that PCR values received belongs to the set of accepted values and then the agency B configuration is secure. From this point trustworthy between agencies A & B exists.
18: Then Agency A allows to the agent Ag the migration to agency B.

We can observe that the protocol fulfills the five main characteristics we mentioned previously. Hence, we have a clear idea of the different components of the system as well as the interaction between them to provide the security in the migration. Secure Migration protocol described is the basis of this research. Thus we want to build a robust solution, for this purpose a validation of this protocol is the next step in this research. Among different alternatives we selected a model checking tool called AVISPA. However a further description about the demonstration of this protocol is out of the scope of this paper.

5 Design and Deployment of the Library

This section analyses the deployment and design of SMLA, for this purpose we study the architecture of this in details as well as we show the components and their related functionalities.

Main use case consists on a user that uses SMLA to develop a regular, but secure multi-agent system. Traditionally in these kind of systems, the user defines the set of agents that compound the system. Concretely JADE defines an agent by means of a class that inherits from Agent class, then throughout this the new agent created is provided of the basic behavior of an agent, then user defines the specific behavior of this agent. Among the basic functionality of a JADE Agent we found the compatibility with inter-containers migration. Concerning the main migration methods we notice; doMove(Location1); to Move the agent from a source container to a destination one. doClone(Location1, String newName); to clone the agent in container1 using newName as the name. Inter-platform migration methods an intra-platform migration methods are shared but the most relevant difference between them is the dynamic type of param1.

Two main services are provided by SMLA that use AgentMobility Service to perform a secure inter-platform migration in the same platform, and SecureInterPlatformMobility service that use InterPlatformMobility service to perform the secure intra-platform migration. Previously we mentioned that JADE Agent class provides two "insecure" migration methods, then we created a new class that inherits from this class and redefines migration methods to perform them securely.

6 Underlying Technology: TPM4Java

TPM4Java is a java library for accessing a trusted platform module in your java applications. This consists on a library deployed by two students from Darmstadt Technique University, Martin Hermanowski and Eric Tews. This is full java developed and provide access to TPM functionality. Briefly TPM4Java architecture is deployed on three main layers.

Concerning the architecture of tpm4java it is built on three layers:

- A high level layer that provides access to common TPM functionality (this is defined in TssHighLevel layer).
- A low level layer that provides access to almost full TPM functionality (this is defined in TssLowLevel layer).
- A backend used by the library that sends commands to TPM.

Therefore more relevant classes used in the delevopment of our library are:

- TssLowLevel. Interface that provides access to high level layer.
- TssHighLevel. Interface that provides access to low level layer.
- PcrSelection. Class that provides a type of data to manage the information of a set of PCR records.
- SimplePrivacyCA. A class that provides an implementation of basic functionality of a CA.
- TCPAIdentityRequest. Creating an AIK a class of this type is generated with needed data by CA generates the needed information to active the AIK.
- TCPAIdentityCredential. This class contains the information generated by CA to activate the AIK.

- TPMKeyWrapper. This class contains the information of a key, with its public part and its private one. Private part is encrypted in such a way that is in clear only in TPM.

Despite of the interesting functionalities provided by TPM4Java some features where missed in its current version. While developing Services Provided by Library some classes and interfaces from TPM4Java library have been modified, which will be included in news versions of TPM4Java. Concretely more relevant improvements are:

- TssLowLevel interface. TPM_OwnerReadPubek method is added because this is not included in TPM4Java but TCG 1.2 specification includes that.
- TssHighLevel interface. GenerateAIK method is modified to returns a TPMKey-Wrapper class corresponding to generated AIK instead of key handler in TPM. Therefore AIK data are needed to store it in the hard disc.
- TssCoreService class implements TssLowLevel interface. TPM_OwnerReadPubek method has been added.
- TssHighLevelImpl class implements TssHighLevel interface. GenerateAIK method has been modified in such a way that a TPMKeyWrapper class is returned.
- PrivacyCA class is added, this is an altered version of SimplePrivacyCA class provided by TPM4Java.

7 Conclusions and Future Work

Concerning next work that we consider interesting to do. We propose the development of a migration library that implements the anonymous direct attestation. We developed a library based on an attestation protocol which uses a Certification Authority this concerns some disadvantages. Some of these disadvantages are that is needed to generate a certificate for each key used for the attestation, this implies that many request are performed to the CA and this is a bottleneck of the system. Also we found that if verification authority and CA act together the security of the attestation protocol can be violated.

The use of the anonymous direct attestation protocol provides some advantages. Certify issuer entity and verifier entity can not act together to violate the system security, then one unique entity can be verifier and issuer of certificates. Last but not least of the advantages is that certificates only need to be issued once which solves the aforementioned bottleneck. These advantages shows that anonymous attestation protocol is an interesting option for attestation.

Improve the keys management system of the library. Our library uses RSA keys for attestation protocol that must be loaded in TPM to be used. However the space for available keys in the TPM is very limited, them it must be carefully managed to avoid arisen space problems. Our library handle the keys in such a way that only one key is loaded in TPM simultaneously, then keys are loaded when will use and downloaded when used. This procedure is not of maximum efficiency due to constantly are keys loading and downloading, in fact there is the possibility that we download the same key that we will use in next step. A possible improvement in key management is establish a cache in such a way that several keys can be loaded simultaneously in TPM and that these will be downloaded when we need more available space for new keys. For this we need to develop a keys replace policy to determine which key delete from TPM to load a new key in the cache.

Another future work is to integrate new functionalities to secure migration services to manage concurrent requests. Secure migration services implemented in the library provide secure migration to a remote container. However they have capacity to handle one unique request simultaneously, then migration requests arriving while migration is performed are refused. This happens due to TPM management of the problem together with the aforementioned key management problem. A possible improvement of the library is provide to secure migration services with the capability to accept and handle several migration requests simultaneously. For this is necessary to develop the previous described capability of several keys simultaneously loaded in the TPM. This improvement implies several difficulties related to key management. There is a possibility that the service must attend several requests which use different keys, forcing to manage the keys in such a way that these are available when are needed. Also we can find the possibility in which any key can be replaced because all are being used, in this case the request might wait until its key is loaded.

References

1. General Magic, Inc. The Telescript Language Reference (1996), http://www.genmagic.com-Telescript/TDE/TDEDOCS_HTML/telescript.html
2. http://jade.tilab.com
3. Maña, A.: Protección de Software Basada en Tarjetas Inteligentes. Ph.D Thesis. University of Málaga (2003)
4. Hachez, G.: A Comparative Study of Software Protection Tools Suited for E-Commerce with Contributions to Software Watermarking and Smart Cards. Ph.D Thesis. Universite Catholique de Louvain (2003), http://www.dice.ucl.ac.be/~hachez/thesis_gael_hachez.pdf
5. Gosling, J., Joy, B., Steele, G.: The Java Language Specification. Addison-Wesley, Reading (1996)
6. Necula, G.: Proof-Carrying Code. In: Proceedings of 24th Annual Symposium on Principles of Programming Languages (1997)
7. Gunter Carl, A., Peter, H., Scott, N.: Infrastructure for Proof-Referencing Code. In: Proceedings, Workshop on Foundations of Secure Mobile Code (March 1997)
8. Yee, B.S.: A Sanctuary for Mobile Agents. Secure Internet Programming (1999)
9. Trusted Computing Group: TCG Specifications (2005), https://www.trustedcomputinggroup.org/specs/
10. Wooldrigde, M.: Agent-based Software Engineering. IEE Proceedings on Software Engineering 144(1), 26–37 (1997)
11. Stern, J.P., Hachez, G., Koeune, F., Quisquater, J.J.: Robust Object Watermarking: Application to Code. In: Pfitzmann, A. (ed.) IH 1999. LNCS, vol. 1768, pp. 368–378. Springer, Heidelberg (2000), http://www.dice.ucl.ac.be/crypto/publications/1999/codemark.pdf
12. Collberg, C., Thomborson, C.: Watermarking, Tamper-Proofing, and Obfuscation - Tools for Software Protection. University of Auckland Technical Report 170 (2000)
13. Maña, A., Muñoz, A., Serrano, D.: Towards Secure Agent Computing for Ubiquitous Computing and Ambient Intelligence. In: Fourth International Conference, Ubiquitous Intelligence and Computing, Hong Kong, China. LNCS. Springer, Heidelberg (2007)

Performance Evaluation of J2ME and Symbian Applications in Smart Camera Phones

Celeste Campo, Carlos García-Rubio, and Alberto Cortés

Departamento de Ingeniería Telemática
Universidad Carlos III de Madrid
Av. Universidad 30, 28911 Madrid, Spain
celeste@it.uc3m.es, cgr@it.uc3m.es, alcortes@it.uc3m.es

Abstract. Nowadays, many smart phones have a built-in camera that is often used as a conventional digital camera. With some image processing in the mobile phone, new applications can be offered to the user, such as augmented reality games or helper applications for visually impaired people. However, image processing applications may require high processor speed and massive dynamic memory, which are scarce resources in smart phones. Image processing algorithms usually consists of long loops iterating some operations. In this paper we evaluate the performance of smart phones executing these operations. We consider both native and J2ME applications. The results we present can help to decide what we can and we cannot do in a mobile phone, and to quantify the performance penalty when using J2ME.

Keywords: Camera Smart Phones, Image Processing, J2ME, Symbian.

1 Introduction

Nowadays, many smart phones have a built-in camera that is used as a conventional digital camera. With the processing capability of current smart phones, we can process the captured images and offer new applications to the user, such as movement detection, pattern recognition, color detection, and so on. These programs allow smart phones to be used for such different applications as, just to cite some examples, for assistance to visually impaired people, for surveillance, or for augmented reality games.

This kind of image processing applications may require high speed and massive dynamic memory, which are bottle-necks in smart phones. In a previous paper [1] we evaluated the performance and the possibilities of phones' cameras in terms of resolution, file size, bit depth and acquisition time, when J2ME (Java 2 Micro Edition [2]) is used, and we measured the speed and memory consumption for basic operations that are frequently repeated in any image processing application.

However, native applications that make a direct use of the operating system API, can be faster and more efficient when doing image processing in a phone. In this paper, we compare the results obtained using native applications with the ones using J2ME. In our new experiments we have used four different Symbian OS [3] mobile phones[1]. Their characteristics are summarized in Table 1.

[1] Our work has been carried out using four typical European high-end GSM smart phones: Nokia 6600, Nokia 6630, Nokia N70, and Nokia N90. We will identify them in the paper as phone 1, phone 2, phone 3, and phone 4, respectively.

J.M. Corchado, D.I. Tapia, and J. Bravo (Eds.): UCAMI 2008, ASC 51, pp. 48–56, 2009.
springerlink.com

Table 1. Technical specifications of the four smart phones used in our performance evaluation study

Phone 1	Operating System	Symbian OS 7.0s
	Platform	Series 60 2nd Ed
	Camera	VGA w/ 2x digital zoom
	Processor	ARM9TDMI
	CPU speed	104MHz
Phone 2	Operating System	Symbian OS 8.0a
	Platform	Series 60 2nd Ed FP2
	Camera	1.3 megapixel w/ 6x digital zoom
	Processor	ARM9EJ-S
	CPU speed	220MHz
Phone 3	Operating System	Symbian OS 8.1a
	Platform	Series 60 2nd Ed FP3
	Camera	2 megapixel w/ 20x digital zoom
	Processor	ARM9EJ-S
	CPU speed	220MHz
Phone 4	Operating System	Symbian OS 8.1a
	Platform	Series 60 2nd Ed FP3
	Camera	2 megapixel w/ 20x digital zoom
	Processor	ARM9EJ-S
	CPU speed	220MHz

2 Camera Access Performance

In this section, we present the results regarding the image acquisition time in a typical camera phone both in J2ME and in Symbian native applications.

J2ME grants access to a phone's camera through the optional library MMAPI [4], which is used mainly to retrieve, open, and play multimedia contents from any source—local or remote. The MMAPI manages the camera just like any other multimedia source, in a straightforward manner. The camera is considered a multimedia source, so the locator *capture://video* can identify it. With the *System.getProperty()* call, we can determine whether the phone supports video and audio recording and in what formats. In J2ME, capturing an image is equivalent to recording a video (the same API is used), and the programmer manages it using a Player created with the *createPlayer()* call in the *Manager* class. Using the *VideoControl.getSnapshot()* method, we can shot a photograph —in this case, an individual video frame—and present it in a range of formats, sizes, and bit depths. To process data, we need to take the RGB pixel values from the image in a standard image format. The most straightforward way to accomplish this is to use the *Image.getRGB()* method, which stores the RGB values as an array of integers.

Regarding native Symbian applications [3], access to the camera is achieved in terms of independent images and not video, as in the MMAPI. The *viewfinder* is not provided, so the application must take photographs and show them to the user periodically. The images are obtained using an active object, and they are offered in a Symbian-specific format, which afterwards can be converted to other formats. One of the parameters is the image quality, which can be high or low. A low quality image (called snapshot), is QQVGA, that is, 160x120 pixels, and 4096 colors (12 bits). A high quality image is VGA, (640x480) and 16 million colors (24 bits).

Typically, snapshots are used for video and *viewfinder*, and high quality images for the actual photograph. On the other hand, it is possible to switch between illumination profiles, night and day, which modify the image brightness by means of the white balance.

In Symbian we can use the *MCameraObserver::ImageReady()* method to capture images. Bitmaps are managed with the *CFbsBitmap* class; to obtain the RGB values of the pixels, which is usually needed for doing any kind of image processing, we can use the *CFbsBitmap::GetPixel()* method.

When measuring the acquisition times using J2ME and Symbian, we observe that Symbian is about 10 times faster than J2ME, and that the acquisition time is independent of the image format used (4K BMP, 64K BMP, and 16M BMP; see section 3 for an explanation of these image formats). The acquisition time increases with image resolution, and this effect is stronger in J2ME (from 969 ms to 6406 ms) than in Symbian.

The slower acquisition time of J2ME can be explained because it is continually building a video stream, from which the images are captured. This process is very costly, and limits the performance of J2ME based image processing applications.

3 Image File Size

Now, we will compare the memory use of Symbian and J2ME camera applications. To do this, we will represent the captured file sizes against image pixel resolution.

J2ME needs less memory than any of the BMP Symbian formats. This is because J2ME image quality is much worst. The J2ME BMP format uses a 256 colors palette, in which just one byte is used to represent each pixel. On the other hand, Symbian captured images can represent much more colors:

- 4K BMP uses 12 bits, i.e., 4,096 (2^{12}) colors.
- 64K BMP uses 16 bits, i.e., 65,536 (2^{16}) colors.
- 16M BMP uses 24 bits, i.e., 16,777,216 (2^{24}) colors.

Despite using a different number of bits per pixel, Symbian 4K BMP and 64K BMP images use both 2 bytes per pixel, so for the same resolution, the file size is the same. This is why 4K BMP and 64K BMP plots in Fig. 2 fit in the same line. Symbian 16M BMP images take up a 50% more memory space than 4K and 64K BMP, since 3 bytes per pixel are used to store the RGB values.

We recall from section 2 that Symbian acquisition time is independent with the image format; so we recommend using the 64K BMP format over the 4K BMP format, since it has more resolution and the requirements regarding memory and speed are the same. 16M BMP images will be used when a higher color resolution is required by our application.

4 Processing Speed

Once captured the image, the application will usually make some processing in the image to detect patterns, movements, colors, etc. If we want to implement applications which perform image processing, our device must be able to run heavy algorithms very fast, sometimes even in real-time.

We will measure the speed of basic operations of the processor, to see the overall speed of the operations, and also which ones are faster and which ones should be avoided. We will estimate the time consumed in the processor for these operations; we will present experimental results measured in real devices; and finally we will discuss these results.

The operations we consider in our study are: increment of a variable, addition of two variables, shift of a variable, multiplication by a constant, division by a constant, multiplication of two variables, division of two variables, "less than" comparison of two variables, "less or equal than" comparison of two variables, equality comparison of two variables.

4.1 Clock Cycles Estimation

For native Symbian applications, we can calculate the time consumed by each of these operations from the number of clock cycles consumed by the assembler code necessary to perform the operation, divided by the CPU frequency. However, as we will see now, the number of clock cycles consumed cannot be accurately calculated, we can just estimate some upper and lower limits.

First, we have to obtain the assembler code the compiler generates for each of the above operations. When compiling a Symbian C++ code, you can specify two compilation options: with or without code optimization. Optimized code is faster but less portable to other processors of the same family. We used the *–S* option of the *gcc* compiler to obtain it.

Once we know the instruction set that correspond to each of the basic operations, we can estimate the time it takes to execute them.

The CPUs used by the phones we consider in our study are the ARM9TDMI [5] and the ARM9EJ-S [6]. These CPUs use a segmented architecture, i.e., the execution of two consecutive instructions may overlap during a variable number of cycles, depending for example on whether the second instruction makes use or not of some result obtained from the first one.

In fact, it is a little bit more complicated. In ARM9 processors, there are four types of CPU cycles: N, S, I, and C. The *ldr* instruction, for instance, that loads a 32-bit register from a 32-bit memory location, takes three cycles to execute: S+N+I (i.e., one S cycle, one N cycle, and one I cycle). I and C cycles of one instruction can be multiplexed only with N or S cycles of other instruction. So, depending on the type of the first cycle of the next instruction, the processor will overlap their execution or not.

As an example, for the increment operation compiled without optimization, three instructions are executed:

```
ldr r3, [sp, #372]
add r2, r3, #
str r2, [sp, #372]
```

Consulting the documentation, the first instruction (*ldr*) can take 2 or 3 clock cycles, the second (*add*) 1 clock cycle, and the third (*str*) 1 or 2. So, the number of CPU cycles required for the whole increment operation is between 4 and 6 CPU cycles. We must divide this figures by the CPU clock frequency (see Table 1) to obtain the execution time of the operation.

Similarly, we can obtain the minimum and maximum number of CPU cycles for each of the operations we are considering in this section. The results are presented in Tables 2 and 3.

As we can see, the variation between minimum and maximum estimated cycles is rather big. Besides that, this method gives as results just for native Symbian code, not for J2ME. For these reasons, we decided to take experimental measurements of execution time.

4.2 Measurements

In this section we show the time cost of the basic operations listed in Section 4.1, both in J2ME and in native Symbian applications.

The maximum number of cycles for the division operation cannot be estimated because it calls to a library function, *__divsi3*, whose execution time we could not estimate.

The programs for doing this testing must be carefully coded. If the compiler detects that the operation we are performing inside the loop is "unnecessary", it optimizes the resulting assembler code. For example, if we add two variables but the result is not assigned to any other variable, the compiler does not translate into assembler that operation; or if we repeat the same operation with the same operands *n* times in a for loop, the compiler gets the operation out of the loop in the resulting assembler code, and executes it just one time. These problems forced us to make inside the loop more than just one operation.

Table 4 describes how we initialize the variables before the loop, and which are the operations we make in each iteration of the loop in our measurement program.

Another problem is that our measurement application, as will happen with an image processing application, is not the only one being executed in the mobile phone. The simultaneous execution of certain operating system activities may affect our measurements, distorting our results.

We have executed each experiment in Table 4 ten times. From the minimum of the obtained measurements, we subtract the time consumed by operations in the loop that were not the one under study. For example, from the result obtained in the measurement of the addition operation, we subtract the time we have measured through the first test that it takes to make an increment, and also the time we have measured for an assignment.

Tables 5, 6 and 7 present the results obtained for Symbian applications with compiler optimization, without compiler optimization, and J2ME applications. The results are normalized to CPU clock cycles.

4.3 Discussion of Results

We must note that both J2ME and Symbian present a similar behavior, with the division as the most costly operation. In both cases, we will try to the use of divisions, which are the slower operations. This result is as expected.

This is only partially true. The J2ME virtual machine (VM) features a dynamic compilation scheme [7], based on the Java Hotspot Virtual Machine for J2SE. This means that the virtual machine has both an interpreter, which at runtime executes Java bytecodes, and a compiler, which also at runtime turns bytecodes into native code. Native code is an order of magnitude faster than interpreted code, but has the disadvantage of taking up additional memory space.

Table 2. Number of clock cycles with optimization

Operation	Min.	Max.
Increment	1	3
Addition	1	3
Shift	1	3
Multiplication by a constant	1	3
Division by a constant	4	12
Multiplication	2	3
Division	3	-
Less than	3	7
Less or equal than	3	7
Equal than	3	7

Table 3. Number of clock cycles without optimization

Operation	Min.	Max.
Increment	4	6
Addition	5	8
Shift	4	6
Multiplication by a constant	6	8
Division by a constant	12	14
Multiplication	6	11
Division	4	-
Less than	7	13
Less or equal than	7	13
Equal than	7	13

Table 4. Initialization and body of the loop in the measurement programs

Operation	Initialization	Body of the loop
Increment	j=0	j++
Addition	var1 = 459; var2 = 396	r=var1+var2; var2++
Shift	var1 = 519	var1=var1>>1;
Multiplication by a constant	var1 = 519	r = var1*5; var1++
Division by a constant	var1 = 519	r = var1/342; var1++
Multiplication	var1 = 459; var2 = 396	r = var1*var2; var1++
Division	var1 = 459; var2 = 396	r = var1/var2; var1++
Less than	var1 = 459; var2 = 396	r = (var1 < var2)
Less or equal than	var1 = 459; var2 = 396	r = (var1 <= var2)
Equal than	var1 = 459; var2 = 396	r = (var1 == var2)

This VM solves this with a third block of the virtual machine, the profiler. The profiler uses statistic techniques to identify, at runtime, parts of the code that the J2ME application runs repeatedly. These parts of the code are called hotspots. Then the hotspots are compiled, and the rest of the code, executed rarely or only once, is interpreted.

Table 5. Number of clock cycles in Symbian with optimization

Operation	Phone 1	Phone 2	Phone 3	Phone 4
Increment	1	1	1	1
Addition	1	1	1	1
Shift	1	1	1	1
Multiplication by a constant	2	2	2	2
Division by a constant	10	8	8	8
Multiplication	10	8	8	8
Division	111	123	123	123
Less than	3	3	3	3
3Less or equal than	3	3	3	3
Equal than	3	3	3	3

Table 6. Number of clock cycles in Symbian without optimization

Operation	Phone 1	Phone 2	Phone 3	Phone 4
Increment	5	4	4	4
Addition	6	5	5	5
Shift	5	4	4	4
Multiplication by a constant	7	6	6	6
Division by a constant	16	13	13	13
Multiplication	16	13	13	13
Division	111	129	127	129
Less than	8	7	7	7
Less or equal than	8	7	7	7
Equal than	8	7	7	7

Table 7. Number of clock cycles with J2ME

Operation	Phone 1	Phone 2	Phone 3	Phone 4
Increment	2	1	1	2
Addition	2	2	2	1
Shift	4	3	3	3
Multiplication by a constant	5	4	4	4
Division by a constant	25	25	13	9
Multiplication	5	3	3	6
Division	236	251	259	261
Less than	12	5	5	5
Less or equal than	12	5	6	7
Equal than	13	8	10	11

This enhancement relies on the assumption that an application spends most of its runtime executing a small portion of the code. This assumption is accomplished in the test application, since all it does is to repeat a loop thousands of times. This loop will soon be identified as a hotspot and compiled, running at processor speed. Also, for the applications we are studying in this document this assumption still holds true. Typically, an image processing algorithm will iterate on an array containing pixels of an image, and repeat the process on subsequent images over time or on different parts of an image. Thus we can expect a good behavior of our J2ME image processing application.

Finally, note that the number of cycles for J2ME applications in phone 1 is higher, for most operations, that for phones 2, 3, and 4. The explanation for this is that the processor present in phones 2, 3, and 4, the ARM9EJ-S, has native support for Java code, while the processor present in phone 1, the ARM9TDMI, has not.

5 Conclusions

J2ME applications are easier to develop and more easily portable to a wider range of devices. However, this comes at a penalty in speed, mainly in image acquisition, and no so much in processing time. Detailed measurements are presented, that helps us to decide when J2ME is fast enough for our application, or if native Symbian is required. We also give recommendations to select among the different Symbian BMP formats supported.

To test some of the possibilities of the image processing on a mobile phone, we have implemented two applications, both in Symbian and in J2ME. The first one is a color reader, designed to help the blind people. It detects the colors present on the image and announces them in loud voice. The second one is a very simple movement detector, to check real-time video.

The Color Detector obtains the colors present in an image taken by the camera, and then plays voice files with the names of the detected colors. To achieve this, a K-means clustering algorithm, is used on 50 equally spaced pixels of the image. The resulting three color values are then studied to see how many colors are there in the image, for example comparing them to see whether they represent different colors of the image or just different shades of the same color. The resulting colors (there may be between zero and three colors remaining), are assigned a color category, with an associated sound file with its name.

The movement detector continuously samples the image looking for movement and on detection, plays a sound and shows the image where the movement took place. The method to detect movement is very simple, so that the speed of the algorithm is limited by the camera frame speed. We simply compare the color of a single pixel in two consecutive images. We take the Manhattan distance between the two colors, and if this distance is higher than a given threshold (to take into account the noise present in the image, which is quite harmful), we assume a movement has taken place.

Acknowledgements. The authors would like to thank Jonatan Tierno, José Carlos Perea, and Consolación Navarrete for their help in the development of this work. This work has been partly supported by the projects ITACA (TSI2007-13409-C02-01) and "Plataforma abierta para la integración en el hogar de servicios cooperativos de teleasistencia y telemedicina" (TSI2006-13390-C02-01) of the Science and Education Ministry, Spain.

References

Tierno, J., Campo, C.: Smart camera phones: limits and applications. IEEE Pervasive Computing Magazine 4(2), 84–87 (2005)

Taivalsaari, A., Huopaniemi, J., Patel, M., Van Peursem, J., Uotila, A.: Programming Wireless Devices with the Java 2 Platform, Micro Second Edition. In: Riggs, R. (ed.). Addison-Wesley Pub. Co., Reading (June 2003)

Edwards, L., Barker, R.: Developing Series 60 Applications: A Guide for Symbian OS C++ Developers. Addison-Wesley, Reading (2004)

Mobile Media API, tech. report JSR 135, Java Community Process, http://jcp.org/en/jsr/detail?id=135

ARM9TDMI (Rev 3) Technical Reference Manual. Ref: DDI0180A, Issued: March 08, 2000, http://www.arm.com/pdfs/DDI0180A.zip

ARM9EJ-S Revision r1p2 Technical Reference Manual. Ref: DDI0222B, Issued: September 30, 2002, http://www.arm.com/pdfs/DDI0222B_9EJS_r1p2.pdf

The Project Monty Virtual Machine. Sun Microsystems White Paper (2002)

Extended Bluetooth Naming for Empowered Presence and Situated Interaction with Public Displays

Rui José[1] and Francisco Bernardo[2]

[1] Departamento de Sistemas de Informação
Universidade do Minho. Campus de Azurem,
4800-058 Guimarães, Portugal
rui@dsi.uminho.pt
[2] Ubisign Tecnologias de Informação, Lda.
Edifício Olympus II. Rua Cidade do Porto, 79
4709-003 Braga, Portugal
info@ubisign.com

Abstract. The paradigm of proximity-based discovery and communication enabled by Bluetooth technology can be very relevant in Ambient Intelligence as an enabler for situated interaction. In this work, we explore the use of Bluetooth naming as a key driver for situated interaction around public displays. Our approach to the use of Bluetooth naming extends beyond self-exposure and introduces support for simple commands in the name that can trigger actions on the displays. Our specific objective is to evaluate the usability of this interaction technique and uncover any guidelines for its usage. We have conducted a study combining a trial in a public bar and a set of usability interviews. The results obtained confirm Bluetooth Extended Naming as an easily adoptable technique for situated interaction and suggest some recommendations to improve its effectiveness.

Keywords: Situated interaction, Bluetooth, Presence, Public displays.

1 Introduction

The paradigm of proximity-based discovery and communication enabled by Bluetooth technology is particularly well suited for situated interaction. In this work, we explore the use of Bluetooth naming as a key driver for situated interaction around public displays. Bluetooth devices have a user-defined name, created primarily for defining how Bluetooth devices present to each other, but which can be set and changed quickly. In the discovery process, these names become visible to nearby devices, enabling a simple proximate self-exposure mechanism, which has been enabling an increasingly strong culture around the social uses of Bluetooth naming [1]. Our approach to the use of Bluetooth naming extends beyond self-exposure and introduces Bluetooth Extended Naming as a technique in which the system can recognise parts of the Bluetooth device name as explicit instructions to trigger actions on the displays.

Bluetooth Extended Naming is necessarily a limited technique that does not aim to support any complex interaction dialogues with situated displays. There are several obvious limitations that may undermine its potential, such as the short size of Bluetooth

J.M. Corchado, D.I. Tapia, and J. Bravo (Eds.): UCAMI 2008, ASC 51, pp. 57–65, 2009.
springerlink.com

names, limitations in text entry, and also the delay in the detection of name change updates introduced by the timings of the discovery process. It is however, in its technical simplicity, a promising technique for situated interaction. Firstly, because it has an extremely low entry barrier. Bluetooth is a widely available technology and setting new Bluetooth device names is normally a relatively simple task that can be accomplished with the base functionality of any mobile phone and without the need for any specialist software. This easy availability is of an huge importance in enabling social practices around the technology and represents a major difference to other sensing and interaction approaches that, albeit more sophisticated, require specific hardware or the installation of specialist software in personal devices. Secondly, the use of Bluetooth presence for situated interaction combines very well implicit and explicit forms of interaction, in fact blurring the distinction between them. Simply by having a discoverable Bluetooth device, people are already part of the situation and implicitly engaging with the system. This low bandwidth, but continuous, flow of presence information can be fundamental in the aggregation of situated content for the display and may act as an important catalyst for more explicit forms of interaction. This is what mainly differentiates interactions based on Bluetooth presence from interactions based on SMS, another widely available technology that is also very relevant for situated interaction with public displays.

1.2 Research Goals and Overview

Our research objective is to evaluate the effectiveness of Bluetooth extended naming as a technique for communicating commands for situated interaction around public displays. In the process, we also expect to uncover guidelines that may help to maximise the efficiency and potential uses of Bluetooth Extended Naming. In this study, we do not address the motivations and practices associated with the use of Bluetooth Extended Naming.

The methodology we chose combines a trial in a public setting and usability interviews. The trial provides important insight on the practical implementation of the technique and on its usability within the complex set of social phenomena that characterize situated interaction in a public setting. The interviews aim to inform the design with insight on the procedural mechanisms associated with this particular use of Bluetooth naming. The results obtained from this study confirm Bluetooth Extended Naming as an easily adoptable technique for situated interaction and suggest some recommendations to improve its effectiveness.

In the remainder of this paper, we will start by reviewing in Section 2, the most relevant pieces of previous research that relate with our own work. In Section 3 we describe instant places, the system that served as an infrastructure for presence-based situated displays. In Section 4, we describe the trial we deployed in a bar and its results. In Section 5 we outline the main results from the usability interviews, and finally, in Section 6 we summarise our conclusions.

2 Related Work

The use of Bluetooth scanning has been extensively explored as a mechanism for sensing presence and uncovering all sorts of patterns, such as the familiarity level of the surrounding environment [2], the social situation [3], and more general large-scale reality mining [4]. In this work, we also build on the sensing possibilities enabled by Bluetooth discovery, but we take a different direction. Instead of intending to uncover information about an existing reality, our focus is on the use of Bluetooth naming as a means for empowered self-expression and as an enabler for situated interaction.

The work by Kostakos [5] in the cityware project is based on a platform for capturing mobility traces via Bluetooth scanning and has explored several ways of leveraging that information, including a set of in-situ visualizations providing people with information about current or recent Bluetooth presences. The system uses in-situ presence information as a way to generate content for a Facebook application that lets people associate physical co-presence information with their social network. We explore the opposite direction in that we aim to build on Bluetooth presence to generate situated content for local displays.

The Proactive displays [6] system explores the use of presence as a driver for situated interaction around public displays. The detection of nearby RFID tags triggers the display of profile information about the owner of the tag, promoting occasional encounters between people around the display. However, this type of approach requires a priori definition of individual profiles with associated data and assumes that everyone will be using a particular type of tag. Furthermore, people have a very limited role in the system, which is basically to move around and be detected.

The use of commands in short text messages is used by several on-line services that support SMS interfaces. The use of a picoformat [7] enables the codification of simple commands while satisfying the restrictions imposed by text input on mobile phones. Even though there are no standards for this, there some emerging SMS-based dialects to support interaction with mobile services, such as twitter nanoformats [8] and Dodgeball [9] .

3 Instant Places

The instant places system displays on a public screen content that is directly or indirectly derived from Bluetooth presence information, including content derived from the presence of commands in Bluetooth names.

The system is composed by one or more Bluetooth enabled computers each connected to a public screen and linked to a central repository, as shown in Fig.1.

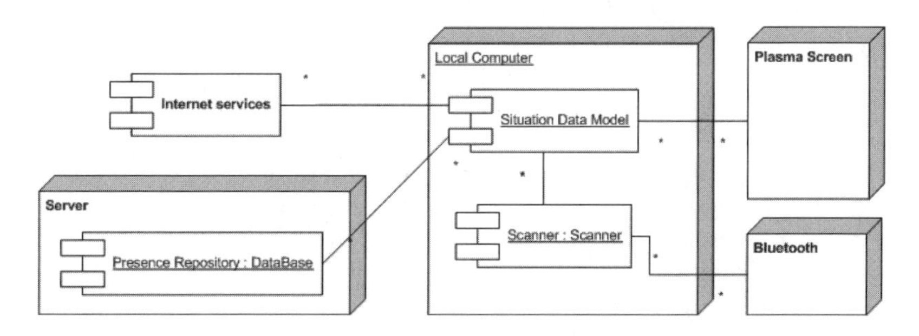

Fig. 1. Instant Places architecture

Information about nearby devices is periodically collected by a Bluetooth scanner and consumed by a situation data model that manages data about the place and present identities. A central repository is used to allow persistent identity information to be recognised across multiple sessions and also to combine information from pervasively distributed data sources, allowing for multiple screens in a large space to share the same presence view. The system does not need any a priori information about people, their profiles, permissions or groups, as all the information in the repository is entirely created from the history of presences.

Support for Bluetooth Extended Naming is an integral part of instant places. With every scanning operation, Bluetooth devices names are parsed in search for commands and their parameters. The syntax used is necessarily simple to comply with the constraints associated with Bluetooth device names. According to specifications, a Bluetooth device name can be up to 248 bytes and should be encoded based on UTF-8. This means that the maximum number of characters entered at the user interface level will actually vary with the type of characters being used and could be down to 82 characters. However, not all devices can be expected to handle more than the first 40 characters of the Bluetooth device name, and in many cases this may actually come down to the first 20 characters or even less. Additionally, text input on mobile devices suffers from restrictions associated with characters sets that may further restrain the syntax possibilities.

In the version of instant places that was used for this study, there were two types of commands, both defined as a command word followed by ":" and a set of comma separated parameters. The tag command associates multiple tags with the device, and is activated by including in the name the expression "tag:" followed by a comma separated list of tags, as in the following example "my device tag:punk,rock". The Flickr command associate a Flickr user name with the device and is activated by including in the device name the expression "flk:" followed by the flickr user name, as in the following example "my device flk:JohnSmith". In both cases, these commands were intended to serve as hints for the display of photo streams obtained from the photo sharing website Flickr.

The central functionality supported by instant places is the visualisation on a public display of content that is situationally relevant and mainly driven by Bluetooth presence information, as shown in Fig. 2

The main driver for the visualisation is the real-time information about currently present devices and their names. The periodic scanning of Bluetooth devices generates a continuously changing flow of presence patterns that is visualised on the public display. Each discovered device is represented as a multivariate icon, in which the device name is clearly displayed. If the device name includes any commands, they are removed before the name is displayed. This use of Bluetooth names can be classified as an implicit form of interaction in the case of the person who unexpectedly finds his or her name on the display. However, it can quickly turn into an explicit form of interaction when that person changes the device name based on the new meaning created by the visualization on the screen.

These elements, per se, guarantee an important base level of content generation and implicit interaction. The dynamic patterns of Bluetooth presence provide an interesting and continuously changing element of situation awareness, and the public display of Bluetooth device names gives them a new meaning and empowers their use.

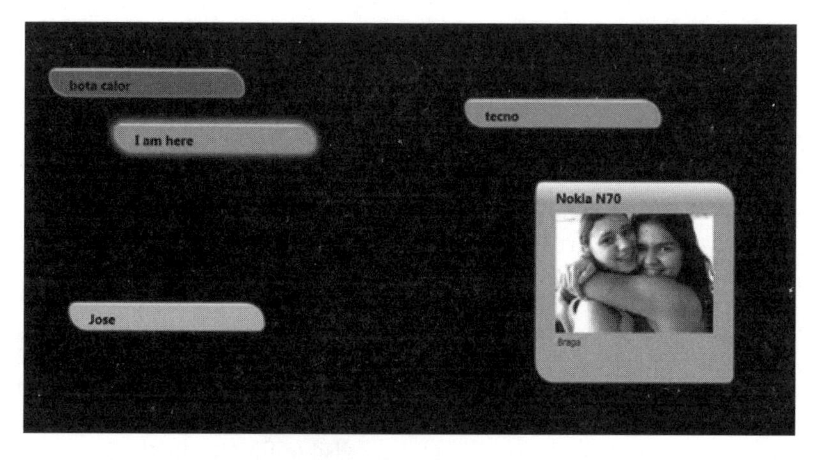

Fig. 2. Instant Places visualisation

However, they are limited in their ability to produce a continuous flow of enticing content, and that is where extended naming provides the extra level of functionality that invites people to more explicit and engaging forms of interaction. By including *tag* or *flk* commands in their device names, people can easily provide seeds that the system uses for selecting further content from Flickr. If a tag is used, the display presents photos tagged with that word. If a flickr user name is used, the display presents the respective photo stream. In both cases, photos are displayed directly on the device icon, which expands itself to create space for the photo display. This "situated mashup" manages to combine situated interaction with a global service to generate a continuously changing display of situationally relevant content.

4 The Trial at the Campus Bar

The first part of this study was a public deployment of instant places. We chose a bar at the campus of University of Minho that is visited every day by several hundred people that come in for coffee or a quick snack, normally in small groups. There are several peak periods, with the busiest moment being at lunch time, when small meals are served Instant places visualisations were displayed using a 42" LCD screen that was already in the bar and is normally used for watching TV.

Prior to the public deployment of the system, we conducted a silent Bluetooth scanning for four weeks with the sole purpose of obtaining a neutral perspective of the local Bluetooth environment. When the system went public, we distributed leaflets with information about the project and instructions about the use of tags in Bluetooth names. The system used in this study was operational for 3 weeks.

The overall results of the trial indicate that the proposed techniques were easily and widely adopted as part of situated interactions around a public display. Table 1 compares Bluetooth usage parameters before and after the system was made public. We estimated the total number of visits to the bar based on sales numbers, and we collected from the logs information about how many unique device addresses and unique devices names were seen during these two periods.

Table 1. General Bluetooth usage patterns before and after instant places

	Silent Scanning	Instant Places
Estimated visits	7625	6526
Unique devices	365	460
Unique names	317	685
% visits with BT visible	4.7%	7.0%
Names per device	0.9	1.5
Names with commands	n.a	112

Even though no effort was made to recruit users, these numbers show a strong effect of the system on Bluetooth presence and naming patterns. There was a significant increase in the percentage of visitors that were visible for Bluetooth discovery (from 4.7% to 7.0%), suggesting that a significant number of people made their device visible specifically for this purpose. Prior to instant places, there were more devices than device names, which can be justified by the existence of many devices using their default names. During the operation of the system, the average number of names per device raised to 1.5. Since there were no name changes during the initial period of silent scanning, it is sensible to conclude that all those changes were induced by the system. Regarding extended naming, there were 112 names that included tag commands. We analysed those names where tags were used to uncover any difficulties in the use of commands. The results are summarised in Table 2.

Table 2. Problems with the use of tag commands

Problem	Type	Occurrences
Names with space after tag:	Syntax error	8
Names with space between multiple tags	Syntax error	5
Names without space before tag:	Syntax error	3
Names only with tags	Ambiguous	6
Names with tag: but no tag specified	Ambiguous	2
Names with tag: repeated	Ambiguous	1

There were 16 cases (14.2% of those who tried to add tags to their name) with at least one syntax error that prevented the successful indication of tags. In 5 of those errors, the same name was observed with the correct syntax, indicating that after a first failed attempt, 5 people managed to somehow correct their mistakes and produce valid tags, lowering the failure rate to 9.8%.There were also 9 ambiguous names that albeit not necessarily incorrect were not anticipated in our definition of the name rules. In 6 of them, the tag command was the name. This was not properly parsed because we were using the first word of the name as an identifier that was not considered for command parsing. There were 2 names without any tag specified, even though the "tag:" expression was present. Interestingly, this may have not been an error, but an emerging strategy for facilitating at a later point the recollection of the tag syntax, or reducing the input needed for adding new tags. One name had the "tag:" expression repeated, which in our current parser would ignore the second expression.

5 Usability Interviews

The second part of this study was a series of interviews with mobile phone owners to evaluate the performance in the execution of tasks related with Bluetooth naming and uncover any limitations or suggestions related with the use of Bluetooth Extended Naming. We conducted interviews with 40 participants (26 male and 14 female), aged between 15 and 28 who said to have had some type of Bluetooth usage before. The interview was divided in two distinct parts: a survey on Bluetooth naming practices that was answered by all the participants; and a set of 5 tasks related with Bluetooth naming that were executed by 12 of the participants. The tasks were executed with the participants own phones, and the objective was to measure performance and highlight any factors that could affect their execution.

The first group of questions was about the visibility mode used in the Bluetooth device. 47% of the responds indicated always having their Bluetooth inactive or invisible. 41% responded that they occasionally had their device visible, 8% that they frequently had their device visible and 3% that they always had their device visible. The reasons for not being visible were most of all related with fears of attacks on the personal device, such as the possibility of infection by virus. Many people also mentioned a possible loss of privacy and saving energy (many respondents seemed to be aware of the consequence of Bluetooth on battery life). The reasons for activating Bluetooth and making it visible, were essentially sharing content (72%) and device pairing (20%). Some people specifically mentioned making use of "temporary visibility", a feature available in some devices that allows visibility to be activated only for a limited period of time.

Regarding Bluetooth naming practices, participants were asked how often did they change their Bluetooth device name. The vast majority answered that they rarely changed their name (61%) or even that they had never changed their device name (31%). Only 8% reported changing their device name occasionally or frequently. On average, participants indicated being able to write device names with 31 characters on their mobile phone, with 16 being the minimum and 65 the maximum.

In the second part of the interview, participants were asked to perform the following 5 tasks:

- T1– Change the current activation state of their Bluetooth device.
- T2– Change their device name (new name had to be realistic)
- T3– Introduce a command in the device name (CMD:activate)
- T4–Introduce a command with two parameters (CMD:activate,num)
- T5–Introduce 2 commands, each with a parameter (CMD:activate CMD:num)

The graphics in Fig. 3 summarises the performance results.

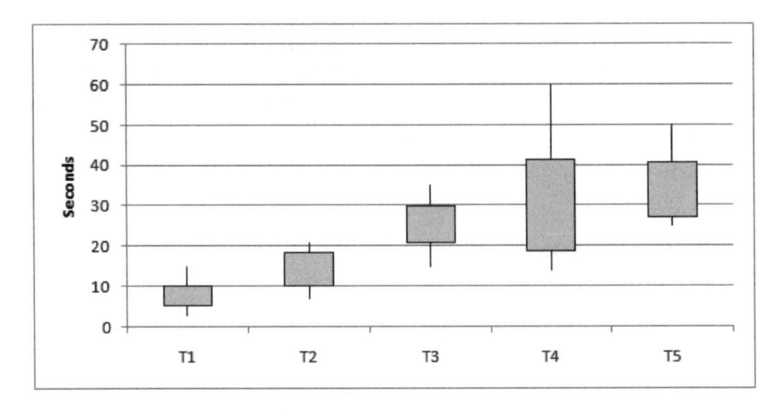

Fig. 3. Min, 1st Quartile, 3rd Quartile and Max times for task execution

Results for tasks 1 and 2 indicate a generally good performance in the most basic operations of activating/deactivating Bluetooth (T1) and changing device name (T2). Particularly good results were obtained by participants with speed dial functionality already associated with Bluetooth activation/deactivation. These are tasks that can be easily and quickly achieved, even while on the move. The introduction of commands in device names (T3,T4 and T5) takes too long to be something that the average user would do without some disruption to other tasks. However, the average duration (about 30 seconds) is perfectly suitable for someone not immediately involved with other tasks, and these were tasks that people were doing for the first time. One of the participants was not able to complete tasks 3, 4 and 5 because his device did not support whitespaces as part of the name.

Participants also made some suggestions on how to improve the performance of these tasks, such as using only lower case letters and replacing the use of the ":" character by the dot character ".", more easily accessible on most mobile phones.

6 Conclusions

The overall results from our study show that there are no significant limitations for a widespread adoption of Bluetooth Extended Naming as a technique for situated interaction. The deployment of the trial has highlighted the high adoption rates that these techniques can have. Even without any active recruitment, a very significant number of people decided to try it. From those who have tried it, less than 10% seem to have been unable to achieve their purpose due to syntax errors. The results from the interviews also show that for most devices there are no significant limitations and that for most people, even without any previous training, the proposed operations can be executed within a periods of time that is appropriate for not too frequent operations.

We also intended to uncover guidelines that should be followed to maximise the efficiency and potential uses of Bluetooth Extended Naming. The results indicate that the major problem with the proposed syntax was the use of white spaces, suggesting the use of an alternative syntax that is more tolerant with the ambiguity that is normally associated with the use of white spaces. They also suggested the use of a character set that maximises the efficiency of text writing on mobile phones, such as a case insensitive syntax and the use of more commonly used characters such as "." as separator.

Regardless of these overall results, the individual experience will be largely determined by the mobile phone being used. Some devices still treat the Bluetooth name uniquely as a technical feature, resulting in cumbersome procedures for changing name and limitation on the length and format of the name itself that may restrict the appropriation of Bluetooth naming for other purposes, such as those studied in this work. However, severe limitations are not common and, what is more important, are not inherent to the Bluetooth standard. As the growing culture around Bluetooth naming continues to evolve these limitations are likely to disappear. These results should be considered by the manufacturers of Bluetooth devices that should aim to facilitate name writing and name change. The existence as part of the factory settings of speed dial options for Bluetooth configuration is a good indication that some device manufactures are attentive, but additional functionality, such as storing pre-defined names for easy selection (very much like changing profile) would be welcome in facilitating the use of these techniques.

References

[1] Kindberg, T., Jones, T.: "Merolyn the Phone": A study of Bluetooth naming practices. In: Krumm, J., Abowd, G.D., Seneviratne, A., Strang, T. (eds.) UbiComp 2007. LNCS, vol. 4717, pp. 318–335. Springer, Heidelberg (2007)

[2] Paulos, E., Goodman, E.: The familiar stranger: anxiety, comfort, and play in public places. In: Proceedings of the SIGCHI conference on Human factors in computing systems. ACM, Vienna (2004)

[3] Nicolai, T., Yoneki, E., Behrens, N., Kenn, H.: Exploring Social Context with the Wireless Rope. In: Presented at 1st International Workshop on MObile and NEtworking Technologies for social applications (MONET 2006), Montpellier, France (2006)

[4] Eagle, N., Pentland, A.: Reality Mining: Sensing Complex Social Systems. Personal and Ubiquitous Computing 10(4) (2006)

[5] Kostakos, V., O'Neill, E.: Capturing and visualising Bluetooth encounters. In: Presented at CHI 2008, workshop on Social Data Analysis, Florence, Italy (2008)

[6] McCarthy, J.F., Nguyen, D.H., Rashid, A.M., Soroczak, S.: Proactive Displays & The Experience UbiComp Project. In: Presented at First International workshop on Ubiquitous Systems for Supporting Social Interaction and Face-to-Face Communication in Public Places, Seattle, Washington, USA (2003)

[7] Mabbett, A., Messina, C., Andy Stack, M., Turner, A.: Picoformats - Microformats (June 2008), http://microformats.org/wiki/picoformats

[8] Julio, G.: Microblogging-nanoformats - Microformats (June 2008), http://microformats.org/wiki/twitter-nanoformats

[9] Google Inc., dodgeball. com:: mobile social software (June 2008), http://www.dodgeball.com/glossary

Infrastructural Support for Ambient Assisted Living

Diego López-de-Ipiña[2], Xabier Laiseca[1], Ander Barbier[2], Unai Aguilera[1], Aitor Almeida[1], Pablo Orduña[1], and Juan Ignacio Vazquez[2]

[1] Tecnológico Fundación Deusto
{xlaiseca,uaguiler,aalmeida,porduna}@tecnologico.deusto.es
[2] Universidad de Deusto. Avda. de las Universidades, 24 - 48007, Bilbao, Spain
{dipina,barbier,ivazquez}@eside.deusto.es

Abstract. This work describes several infrastructure contributions aimed to simplify the deployment of Ambient Assisted Living (AAL) environments so that elderly people can maximize the time they live independently, through the help of ICT, at their own homes. Three core contributions are reviewed: a) a multi-layered OSGi-based middleware architecture which enables adding new environment monitoring and actuating devices seamlessly, b) an easy-to-use elderly-accessible front-end to comfortably control from a touch screen environment services together with a custom-built alert bracelet to seek assistance anywhere at any time and c) a rule-based engine which allows the configuration of the reactive behaviour of an environment as a set of rules.

1 Introduction

By 2020, 25% of the EU's population will be over 65 [1][4]. Spending on pensions, health and long-term care is expected to increase by 4-8% of GDP in coming decades, with total expenditures tripling by 2050. However, older Europeans are also important consumers with a combined wealth of over €3000 billion.

Ambient Assisted Living (AAL) [10] is an initiative from the European Union to emphasize the importance of addressing the needs of the ageing European population by reducing innovation barriers on ICT with the goal to lower future social security costs. The program intends to extend the time the elderly can live in their home environment by increasing the autonomy of people and assisting them in carrying out their daily activities. Assisted Living solutions for elderly people using ambient intelligence technology can help to cope with this trend, by providing some proactive and situation aware assistance to sustain the autonomy of the elderly, be helpful in limiting the increasing costs while concurrently providing advantages for the affected people by increasing quality of life. The goal is to enable elderly people to live longer in their preferred environment, to enhance the quality of their lives and to reduce costs for society and public health systems.

Ambient Assisted Living fosters the provision of equipment and services for the independent living of elderly people, via the seamless integration of info-communication technologies within homes and extended homes, thus increasing their quality of life and autonomy and reducing the need for being institutionalised. These include assistance to carry out their daily activities through smart objects, health and

J.M. Corchado, D.I. Tapia, and J. Bravo (Eds.): UCAMI 2008, ASC 51, pp. 66–75, 2009.
springerlink.com

activity monitoring systems including wearables as well as context-aware services, enhancing safety and security, getting access to social, medical and emergency systems, and facilitating social contacts, in addition to context-based infotainment and entertainment.

The ZAINGUNE research project [17] is our approach to address the objectives of the AAL initiative. This paper describes our efforts devising new middleware and hardware infrastructure in order to make the AAL vision reality.

The structure of this paper is as follows. Section 2 reviews some related work on earlier research efforts on ICT innovation for AAL. Section 3 details the multi-layered OSGi-based middleware infrastructure proposed by ZAINGUNE. Section 4 explains the interaction methods offered by our platform. Section 5 reviews the intelligence behaviour though reasoning offered by ZAINGUNE. Finally, section 6 concludes the paper and details some future work plans.

2 Related Work

There have been quite a few attempts [3][6][12] to create middleware which aims to simplify intelligent environment deployment, configuration and management. Among them, quite a few have considered using OSGi [13], since it is a software infrastructure which ensures platform and producer independence and easy programmability. It provides an execution environment and service interfaces to allow the discovery and dynamic cooperation of heterogeneous devices and services to guarantee evolution and external connectivity, allowing us remote control, diagnosis and management. This feature set explains why OSGi is emerging as a de facto industry standard for gateways that monitor, control and coordinate the heterogeneous devices (sensor and actuators) dynamically deployed in any kind of environment (house, factory or car). For all these reasons, we have also decided to base our middleware contributions for AAL in OSGi.

Other researchers have noticed the importance of adding reasoning capabilities to OSGi servers in order to make them more suitable for AmI environment management. A current trend of this regard is to adopt OWL-based [15] semantic ontologies [2] as knowledge repositories upon which reasoning takes place. We have taken this very same approach in a project related to ZAINGUNE named SmartLab [18], concluding that adopting semantic technologies to model and reason upon context is a very powerful mechanisms, however, it does impose high processing power demands and does not adjust, for the time being, to the real-time response requirements of AAL environments. For that, in ZAINGUNE we have adopted a traditional non-semantic more responsive and efficient-computationally rule-based engine, i.e. JBoss Rules [16], appropriate for the reactive intelligence of our system targeted to be deployed in a real house in summer 2008.

Asterisk [5] is an open source/free software implementation of a telephone private branch exchange (PBX) originally created in 1999 by Mark Spencer of Digium. Like any PBX, it allows a number of attached telephones to make calls to one another, and to connect to other telephone services including the public switched telephone network (PSTN). The basic Asterisk software includes many features available in proprietary PBX systems: voice mail, conference calling, interactive voice response (phone menus), and automatic call distribution. Users can create new functionality by writing dial plan. To attach ordinary telephones to a Linux server running Asterisk, or to connect to PSTN trunk lines, the server must be fitted with special hardware. Digium and a number of other firms sell PCI cards to attach telephones, telephone lines, T1 and E1 lines, and other analogue and digital phone services to a server. Asterisk supports a wide range of Voice over IP protocols, including SIP, MGCP and H.323. Asterisk can interoperate with most SIP telephones, acting both as registrar and as a gateway between IP phones and the PSTN. In ZAINGUNE we use the Asterisk technology to offer phone-mediated communication among home inhabitants and external users and phone-mediated interaction between home inhabitants and their instrumented intelligent environment.

Research activity in AAL has not reached its peak yet, however, it is envisaged it will do so in the forthcoming years since the European Union has just launched the 1st call for proposals under the Ambient Assisted Living Joint Programme [1]. Next we review some past AAL-related projects.

In the Gator Tech House [9], a work carried out by Florida University, a whole flat is instrumented with an assortment of coordinated, through an OSGi central server, assistive smart objects such as a smart mailbox which notifies letter arrival, a smart front door which enables to check who is outside the door and to remotely open it or a smart bathroom with a toilet paper sensor, a flush detector or a temperature regulating shower. The overall goal is to define a set of smart objects and services which populate an assistive home for elderly people.

The PAUL (Personal Assistant Unit for Living) system from University of Kaiserslautern [7] collects signals from motion detectors, wall switches or body signals, and interprets them to assist the user in his daily life but also to monitor his health condition and to safeguard him. The data is interpreted using fuzzy logic, automata, pattern recognition and neural networks. It is a good example of the application of artificial intelligence to create proactive assistive environments.

Some relevant projects funded by EU FP 6 and which cover areas related to ZAINGUNE were PERSONA (http://www.aal-persona.org/), CAALYX (http://www.caalyx.eu/), netcarity (http://www.netcarity.org/) or Soprano (http://www.soprano-ip.org/).

3 A Platform for AAL

The ZAINGUNE project aims to provide the software and hardware infrastructure necessary to intelligently control the automation elements, sensors and actuators placed at a home in order to provide assistive services for the elderly or disabled inhabitants of them in their daily activities. A multi-disciplinary team has been created to carry out this work composed of: a) a building automation engineering company expert on EIB/KNX [11] bus installations, namely TECDOA, b) a company expert on undertaking Asterisk-based VoIP deployments, namely IRONTEC, c) a public housing governmental company belonging to the Basque Government, namely VISESA, and d) a research centre specialized on defining context-aware intelligent middleware for AmI environments, namely Tecnológico Fundación Deusto.

ZAINGUNE proposes an OSGi-based middleware infrastructure powered by a rule-based reasoning engine which allows the coordination and cooperation of the sensing (presence detectors, door and window opening or flooding sensors) and actuation (lighting and environment control) devices attached to a KNX-EIB bus with the sensing mechanisms offered by a set of IP surveillance cameras and the control and alert VoIP-based telephony and messaging actuation mechanisms provided through an Asterisk installation. All these heterogeneous sensing and actuation tools have been integrated into a real environment, in this case a public housing flat targeted for renting to elderly or disabled people provided by VISESA in the city of Vitoria-Gasteiz.

The following sections describe the most remarkable ICT innovations for AAL brought forward by this project, which can be summarised as:

- A multi-layered middleware infrastructure offering an easy to program web service-based API and enabling the easy integration of heterogeneous sensing and actuation mechanisms, based on the OSGi standard.
- A rule-based engine which embodies the intelligence offered by an environment instrumented by the ZAINGUNE infrastructure. It enables the configuration of the services offered by an environment as a set of IF-THEN rules, rather than having to reprogram the logic for every specific environment or every change in such environments.
- An advanced control dashboard over a tactile screen based on the web-gadget metaphor easily accessible even by computer-illiterate users, such as most of current elderly people, which facilitates triggering of services both locally (in-house) or remotely (by family members, care centers or own users when they are away). Furthermore, a subset of the functionality accessible is also available through voice commands by means of Asterisk VoIP technology.
- A custom-built alert bracelet designed to accompany the user anywhere at any time within a house which offers an alert button through which emergency help can be sought and a screen through which relevant messages can be presented.

The ZAINGUNE OSGi-based infrastructure allows for the provision of intelligent home assistive services in areas such as security, prevention or energy efficiency, which are activated whenever the rules that model such behaviours trigger. Such environment behaviour rules encapsulate the intelligent reactivity of the environment and the conditions upon which such reactivity takes place.

4 A Middleware for AAL

The AAL-enabling middleware infrastructure devised in ZAINGUNE is set to meet four different goals:

- *Heterogeneous device support.* Enable the integration of heterogeneous sensing and actuating devices dynamically and in the most fault-tolerant manner.
- *Assistive Environment as a set of Cooperating Services.* Abstract away the functionality offered by different devices using different protocols and offering very diverse functionality in the form of services which are easily combined and composed to give place to higher level more complex services.
- *Programmability through a SOA-based approach.* Offer a web service-based API which is easily consumable by third parties ignorant of the inner hardware and communication details of the infrastructure deployed within a home.

- *Accessibility through an easy to user web gadget-based and secure front-end.* Enable the dynamic generation of web gadget front-ends representing the environmental services offered either by the hardware devices deployed or by the combination of the services offered by such devices. Authorization is required to access such front-end. RFID-based authentication is used locally whilst traditional user/password-based login, remotely, in order to access the ZAINGUNE-instrumented web front-end. The flows of data are communicated securely by enforcing the use of an SSL channel.

Internally, this middleware, as it is depicted by Fig. 1, is composed of the following three layers: a) hardware device layer, b) software platform layer and c) application environment layer.

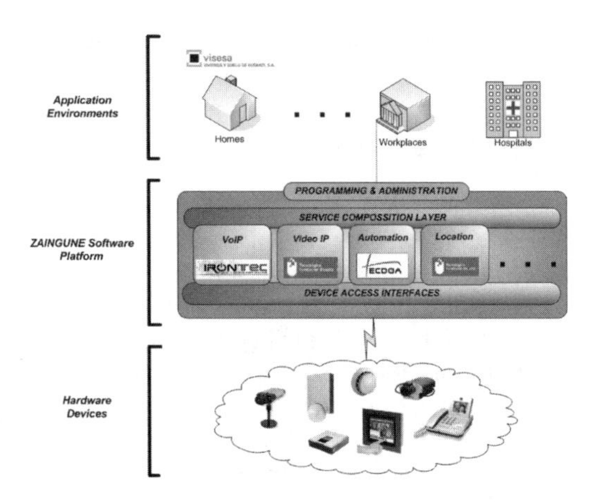

Fig. 1. Architecture of ZAINGUNE middleware infrastructure

The *hardware device layer* is composed of the sensors and actuators which populate an environment. ZAINGUNE combines the capabilities of all the devices in Layer 1 in order to offer intelligent services to the end user. Examples of the standard and proprietary (custom-built) devices used are: a) EIB/KNX devices such as flooding sensors, door/window opening sensors, light intensity actuators and so on; b) VoIP Asterisk-compatible handsets and telephony PC cards and PTZ-enabled PTZ cameras; c) indoor location systems such as Ubisense, KNX presence sensors or RFID readers; and d) our custom-built alert bracelet device which combines an organic screen (μOLED-96-G1 of 4D Systems) with a WSN mote based on Mica2DOT capable of displaying messages broadcasted by other nearby sensing motes.

The *software platform layer* transforms the functionality provided by the devices managed by Layer 1 into software services which can be combined to give place to more advanced services. It is hosted in an OSGi server which controls the heterogeneous sensing and actuating infrastructure mentioned. Every device within an AAL environment (home, residence, hospital) is encapsulated as a bundle or execution unit within our OSGi environment. These bundles are in charge of intermediating among the applications layer and the service composition service and the hardware devices to which commands are issued through communication interfaces. Examples of the bundles created are: a) the EIB/KNX bundle which communicates through a BCU (Bus Coupling Unit) USB interface and using the KNXnet IP to KNX tunnelling server with an environment's KNX devices; b) the Video IP bundle which supports the control protocol used by D-Link DCS-5300G wireless PTZ surveillance cameras; c) the VoIP bundle which through an Ambient MD3200 card uses Asterisk software to communicate with the telephony system; d) the Ubisense bundle which connects with the Ubisense web service deployed in the machine where the Ubisense software server is located; e) the RFID bundle which enables RFID tag reading from OBID classic-pro 13.56 MHz mifare® RFID; and f) the alert bracelet bundle which controls an IP to WSN gateway in order to receive and send from/to data to our own users' alert bracelets.

Apart from the mentioned modules, several high level services have been defined which combine the functionality in the form of services associated directly to devices in the environment. Thus, high level services resulting from the combination of several basic services are produced. Examples of such are our hybrid location service merging RFID-based location details with the ubisense accurate location system or the voice synthesis and call management Asterisk-based service.

Finally, the software platform layer includes two essential components responsible of both managing the behaviour of all the deployed OSGi services and giving HTTP access to such functionality.

- **ZainguneController** – it is the core component of the internal ZAINGUNE server architecture. It manages and controls access to the components in the form of OSGi bundles which are supported by ZAINGUNE. The brain of such component is the Zaingune Rule Reasoner, based on the JBoss Rules open source system.
- **ZainguneServlet –** it behaves as an Web Service/OSGi gateway exporting the OSGi bundle functionality through Web Services. Furthermore, this component does not only allow third party applications to access ZAINGUNE internal functionality through SOAP but it also generates advanced web gadgets associated both to the basic (directly-mapped to devices) and advanced services (composition of several basic and advanced services). Fig. 2 shows the ZAINGUNE dashboard application depicting the gadgets generated by this component. It makes use of small size web gadgets in order to be accessible from different clients such as web browsers, tactile screens or mobile devices. Developed over the JavaScript library X (http://www.cross-browser.com/), it is completely modular, easing the creation and incorporation of new user interfaces for the management and control of new devices connected to the system.

The last layer, *applications environment layer*, includes all the possible application scenarios for the ZAINGUNE infrastructure. In our case, this infrastructure has been applied in offering assistive services in a public housing flat targeted to elderly or disabled people. However, the web service API offered by ZAINGUNE could be used to provide AAL services in hospitals, care centres or residences.

5 Multi-modal Environment Interaction

ZAINGUNE offers several mechanisms for users (and administrators) to act both locally and remotely over the environment:

- **Web gadget-based interaction.** This is the default recommended method of interacting with a ZAINGUNE-enhanced environment. An easy to use interface, namely Environment Controller, based on web gadgets with meaningful icons and big buttons enables the selection of the service category on which operations want to be carried out, namely *Help*, *Communications*, *Home* and *Surveillance*, and the services associated to such categories. Next, we explain the functionality included in each of those categories:
  - *Help*: a simple icon can be pressed by a user in order to seek help both from his family or a care centre. As a result of such interaction an automatic phone call is performed to a set of preconfigured phone numbers to which a prerecorded risk alert message is delivered.
  - *Communication Management*. If offers the user several mechanisms to communicate with his relatives and friends. The default selected option if "call by photo" (see right hand side of Fig. 2), where an elderly person does not have to remember any phone number, it only needs to click on the photo of the person she wants to talk to in order to initiate a call to that person. On the other hand, more computer literate users may also check their email or send SMS message through this service category.
  - *Home control*. Selecting this category the user is presented with a list of the different rooms of her home. Selecting a room, all the devices and services offered within that room are offered to the user. As Fig. 2 shows, aspects such as obtaining the current temperature, video surveillance of the room, door or window opening control or lighting luminance control can be undertaken from this section.
  - *Surveillance control*. Through this category a user can have a look to the images captured by all the IP cameras deployed within a home. This can be useful for an inhabitant to check who is outside her flat and decide whether to open the door or not. Besides, it can also be useful for relatives or care centres authorized to supervise the house inhabitants.
- **Phone touchpad-based and voice-based interaction.** Both a remote and local user can interact with the environment by using preset keystroke configurations to control the environment elements. The same commands can also be issued through voice. This interaction mechanism is possible thanks to the integration of the Asterisk VoIP system with ZAINGUNE infrastructure. Besides, as a side

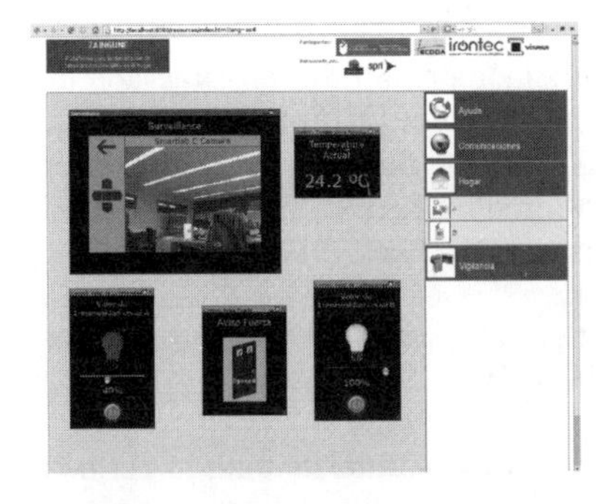

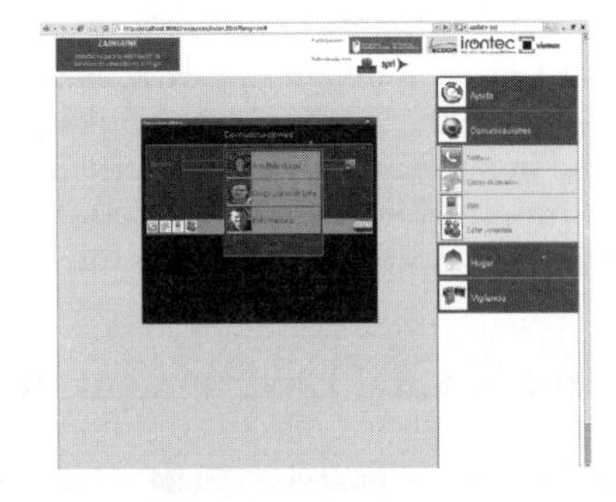

Fig. 2. ZAINGUNE Environment Controller (left) and "call by photo" (right)

effect of the wide deployment of VoIP phones over the house, we use their speakers to offer vocal feedback to users when alert situations take place.

- **Alert bracelet-based interaction.** Every inhabitant of a ZAINGUNE instrumented environment may carry an alert bracelet as the one shown in Fig. 3. This rather restricted custom-built device has been designed for only one purpose, i.e. assistance seeking or alert notification. Such bracelet could be improved by adding living signal monitoring sensors.

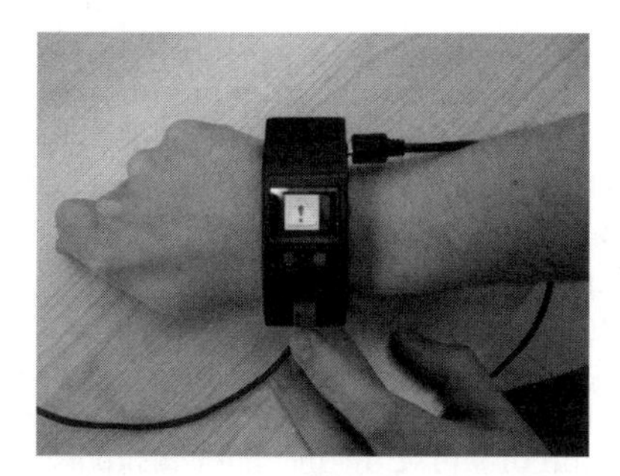

Fig. 3. ZAINGUNE Alert Bracelet

6 Intelligence through Rule-Based Reasoning

One of the most remarkable features of ZAINGUNE's middleware infrastructure is the provision of a rule-based reasoning engine. The adoption of this engine offers two main advantages: a) it decouples environment configuration from middleware programmability. A user can simply reconfigure the behaviour of the environment by defining new rules, without having to change the source code, and b) it enables environment-initiated proactive reactions, giving answer to the user's current needs without always requiring their explicit request by any of above three mentioned interaction mechanisms.

Environment intelligence is encapsulated as a set of rules which trigger when certain sensorial situations are matched. For instance, if the flooding sensor deployed in the house bathroom is activated, the water source of the home could automatically be closed. In essence, in the rule-based paradigm used the left-hand-side part of the rules represents sensing conditions whilst the right-hand-side depicts actions to be undertaken when the left-hand-side situations are matched. These rules consist on the correlation of events coming from the sensorial infrastructure connected to the EIB/KNX bus and the set of IP cameras and VoIP telephones distributed though a home to determine the occurrence of certain situations and the activation, as a consequence, of a set of actions over the EIB or VoIP-based actuators deployed at home.

This rule-based paradigm is employed to configure the reactive behaviour of a ZAINGUNE-controlled environment. Some examples of the assistive reactions possible through ZAINGUNE, in our AAL application domain are: a) *efficient management of energy resources*, b) *security at home* or c) *danger situation prevention.*

7 Conclusions and Further Work

This paper has described several ICT infrastructure contributions to enable an Ambient Assisted Living (AAL) environment. Research efforts such as ZAINGUNE should lead to the progressive adoption of ICT at elderly people's homes so that they can stay active and productive for longer; to continue to engage in society with more accessible online services; and to enjoy a healthier and higher quality of life for longer.

The main outcome of this work is an OSGi application server powered by a rule-based reasoning engine which integrates a KNX/EIB automation bus together with VoIP (through Asterisk) and VideoIP infrastructure in order to configure home environments which are aware and reactive to the special needs of people inhabiting them. Furthermore, different interaction mechanisms to request services from such application server are proposed such as a touch screen-based web gadget-based dashboard, an alert bracelet or VoIP phone-mediated interaction.

Future areas of work in ZAINGUNE will be: a) the development of a new low-cost easily deployable indoor location system based on WSN technology; b) the creation of a wizard through which the ZAINGUNE server can be parameterised for a new environment, c) the integration of workflow mechanisms to assemble complex advanced services from basic ones without requiring code modification and d) the evaluation of our infrastructure in a real home deployment.

Acknowledgments. Thanks to the Industry, Commerce and Tourism Department of Basque Government for sponsoring this work through grant IG-2007/00211.

References

1. Ambient Assisted Living Joint Programme (2008), `http://www.aal-europe.eu`
2. Chen, H.: An Intelligent Broker Architecture for Pervasive Context-Aware Systems. Ph.D thesis, University of Maryland, Baltimore County (2004)
3. Dey, A.K.: Providing Architectural Support for Building Context-Aware Applications, Ph.D thesis, Georgia Institute of Technology (2000)
4. Communication from the Commission to the European Parliament, the Council, the European Economic and Social Committee and the Committee of the Regions.: Ageing well in the Information Society - an i2010 Initiative - Action Plan on Information and Communication Technologies and Ageing {SEC(2007)811}
5. Digium Inc.: Asterisk: The Open Source PBX & Telephony Platform (2008), `http://www.asterisk.org/`
6. Edwards, W.K.: Discovery Systems in Ubiquitous Computing. IEEE Pervasive Computing 5, 70–77 (2006)
7. Floeck, M., Litz, L.: Integration of Home Automation Technology into an Assisted Living Concept. Assisted Living Systems - Models, Architectures and Engineering Approaches (2007)
8. Gu, T., Pung, H.K., Zhang, D.Q.: Toward an OSGi-Based Infrastructure for Context-Aware Applications. IEEE Pervasive Computing 3(4), 66–74 (2004)
9. Helal, A., Mann, W., Elzabadani, H., King, J., Kaddourah, Y., Jansen, E.: Gator Tech. Smart House: A Programmable Pervasive Space. IEEE Computer magazine, 64–74 (2005)

10. International newsletter on on micro-nano integration.: Ambient Assisted Living no. 6/07 (2007), http://mstnews.de
11. KNX Association.: KNX – Open Standard for Home and Building Control (2008), http://www.knx.org/
12. Lee, C., Nordstedt, D., Helal, S.: Enabling Smart Spaces with OSGi. IEEE Pervasive Computing 2(3), 89–94 (2003)
13. OSGi Alliance.: OSGi Alliance Home Site (2008), http://www.osgi.org/Main/HomePage
14. OSGi Event Admin. (2008), http://www2.osgi.org/javadoc/r4/org/osgi/service/event/EventAdmin.html
15. OWL Model Theory (2008), http://www.w3.org/TR/2002/WD-owl-semantics-20021108/
16. RedHat Inc.: JBoss Rules Home Site (2008), http://www.jboss.com/products/rules
17. Tecnológico Fundación Deusto.: ZAINGUNE Project Web Page (2008), http://www.tecnologico.deusto.es/projects/zaingune
18. Tecnológico Fundación Deusto.: SmartLab Project Web Page (2008), http://www.tecnologico.deusto.es/projects/smartlab

ALZ-MAS 2.0; A Distributed Approach for Alzheimer Health Care

Óscar García, Dante I. Tapia, Alberto Saavedra, Ricardo S. Alonso, and Israel García

R&D Department, Tulecom Group S.L.
Hoces del Duratón, 57, 37008, Salamanca, Spain
{oscar.garcia,dante.tapia,alberto.saavedra,ricardo.alonso,
israel.garcia}@tulecom.com

Abstract. This paper presents ALZ-MAS 2.0; an Ambient Intelligence based multi-agent system aimed at enhancing the assistance and health care for Alzheimer patients living in geriatric residences. The system makes use of FUSION@[1], a multi-agent architecture which facilitates the integration of distributed services and applications to optimize the construction of Ambient Intelligence environments. The architecture optimizes the development of distributed multi-agent systems, where applications and services can communicate in a distributed way, even from mobile devices, independent of a specific programming language or operating system. The results obtained demonstrate that ALZ-MAS 2.0 is far more robust and has better performance than the previous version.

Keywords: Multi-Agent Systems, Ambient Intelligence, Services Oriented Architectures, Health Care.

1 Introduction

The continuous technological advances have gradually surrounded people with devices and technology. It is necessary to develop intuitive interfaces and systems with some degree of intelligence, with the ability to recognize and respond to the needs of individuals in a discrete and often invisible way, considering people in the centre of the development to create technologically complex and intelligent environments.

Ambient Intelligence (AmI) is an emerging multidisciplinary area based on ubiquitous computing, which influences the design of protocols, communications, systems, devices, etc., proposing new ways of interaction between people and technology, adapting them to the needs of individuals and their environment [14]. It offers a great potential to improve quality of life and simplify the use of technology by offering a wider range of personalized services and providing users with easier and more efficient ways to communicate and interact with other people and systems [14] [5]. However, the development of systems that clearly fulfil the needs of AmI is difficult and not always satisfactory. It requires a joint development of models, techniques and technologies based on services. An AmI-based system consists on a set of human

[1] FUSION@ is an experimental multi-agent architecture developed by the BISITE Research Group at the University of Salamanca, Spain. For more information about FUSION@, visit http://bisite.usal.es/

J.M. Corchado, D.I. Tapia, and J. Bravo (Eds.): UCAMI 2008, ASC 51, pp. 76–85, 2009.
springerlink.com

actors and adaptive mechanisms which work together in a distributed way. Those mechanisms provide on demand personalized services and stimulate users through their environment according specific situation characteristics [14].

This paper describes ALZ-MAS 2.0 an Ambient Intelligence based multi-agent system aimed at enhancing the assistance and health care for Alzheimer patients living in geriatric residences. Unlike previous versions, ALZ-MAS 2.0 makes use of a *Flexible User and ServIces Oriented multi-ageNt Architecture* (FUSION@). This architecture presents important improvements in the area of Ambient Intelligence (AmI) [14]. One of the most important characteristics is the use of intelligent agents as the main components in employing a service oriented approach, focusing on distributing the majority of the systems' functionalities into remote and local services and applications. The architecture proposes a new and easier method of building distributed multi-agent systems, where the functionalities of the systems are not integrated into the structure of the agents, rather they are modelled as distributed services and applications which are invoked by the agents acting as controllers and coordinators.

Agents have a set of characteristics, such as autonomy, reasoning, reactivity, social abilities, pro-activity, mobility, organization, etc. which allow them to cover several needs for Ambient Intelligence environments, especially ubiquitous communication and computing and adaptable interfaces. Agent and multi-agent systems have been successfully applied to several Ambient Intelligence scenarios, such as education, culture, entertainment, medicine, robotics, etc. [5] [12] [14]. The characteristics of the agents make them appropriate for developing dynamic and distributed systems based on Ambient Intelligence, as they possess the capability of adapting themselves to the users and environmental characteristics [9]. The continuous advancement in mobile computing makes it possible to obtain information about the context and also to react physically to it in more innovative ways [9]. The agents in this architecture are based on the deliberative Belief, Desire, Intention (BDI) model [10] [3] [11], where the agents' internal structure and capabilities are based on mental aptitudes, using beliefs, desires and intentions [2] [8]. Nevertheless, Ambient Intelligence developments need higher adaptation, learning and autonomy levels than pure BDI model [3]. This is achieved in FUSION@ by modelling the agents' characteristics [15] to provide them with mechanisms that allow solving complex problems and autonomous learning.

In the next section, the problem description and the main characteristics of FUSION@ are briefly presented. Section 3 describes the basic components of ALZ-MAS 2.0 and shows how FUSION@ has been used to distribute its functionalities. Finally section 4 presents the results and conclusions obtained.

2 FUSION@, an Alternative for Distributed Multi-agent Systems

The development of AmI-based software requires creating increasingly complex and flexible applications, so there is a trend toward reusing resources and share compatible platforms or architectures. In some cases, applications require similar functionalities already implemented into other systems which are not always compatible. At this point, developers can face this problem through two options: reuse functionalities already implemented into other systems; or re-deploy the capabilities required, which means more time for development, although this is the easiest and safest option in most cases. While the first option is more adequate in the long run, the second one is most chosen by developers, which leads to have replicated functionalities as well as greater difficulty in migrating systems and applications. Moreover, the absence of a strategy for integrating applications generates multiple points of failure that can affect the systems' performance. This is a poorly scalable and flexible model with reduced response to change, in which applications are designed from the outset as independent software islands.

Ambient Intelligence plays an important role in FUSION@. It has been designed to facilitate the development of distributed multi-agent systems with high levels of human-system-environment interaction, since agents have the ability to dynamically adapt their behaviour at execution time. It also provides an advanced flexibility and customization to easily add, modify or remove applications or services on demand, independently of the programming language.

FUSION@ formalizes four basic blocks: Applications, which represent all the programs that can be used to exploit the system functionalities. They can be executed locally or remotely, even on mobile devices with limited processing capabilities, because computing tasks are largely delegated to the agents and services; An Agents Platform as the core of FUSION@, integrating a set of agents, each one with special characteristics and behaviour. These agents act as controllers and administrators for all applications and services, managing the adequate functioning of the system, from services, applications, communication and performance to reasoning and decision-making; Services, which are the bulk of the functionalities of the system at the processing, delivery and information acquisition levels. Services are designed to be invoked locally or remotely; and finally a Communication Protocol which allows applications and services to communicate directly with the Agents Platform. The protocol is based on SOAP specification and it is completely open and independent of any programming language [4].

These blocks are managed by means of pre-defined agents which provide the basic functionalities of FUSION@: CommApp Agent is responsible for all communications between applications and the platform; CommServ Agent is responsible for all communications between services and the platform; Directory Agent manages the list of services that can be used by the system; Supervisor Agent supervises the correct functioning of the other agents in the system; Security Agent analyzes the structure and syntax of all incoming and outgoing messages; Manager Agent decides which agent must be called by taking into account the services performance and users preferences; Interface Agents are designed to be embedded in users' applications. Interface agents communicate directly with the agents in FUSION@ so there is no need to employ the communication protocol, rather the FIPA ACL specification.

FUSION@ also facilitates the inclusion of context-aware technologies that allow systems to automatically obtain information from users and the environment in an evenly distributed way, focusing on the characteristics of ubiquity, awareness, intelligence, mobility, etc., all of which are concepts defined by Ambient Intelligence. The goal in FUSION@ is not only to distribute services and applications, but to also promote a new way of developing AmI-based systems focusing on ubiquity and simplicity.

In the next section, ALZ-MAS 2.0 is presented, where FUSION@ has helped to distribute most of its functionalities and re-design a completely functional multi-agent system aimed at improving several aspects of dependent people.

3 ALZ-MAS 2.0

Ambient Intelligence based systems aim to improve quality of life, offering more efficient and easy ways to use services and communication tools to interact with other people, systems and environments. Among the general population, those most likely to benefit from the development of these systems are the elderly and dependent persons, whose daily lives, with particular regard to health care, will be most enhanced [5] [6] [13]. Dependent persons can suffer from degenerative diseases, dementia, or loss of cognitive ability [7]. In Spain, dependency is classified into three levels [7]: Level 1 (moderated dependence) refers to all people that need help to perform one or several basic daily life activities, at least once a day; Level 2 (severe dependence) consists of people who need help to perform several daily life activities two or three times a day, but who do not require the support of a permanent caregiver; and finally Level 3 (great dependence) refers to all people who need support to perform several daily life activities numerous times a day and, because of their total loss of mental or physical autonomy, need the continuous and permanent presence of a caregiver.

FUSION@ has been employed to develop an improved version of ALZ-MAS (*ALZheimer Multi-Agent System*) [5] [6], a multi-agent system aimed at enhancing the assistance and health care for Alzheimer patients living in geriatric residences. ALZ-MAS 2.0 is a distributed system designed upon Ambient Intelligence. The main functionalities in the system include reasoning and planning mechanisms that are embedded into deliberative BDI agents, and the use of several context-aware technologies to acquire information from users and their environment. In the remainder of this section, the main characteristics of the previous version of ALZ-MAS are described, followed by a description of the new ALZ-MAS 2.0 system developed by means of the FUSION@ architecture.

As can be seen on Figure 1, ALZ-MAS structure has five different deliberative agents based on the BDI model (BDI Agents), each one with specific roles and capabilities:

- User Agent. This agent manages the users' personal data and behaviour (monitoring, location, daily tasks, and anomalies). The User Agent beliefs and goals applied to every user depend on the plan or plans defined by the super-users. User Agent maintains continuous communication with the rest of the system agents, especially with the ScheduleUser Agent (through which the scheduled-users can communicate the result of their assigned tasks) and with the SuperUser Agent. The User Agent must ensure that all the actions indicated by the SuperUser are carried out, and sends a copy of its memory base (goals and plans) to the Admin Agent in order to maintain backups. There is one agent for each patient registered in the system.
- SuperUser Agent. It also runs on mobile devices (PDA) and inserts new tasks into the Manager Agent to be processed by a Case-Based Reasoning mechanism. It also

needs to interact with the User Agents to impose new tasks and receive periodic reports, and with the ScheduleUser Agents to ascertain the evolution of each plan. There is one agent for each doctor connected to the system.

- ScheduleUser Agent. It is a BDI agent with a Case-Based Planning (CBP) mechanism embedded in its structure. It schedules the users' daily activities and obtains dynamic plans depending on the tasks needed for each user. It manages scheduled-users profiles (preferences, habits, holidays, etc.), tasks, available time and resources. Every agent generates personalized plans depending on the scheduled-user profile. There is one ScheduleUser Agents for each nurse connected to the system.
- Admin Agent. It runs on a Workstation and plays two roles: the security role that monitors the users' location and physical building status (temperature, lights, alarms, etc.) through continuous communication with the Devices Agent; and the manager role that handles the databases and the task assignment. It must provide security for the users and ensure the efficiency of the tasks assignments. There is just one Manager Agent running in the system.
- Devices Agent. This agent controls all the hardware devices. It monitors the users' location (continuously obtaining/updating data from sensors), interacts with sensors and actuators to receive information and control physical services (temperature, lights, door locks, alarms, etc.), and also checks the status of the wireless devices connected to the system (e.g. PDAs). The information obtained is sent to the Manager Agent for processing. This agent runs on a Workstation. There is just one Devices Agent running in the system.

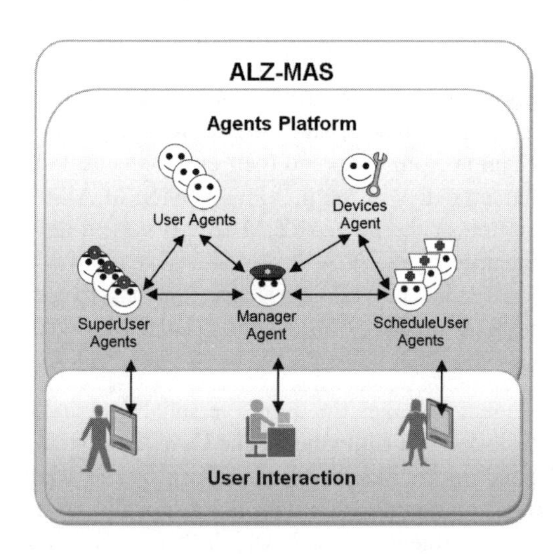

Fig. 1. ALZ-MAS basic structure

In the initial version of ALZ-MAS, each agent integrates its own functionalities into their structure. If an agent needs to perform a task which involves another agent, it must communicate with that agent to request it. So, if the agent is disengaged, all its functionalities will be unavailable to the rest of agents. This has been an important issue in ALZ-MAS, since agents running on PDAs are constantly disconnecting from the platform and consequently crashing, making it necessary to restart (killing and launching new instances) those agents. Another important issue is that robust reasoning mechanisms [1] are integrated into the agents. These mechanisms are busy almost all the time, overloading the respective agents. Because these mechanisms are essential in the system, they must be available at all times. The system depends on these mechanisms to generate all decisions, so it is essential that they have all processing power available in order to increase overall performance. In the new version of ALZ-MAS, these mechanisms have been modelled as services, so any agent can make use of them.

As seen on Figure 2, the entire ALZ-MAS structure has been modified according to FUSION@ model, separating most of the agents' functionalities from those to be modelled as services. However, all functionalities are the same in both approaches, since we have considered it appropriated to compare the performance of both systems to prove the efficiency of FUSION@ model.

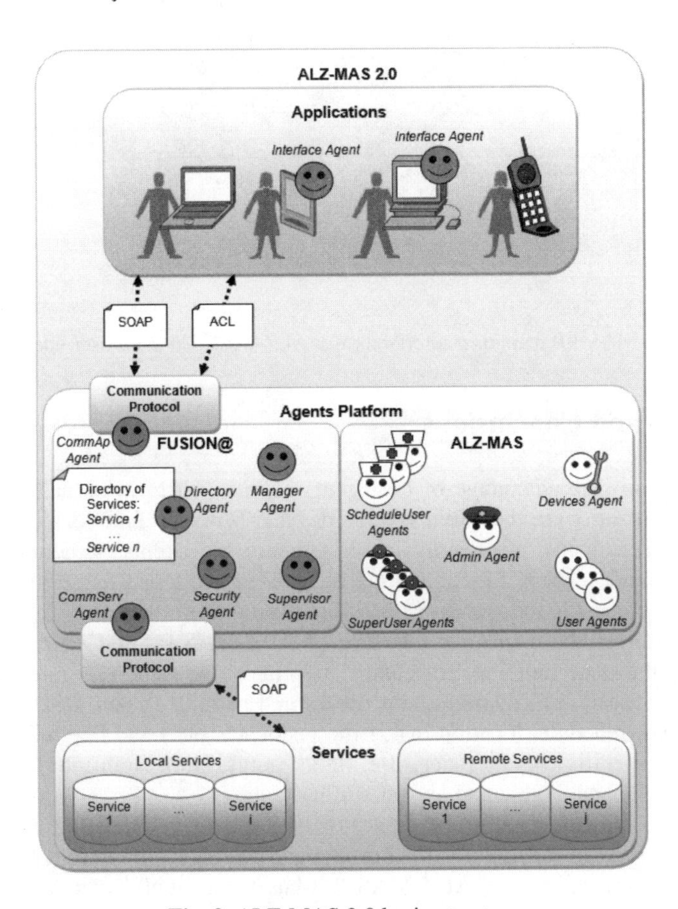

Fig. 2. ALZ-MAS 2.0 basic structure

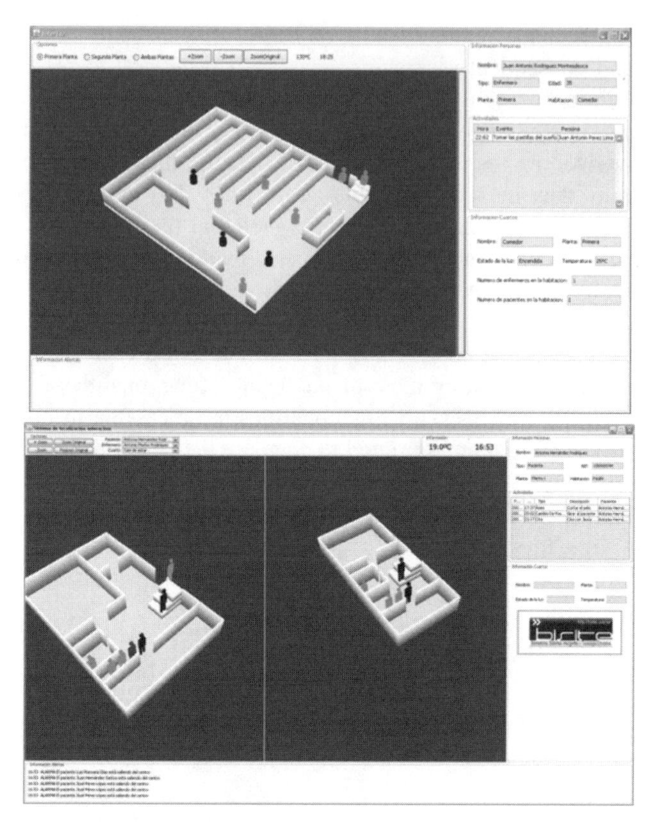

Fig. 3. ALZ-MAS 1.0 main user interface (up); ALZ-MAS 2.0 main user interface (down)

4 Results and Conclusions

FUSION@ can take advantage of the agents characteristics in order to enhance the services organization. In the initial version of ALZ-MAS, all functionalities, including reasoning mechanisms, are deeply integrated into the agents' structure. Following FUSION@ model, these functionalities have been modelled as services linked to agents, thus increasing the system's overall performance.

Figure 3 shows the main user interface of ALZ-MAS (a) and ALZ-MAS 2.0 (b), which have the same functionalities and share almost the same user interface. The interfaces show basic information about nurses and patients (name, tasks that must be accomplished, schedule, location inside the residence, etc.) and the building (outside temperature, specific room temperature, lights status, etc.). Both interfaces are managed by the Manager Agent and appear similar to users.

One of the most demanding processes in ALZ-MAS is a planning mechanism [5] [6]. This mechanism schedules (i.e. plan) the activities for all nurses connected to the system. The performance of ALZ-MAS 2.0 has been highly improved, mainly because most of the functionalities, including the planning mechanism, have been modelled as Services.

To generate a new plan, a ScheduleUser Agent (running on a PDA) sends a request to the Agents Platform. The message is processed by the Manager Agent which decides to use the planner service (PlanServ). The platform invokes the planner service. PlanServ receives the message and starts to generate a new plan. Then, the solution is sent to the platform which delivers the new plan to all ScheduleUser Agents running. The planner service (i.e. planning mechanism) creates optimal paths and scheduling in order to facilitate the completion of all activities defined for the nurses connected to the system.

Several tests have been done to compare the overall performance of ALZ-MAS and ALZ-MAS 2.0, the latter making use of FUSION@. The tests consisted of a set of requests delivered to the planning mechanism which in turn had to generate paths for each set of tasks (i.e. scheduling). A task is a java object that contains a set of parameters (TaskId, MinTime, MaxTime, ScheduleTime, UserId, etc.). ScheduleTime is the time in which a specific task must be accomplished, although the priority level of other tasks needing to be accomplished at the same time is factored in. The planning mechanism increases or decreases ScheduleTime and MaxTime according to the priority of the task: ScheduleTime=ScheduleTime-5min*TaskPriority and MaxTime=MaxTime+5min*TaskPriority.

Once these times have been calculated, the path is generated taking the RoomCoordinates into account. There were 30 defined agendas each with 50 tasks. Tasks had different priorities and orders on each agenda. Tests were carried out on 7 different test groups, with 1, 5, 10, 15, 20, 25 and 30 simultaneous agendas to be processed by the planning mechanism. 50 runs for each test group were performed, all of them on machines with equal characteristics. Several data have been obtained from these tests, notably the average time to accomplish the plans, the number of crashed agents, and the number of crashed services. For ALZ-MAS 2.0 five planner services with exactly the same characteristics were replicated in the same workstation on which the system was running.

Figure 4 (left) shows the average time needed by both systems to generate the paths for a fixed number of simultaneous agendas. The previous version of ALZ-MAS was unable to handle 15 simultaneous agendas and time increases to infinite because it was impossible to perform those requests. However, ALZ-MAS 2.0 had 5 replicated services available, so the workflow was distributed and allowed the system to complete the plans for 30 simultaneous agendas. Another important data is that

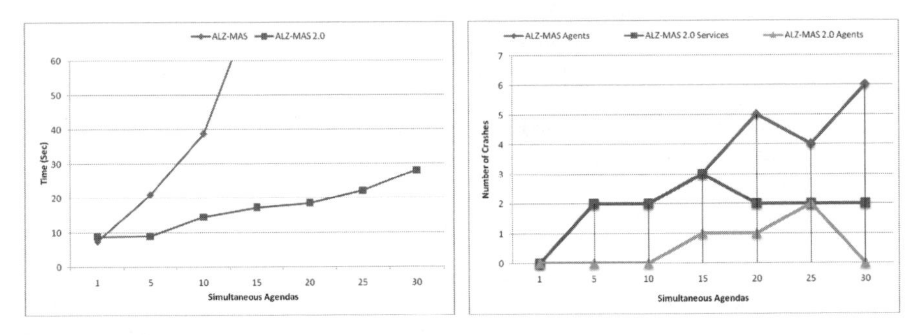

Fig. 4. Time needed for both systems to generate paths for a group of simultaneous agendas (left); Number of agents and services crashed at both versions of ALZ-MAS (right)

although the previous version of ALZ-MAS performed slightly faster when processing a single agenda, performance was constantly reduced when new simultaneous agendas were added. This fact demonstrates that the overall performance of ALZ-MAS 2.0 is better when handling distributed and simultaneous tasks (e.g. agendas), instead of single tasks.

Figure 4 (right) shows the number of crashed agents and services for both versions of ALZ-MAS during tests. None of the tests where agents or services crashed were taken into account to calculate data, so these tests were repeated. As can be seen, the previous version of ALZ-MAS is far more unstable than ALZ-MAS 2.0, especially when comparing the number of crashed agents. These data demonstrate that this approach provides a higher ability to recover from errors.

Although these tests have provided us with very useful data, it is necessary to continue developing and enhancing the system presented. Results also demonstrate that FUSION@ is adequate for building complex systems and exploiting composite services, in this case ALZ-MAS 2.0. Future work consists on improving ALZ-MAS 2.0 functionalities and performance, as well as applying it into a real case scenario.

Acknowledgments. This work has been partially supported by the FIT-350300-2007-84 project.

References

1. Aamodt, A., Plaza, E.: Case-Based Reasoning: foundational Issues, Methodological Variations, and System Approaches. AI Communications 7, 39–59 (1994)
2. Bratman, M.E.: Intentions, plans and practical reason. Harvard University Press, Cambridge (1987)
3. Bratman, M.E., Israel, D., Pollack, M.E.: Plans and resource-bounded practical reasoning. Computational Intelligence 4, 349–355 (1988)
4. Cerami, E.: Web Services Essentials Distributed Applications with XML-RPC, SOAP, UDDI & WSDL, 1st edn. O'Reilly & Associates, Inc., Sebastopol (2002)
5. Corchado, J.M., Bajo, J., De Paz, Y., Tapia, D.I.: Intelligent Environment for Monitoring Alzheimer Patients. In: Agent Technology for Health Care. Decision Support Systems. Eslevier, Amsterdam (in press, 2008)
6. Corchado, J.M., Bajo, J., Abraham, A.: GERAmI: Improving the delivery of health care. IEEE Intelligent Systems, Special Issue on Ambient Intelligence 23(2), 19–25 (2008)
7. Costa-Font, J., Patxot, C.: The design of the long-term care system in Spain: Policy and financial constraints. Social Policy and Society 4(1), 11–20 (2005)
8. Georgeff, M., Rao, A.: Rational software agents: from theory to practice. In: Jennings, N.R., Wooldridge, M.J. (eds.) Agent Technology: Foundations, Applications, and Markets. Springer, New York (1998)
9. Jayaputera, G.T., Zaslavsky, A.B., Loke, S.W.: Enabling run-time composition and support for heterogeneous pervasive multi-agent systems. Journal of Systems and Software 80(12), 2039–2062 (2007)
10. Jennings, N.R., Wooldridge, M.: Applying agent technology. Applied Artificial Intelligence 9(4), 351–361 (1995)

11. Pokahr, A., Braubach, L., Lamersdorf, W.: Jadex: Implementing a BDI-Infrastructure for JADE Agents. In: EXP - in search of innovation (Special Issue on JADE), Department of Informatics, University of Hamburg, Germany, pp. 76–85 (2003)
12. Schön, B., O'Hare, G.M.P., Duffy, B.R., Martin, A.N., Bradley, J.F.: Agent Assistance for 3D World Navigation. LNCS, vol. 1, p. 499. Springer, Heidelberg (1973)
13. van Woerden, K.: Mainstream Developments in ICT: Why are They Important for Assistive Technology? Technology and Disability 18(1), 15–18 (2006)
14. Weber, W., Rabaey, J.M., Aarts, E.: Ambient Intelligence. Springer, New York (2005)
15. Wooldridge, M., Jennings, N.R.: Intelligent Agents: Theory and Practice. The Knowledge Engineering Review 10(2), 115–152 (1995)

Ambient Assisted Living

Ricardo Costa[1], Davide Carneiro[2], Paulo Novais[2], Luís Lima[1], José Machado[2], Alberto Marques[3], and José Neves[2]

[1] College of Management and Technology - Polytechnic of Porto, Rua do Curral – Casa do Curral, 4610-156 Felgueiras, Portugal
[2] Departamento de Informática/CCTC, Universidade do Minho, Campus de Gualtar, 4710-553 Braga, Portugal
[3] Centro Hospitalar do Tâmega e Sousa, EPE, Penafiel, Portugal
rfc@estgf.ipp.pt, pg10906@alunos.uminho.pt, pjon@di.uminho.pt, lcl@estgf.ipp.pt, jmac@di.uminho,pt, marquestris@chts.min-saude.pt, jneves@di.uminho.pt

Abstract. The quality of care practice is difficult to judge. Indeed, support and care provision is very personal, i.e., assessments are individual and lead to specific care packages, involving social services, health workers, care agencies. We expect privacy in our own affairs and confidentially from those to whom we disclose them. Therefore, we are in an urgent need for new, technological and formal approaches to problem solving, as the increase of population with special care requirements. Following this line of thought, it is one's goal to present the VirtualECare framework, an intelligent multi-agent system able to monitor, interact and serve its customers, which are in need of care services, based in open standards, expecting not only to fulfil the objectives referred to above, but also to overcome the problems induced by the use of new technologies and formalisms.

Keywords: Ambient Intelligence, Group Decision Support Systems, Virtual Organizations, Assisted Living.

1 Introduction

Healthcare costs are rising, but the quality-of-service is often poor or simply non-existent, as it is the lack of admission and choice of services. Definitely, some changes have to be imposed, where the use of new technologies and practices is inevitable (e.g. patients with chronic diseases may have to be empowered). Indeed, the use of Information Technology (IT) based systems in disease management make the patients to be in control of the situation, leading to a lowering of the healthcare costs. Under this scenario, their treatment will take place at home, through the use of technologies and appliances that will cater for real time monitoring and evaluation of critical data, triggering alarms and making recommendations, in case of necessity. Family and relatives will also be empowered, as they will have access, in real time, to the collected information [1].

Therefore, it will come not as a surprise, when we perceive that IT healthcare based systems have been developed or are in development, following this line of thought [2-8]. However, some are simple "panic buttons", others use domotics smart

J.M. Corchado, D.I. Tapia, and J. Bravo (Eds.): UCAMI 2008, ASC 51, pp. 86–94, 2009.
springerlink.com

sensors and, the most advanced ones, focus the development on context-aware interfaces for older people or "smart-homes" for aging-in-place. There are also generic frameworks aimed to integrate the systems referred to above, being one the Telecare.

1.1 Ambient Intelligence

Ambient Intelligence (AmI) is a relatively new paradigm in Information Technology (IT), in which people is empowered through a digital environment that is not only aware of their presence, but sensitive, adaptive, and responsive to their needs, habits, gestures and passions [6]. AmI is built on advanced networking technologies, shaped by a broad range of bits and pieces (e.g., mobile devices). Adding adaptive user-system interaction methods, based on new insights in the way people interact with computing devices (e.g., social user interfaces), digital environments may be twisted in order to improve quality-of-living of people by acting on their behalf. These context aware systems combine ubiquitous information, communication, and entertainment, with enhanced personalization, natural interaction and intelligence. The path to pursue, in order to achieve this goal, relies on a mix of different receptiveness from disciplines as Artificial Intelligence, Psychology or Mathematical Logic, coupled with different computational paradigms and methodologies for problem solving, such as the conceptualization of figures (e.g., the one of software agent), and its social counterpart in the form of Group Decision Support Systems environments [10, 11, 12]. Thus, with respect to the interaction of the different actors, the cooperation elements can be established as productions of a kind of common language; i.e., their social relationship, norms, values and institutions, will be set, in this work, in terms of an extension to the logic programming language, being their knowledge bases built as logical theories that found their foundations on this extension [13]. Conclusions are supported by deductive proofs, or by arguments that include conjectures and motivate new topics of inquiry, i.e., if deduction is fruitless, the agent inference engine resorts to abduction, filling in missing pieces of logical arguments with plausible conjectures to obtain answers that are only partly supported by the facts available.

2 Business Integration for Healthcare

Our objective is to present an intelligent multi-agent system not only to monitor and to interact with its costumers (being those elderly people or their relatives), but also to be interconnected to other computing systems running in different healthcare institutions, leisure centers, training facilities or shops. The VirtualECare [9] architecture is a distributed one, being their components unified through a network (e.g., LAN, MAN, WAN), and each one with a different role (Fig. 1).

SupportedUser* –** Elderly people with special healthcare needs, whose critical data is sent to the ***CallCareCenter and forwarded to the ***Group Decision Supported System***;

Home* – *SupportedUser natural premises. The data collected here is sent to the ***Group Decision Supported System*** through the ***CallCareCenter,*** or to the ***CallServiceCenter*** (which speak for themselves);

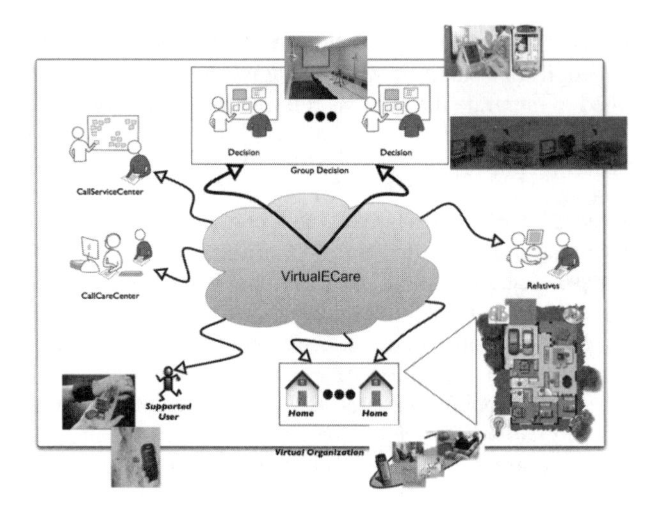

Fig. 1. The System Architecture

Group Decision **–** It is in charge of all the decisions taken at the VirtualECare platform. Our work will be centred on this key module;

CallServiceCenter **–** Entity with all the necessary computational and qualified personal resources, capable of receiving and analyze the miscellaneous data and take the necessary actions according to it;

CallCareCenter **–** Entity in charge of the computational and qualified personal resources (i.e., healthcare professionals and auxiliary personnel), capable of receiving and analyze the clinical data, and to take the necessary actions.

Relatives SupportedUser - Relatives which may and should have an active role in the supervising task of their love ones, providing precious complementary information (e.g., loneliness).

In order to the Group Decision Support System take their decisions, one needs of a digital profile of the SupportedUser, which may provide a better understanding of his/her special needs. In this profile we may have different types of data, ranging from the patient Electronic Health Record to their own personal experiences and preferences (e.g., musical, gastronomic). It will provide tools and methodologies for creating an information-on-demand environment that can improve quality-of-living, safety, and quality-of-patient care.

3 OSGi

OSGi is an initiative that intends to establish standards in Java programming, highly specific, catering for the sharing of Java classes, that may be achieved in terms of a services platform paradigm [14, 15]. Indeed, in order to perform tasks on the user's behalf, we need devices and applications able to detect the user's presence (being it physically or not), which may lead us to the so-called context-aware systems. These kinds of systems take advantage of the emerging pervasive computing solutions to be able to provide "anytime, anywhere" computing, which has attracted much attention from researchers in recent years in order to demonstrate the technology's usefulness. However, building context-aware applications is relatively complex and requires an adequate infrastructure to support a generic, platform independent framework.

When adapting the proposed architecture to fit the OSGi specifications we are faced up to some challenge, being the most obvious "how to make the agents, used in some components of the architecture, OSGi compatible (since some of our components are agent based), and how to use OSGi in our distributed architecture (since OSGi is a service-oriented centralized architecture)". Besides that, agents can be very different among themselves, including the signatures of the methods they declare, so we must ensure that every agent is compatible with each other and with regular OSGi bundles. We address these issues, and present the solutions adopted, in the following sections, describing how we made our architecture to be OSGi compatible.

3.1 Multi-agent Systems and OSGi

Adopting OSGi on each component of our architecture forced us to find a way to make one's agents compatible with OSGi bundles. The aim is to make accessible the functionalities of an agent (e.g., its methods) as services to other bundles. It would not be advisable to convert each agent into an OSGi bundle, since it would increase the development time and will throw away the advantages of Agent or Multi-Agent Systems (MAS) based methodologies for problem solving. Therefore, the verdict was to create an OSGI bundle that could make the bridge between regular bundles and Jade: the MAS bundle. This bundle can deal with one Agent Container (AC) and implement the methods declared in the interface of the agents in that AC as its own services. Moreover, this bundle must be able to start and stop agents, which in practice, correspond to the start and stop of the services they provide. The bundle, upon the reception of an invocation for an offered service from any other bundle, sends the invocation to the correspondent agent and delivers the respective result to the calling bundle. Please note that an agent, when trying to satisfy an invocation, may require the services provided by other bundles currently available. This is possible through the MAS bundle.

We are now ready to make a more detailed description of the MAS bundle. There are two methods for controlling the bundle, either by the client or by the administrator, in order to start or stop the respective bundle. Once this MAS bundle registers the services of the agents it creates, it declares the public methods of the agents on its interface, in order to make them visible to the other bundles as regular services. As for the interface between the MAS bundle and the Jade system, a JadeGateway agent (JGa) is used. The task of this agent is to act as a bridge between Jade and non-Jade code. When the MAS bundle starts, this agent is created along with the other agents. The JGa has the knowledge of which services are provided by each agent that is running. Hence, whenever a request from a service arrives to the MAS bundle, it knows to which agent that request should be forwarded. When the request arrives, a shared object is created in the MAS bundle, i.e., the blackboard. This object has some fields such as the name of the service to be invoked or the content that is the response from the final agent. The MAS bundle simply fills the service name field, which is the name of the ser-vice that was invoked by another bundle. Then, the blackboard is passed to the GWa which, interacting with one or more agents, gets the answer needed to the invocation of the service. That answer is then written to the content field of the shared object and returned to the MAS bundle. The final part consists of the bun-dle resuming the invocation of the service by returning to the calling bundle, the content of the shared object. The bundle that requested the service will never notice all that went on while it waited for the result of the service invocation.

Likewise, if an agent needs to use a service from another bundle, it contacts the MAS bundle, which is responsible for contacting the correct bundle, invoking the service and forwarding the result back to the agent.

The more specific issue of the interaction between agents, inside each platform, is outside the scope of OSGi, and, therefore, must be addressed. Agent communication is indeed a very important subject since it implies directly with the performance and behavior of the whole system. FIPA (Foundation for Intelligent Physical Agents) establishes several agent-related standards, being one of them the Agent Communication Language (FIPA-ACL). This standard defines how to construct a message, syntactically and semantically. It specifies the parameters a message may have (e.g., sender, content, performative), and how to use them. The communication between the agents of our architecture complies with FIPA-ACL standard. By doing so, we solve some drawbacks and enlarge the compatibility of the architecture to foreign agents that follow the same standard. At this point, any agent that complies with FIPA-ACL can run inside a container controlled by a MAS bundle, and may provide its methods as services.

3.2 OSGi Locally

Let us describe in this section the role that OSGi will play in our architecture. What we have is a group of nodes (e.g. the supported user, the CallCareCenter) and we use OSGi inside each of these nodes. Let us take as an example the sup-ported user's home. There is a group of components, which are part of the home and should be connected to the system. In one hand we have the 1-Wire sensors, which give the system the knowledge about the temperature, humidity, luminosity and other factors. In another hand, we have the X10 network, which allow for house equipments, like lights or other electric devices, to be controlled from a computer. Our objective here is to connect a group of heterogeneous devices in an integrated way and that is the OSGi contribution at this level.

Inside the house, the sensors are connected to the central computer through the serial port and a bundle is responsible for constantly reading the values from the sensors and registering them. This bundle exports as a service the values of the sensors of the house, which can then be used by the rest of the bundles. There is also a bundle for each X10 equipment and each of this bundles exports as a ser-vice the commands that can be issued to the equipment it represents (such as ON, OFF, UP, DOWN, etc).

Let us assume that the air conditioning system has enough autonomy to control the temperature based on the client preferences. It would be a hard task for an X10 equipment to interact with a 1-Wire sensor and acquire information from it without intermediary equipment. With OSGi, the autonomy is given to the bundle, which can easily issue X10 commands to the equipment through the serial port based on the information that it obtains from the sensor bundle. With OSGi, entities that are different may be easily integrated and put to work together ranging from sensors and actuators to house equipments, people or software agents.

Another problem we face when adopting OSGi is the compatibility between bundles. This problem arises from the very different bundles that we can have, aggravated by the fact that with OSGi, any bundle written by anyone can be incorporated into the system. As an example, imagine a class that exports a method which signature is a data structure declared within that same method. The bundle that made the call to the service will not "understand" the result. There is, therefore, the need to ensure the compatibility between all bundles. This was achieved by de-fining an ontology to be used by all the bundles. This ontology is a Java package on which the classes relating to all the objects that can be used by the methods on its signatures are declared. If every bundle imports this ontology, the compatibility is assured and, if a bundle is added which implements a new object, only the ontology must be updated. None of this would however be necessary if all the methods only used standard Java classes.

Having addressed these main issues, OSGi can be used locally in the components of our architecture simplifying its implementation. Moreover, OSGi provides a bundle that, off the shelf, allows for UPnP components to be viewed by the other bundles as services, extending its possibilities. Let's look again at the home of the supported user: each bundle can provide local services like the control of the lights or the air conditioning system. If the user has, for example, an UPnP TV, its control will also be provided as a service to the user the moment he connects it. The fact of OSGi supporting UPnP devices has yet the advantages associated do the zero-configuration needed for including new devices, therefore, for in-cluding new services, which is especially useful for the elderly.

Regarding the service providers, like the CallCareCenter, the implementation of OSGi allows for a better local organization. Each qualified health professional or computational resource can, for example, be a local service used by the entity responsible for generating the answer to the user who requested it.

In the next section we will show how OSGi must be expanded so that it interconnects the OSGi cells that are each component of the architecture, enabling the user to access remote services and maintaining the compatibility with the OSGi standard.

3.3 OSGi Remotely

The fact that OSGi has a centralized architecture while our architecture is a distributed one, raises a last challenge that must be overcome. What we have is standard OSGi cells, each one being a component of the architecture. These cells, when interconnected, create a virtual organization, on a service-sharing basis. There must be, therefore, a way of accessing remote services of other cells since OSGi standard does not address this problem. R-OSGi is an extension to OSGi standard, which allows for a centralized OSGi application to be distributed, using proxy bundles. A proxy bundle is a bundle which provides not only a remote ser-vice exactly as if it was a local one, but also allows for local services to be remotely accessed.

The main idea behind R-OSGi is for remote services to be accessed by bundles like they would be if they were local ones, in a completely transparent way. What we do is to add an additional bundle to each cell that provides at least a remote service. This bundle is responsible not only for checking the service registry of the OSGi cell it is on, but also to search for services which should be provided remotely, announcing them on an external port. Remote bundles which want to sub-scribe its services will make a connection to that port and subscribe it. Moreover, each cell should also start a bundle for each service (or bundle of services) it wants to remotely access. This bundle subscribes the remote service as soon as it is needed and registers it on the local OSGi cell, as if it was the bundle providing it. Subsequently, any local bundle can use the service without the need to know if it is accessing a remote or a local one.

What happens is that, when a local bundle in cell A calls for some remote method, that call goes normally to the proxy bundle as if it was the one providing the service. After receiving the method call, the proxy bundle identifies the remote cell which is providing the service; let us call it cell B. It starts a connection with the proxy bundle of cell B and sends the method invocation. The proxy bundle in cell B receives and forwards it to the right local bundle (it may also need to connect to other remote bundles), which returns the result of the invocation to the proxy bundle. The proxy bundle in B then sends the result to the proxy server in A, which forwards it to the original bundle that called the method. The Fig. 2 is an example of a portable device, which connects to a remote server to satisfy a request of the user to plan a trip. The user, as well as the interface service on his device, does not need to distinguish if a remote service was used or not, but to know the results.

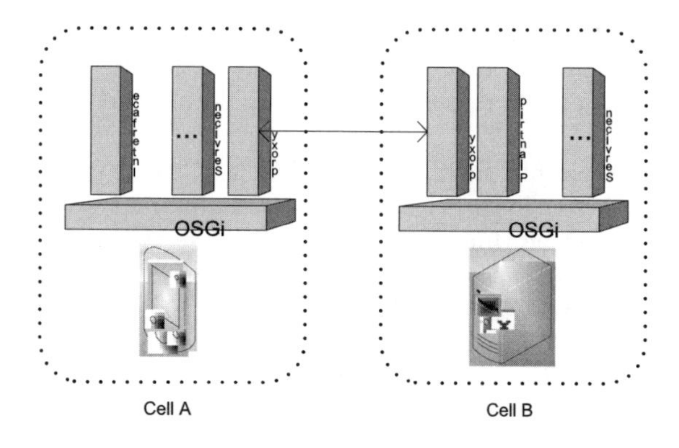

Fig. 2. Using remote services

The main advantage of having a Virtual Organization (VO) implemented with the R-OSGi technology, is on the dynamic configuration of the group. New members willing to join the VO can easily start making part of it and, when no more want to offer a service, they simply stop the bundle which provides that service. Issues raised from stopping services being used remotely are treated locally by the service registries of each OSGi cell, and constitute no new problem to be ad-dressed, since even local services can stop being available, once the service registry already takes care of that.

In Fig. 3 we present an example of the architecture from the OSGi point of view. The OSGi network will be made of standard OSGi artifacts (i.e., the OSGi cells), that may interact with one another. In this example, components in the area of the user body, such as a heart beat monitor, a handheld device or his personal assistant, are an

OSGi cell, the devices and equipments of the user's house are an-other one, and the group decision along with the persons or equipment that make it, create even another OSGi cell. The red proxy bundles provide the local services of their cells as remote services, and the green proxy bundles provide a remote service as local services in their OSGi cells. In this implementation, the Personal Agent uses a remote service to plan the weekend, provided by the Group Decision OSGi, when the user requests it. It also uses the lights service provided by the Home OSGi to turn the lights on or off, based on the location of the user. At last, the Temperature Monitor bundle uses the location service and the preferences of the user provided by the Location and Personal Agent to control the temperature at the home through the Air Conditioning bundle. Note that a service that is provided by an agent and is to be remotely accessed, is registered locally by the MAS bundle and remotely by the proxy bundle.

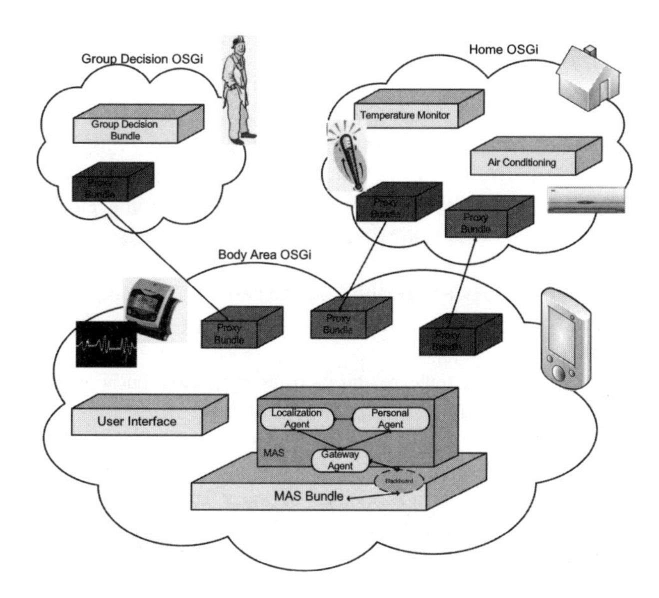

Fig. 3. The architecture from the OSGi point of view

4 Conclusion

Following the decision of using proxies in R-OSGi we were using Java reflection API to dynamically creating the classes of the new unknown bundles that could be added. However, we concluded that the increase in complexity and the decrease of performance in the key components of the architecture were reasons strong enough for us to abandon this idea and let each bundle that needs to connect to a remote service to connect on its own. This option was yet encouraged by the fact that R-OSGi makes remote method invocation very simple to use so there is no reason for us not to stick with it.

Based on open standards, we have presented a framework to an early deployment of a prototype for the VirtualECare system. In future work, we expect to elaborate on real life scenarios and situations, in order to make the necessary's developments to set a working prototype, that could provide to the population in general, and the elderly, in particular, a certain amount of remote services (e.g., healthcare, entertainment), without delocalizing or messing up with their routines, in a more effective and intelligent way.

References

1. Healthcare 2015: Win-win or lose-lose? In: Book Healthcare 2015: Win-win or lose-lose? (IBM Global Business Services) (2006)
2. Nehmer, J., Becker, M., Karshmer, A., Lamm, R.: Living assistance systems: an ambient intelligence approach. In: Book Living assistance systems: an ambient intelligence approach, pp. 43–50 (2006)
3. Giráldez, M., Casal, C.: The Role of Ambient Intelligent in the Social Integration of the Elderly. IOS Press, Amsterdam (2005)
4. Aguilar, J., Cantos, J., Expósito, G., Gómez, P.: Tele-assistance Services to Improve the Quality of Life for Elderly Patients and their Relatives: The Tele-CARE Approach. The Journal on Information Technology in Healthcare 2(2), 109–117 (2004)
5. Sarela, A., Korhonen, I., Lotjonen, J., Sola, M., and Myllymaki, M.: IST Vivago® - an intelligent social and remote wellness monitoring system for the elderly. In: Book IST Vivago® - an intelligent social and remote wellness monitoring system for the elderly, pp. 362–365 (2003)
6. Riva, G.: Ambient Intelligence in Health Care. Cyberpsychology & Behavior 6(3) (2003)
7. Holmlid, S., and Björklind, A.: Ambient Intelligence to Go. In: Book Ambient Intelligence to Go (2003)
8. Brown, S.J.: Next generation telecare and its roles in primary and community care. Health and social care in the community 11(6), 459–462 (2003)
9. Costa, R., Novais, P., Machado, J., Alberto, C., Neves, J.: Inter-organization Cooperation for Care of the Elderly. In: Wang, W., Li, Y., Duan, Z., Yan, L., Li, H., Yang, X. (eds.) Integration and Innovation Orient to E-Society. Springer, Heidelberg (2007)
10. Marreiros, G., Novais, P., Machado, J., Ramos, C., Neves, J.: An Agent-based Approach to Group Decision Simulation using Argumentation. In: Proceedings of the International MultiConference on Computer Science and Information Tecnology, Workshop Agent-Based Computing III, Wisla, Poland, pp. 225–232 (2006) ISSN 1896-7094
11. Dosi, G.: Sources, procedures and microeconomics effects of innovation. Economic Literature 26, 1120–1171 (1998)
12. Malmberg, A.: Industrial Geography: agglomeration and local milieu: Progress in Human Geography, pp. 392–403 (1996)
13. Neves, J.: A Logic Interpreter to Handle Time and Negation in Logic Data Bases. In: Book A Logic Interpreter to Handle Time and Negation in Logic Data Bases, pp. 50–54 (1984)
14. Initiative, O.S.G.: Osgi Service Platform, Release 3. IOS Press, Amsterdam (2003)
15. Chen, K., Gong, L.: Programming Open Service Gateways with Java Embedded Server(TM) Technology. Prentice Hall PTR, Englewood Cliffs (2001)

Quality of Service in Wireless e-Emergency: Main Issues and a Case-Study

Óscar Gama[1], Paulo Carvalho[1], J.A. Afonso[2], and P.M. Mendes[2]

[1] Informatics Dept., University of Minho, Braga, Portugal
{osg,pmc}@di.uminho.pt
[2] Industrial Electronics Dept., University of Minho, Guimarães, Portugal
{jose.afonso,paulo.mendes}@dei.uminho.pt

Abstract. Due to its critical nature, emergency healthcare (e-emergency) systems should be totally reliable, efficient and support real-time traffic. Therefore e-emergency networks must provide proper quality of service (QoS) levels. After assessing the relevance of QoS deployment in different e-health contexts, this paper presents a pragmatic case-study intended to be deployed in a hospital room containing patients with high risk abnormalities, whose vital signals are being monitored by personal wireless body sensor networks. After justifying the unsuitability of ZigBee standard in this e-emergency scenario, the use of Low-Power, Real-Time (LPRT) protocol for wireless sensor networks, is proposed as an adequate candidate for such task. For the present case-study, the protocol is able to fulfill quantitatively the required QoS levels.

Keywords: e-Health, e-Emergency, WSN, QoS, TDMA.

1 Introduction

An e-health system consists of a group of sensors attached non-invasively to a patient in order to sense the physiological parameters. It has been used in hospitals during the last decades using conventional wired equipment, hence not allowing the patient to move around freely. However, recent advances in wireless sensors technology are changing this scenario by permitting mobile and permanent monitoring of patients, even during their normal daily activities [1].

An e-health system should be able to accomplish at least one crucial aim: to monitor a patient and, when an emergency occurs, to trigger immediately an event to alert the patient and/or to warn a remote caregiver. In this way, both the patient and the caregiver can take timely the right procedure in accordance with the clinical episode. The system should also be able to trigger an alert anticipating the case where the patient is unaware of his/her health gravity. When a patient´s clinical state turn from a non-critical situation into a critical one, a context change occurs and consequently the healthcare network should adapt its performance requirements to the new situation. For instance, higher monitoring activity and lower delay transmission of the vital signals might be required when the patient´s clinical situation changes from non-critical to critical. Hence, healthcare networks should provide QoS facilities for e-emergency services, since these clearly demand for high reliability, guaranteed bandwidth, and short delays.

J.M. Corchado, D.I. Tapia, and J. Bravo (Eds.): UCAMI 2008, ASC 51, pp. 95–102, 2009.
springerlink.com

2 Vital Signal Monitoring

e-Health requires the monitoring of several vital signals simultaneously. The electrical characteristics of the vital signals usually used in emergency medical care are presented in Table 1 [2, 3].

At non-emergency medical situations, electrocardiogram (ECG) and blood oxygen saturation (SpO2) signals are usually transmitted in bursts, while signals such as body temperature and blood glucose, are transmitted in single packets to the base-station [4]. In fact, to reduce the traffic load and the power consumption of a body sensor network (BSN), the current trend in telemedicine systems is to enhance sensor node intelligence, available memory, processing power, and enabling on-line solicited requests only for results. In this way, continuous and bulky data transfer is sporadic, occurring only in intermittent occasions [4]. However in emergency cases this should not be the rule, since patient's life is priceless and above any other consideration. Continuous and bulky data transfer in real-time might be prevalent here.

Table 1. Vital Signal Electrical Characteristics

Vital signal (Hz)	**Freq. range** (Hz)	**Sampling rate** (Hz)	**Resolution** (bit)
ECG (per lead)	0.01…60-125	120-250	16
Temperature	0…0.1-1	0.2-2	12
Oximetry	0 … 30	60	12
Arterial pressure	0 … 60	120	12
Respiration rate	0.1 … 10	20	12
Cardiac rate	0.4 … 5	10	12

3 QoS Needs in e-Health

Some authors argue that differentiation based on data priority is inherent to wireless sensor networks (WSN), since it is normal to have sensors to monitor distinct physical parameters simultaneously, just as in BSNs. Here, the importance of the collected information is necessarily distinct, and therefore the network must prioritize the transmission of critical data when occurs a sudden clinical change in the patient. For example, in patients with cardiac diseases, heart activity information is more important than body temperature data. And depending on the patient´s clinical condition, the priority assigned to a vital signal can change dynamically. For instance, glucose data might be assigned a low priority when readings are in the normal range, but a higher priority might be reassigned to it when readings indicate hypo or hyper-glycemia.

Most current BSNs only offer the best-effort service, which is limitative for e-emergency support. In these networks, QoS provision is required to assist critical cases conveniently. This will enable, for instance, guaranteed bandwidth to higher priority streams for an efficient data delivery, even in case of fading or interference.

QoS control mechanisms are usually deployed in networks to guarantee consistent service levels concerning certain parameters, such as, packet loss, delay, jitter, and available bandwidth. These are the traditional end-to-end QoS parameters used to characterize the performance of communication infrastructures, including BSNs. For instance, the total delay of an ECG signal being displayed in the monitor should be less than 3 s for useful real time analysis by the cardiologists [5]; ECG signals require a minimum sampling rate of 250 Hz to guarantee that jitter does not affect the estimation of the R-wave fiducial point, which modifies considerably the spectrum [6]. No significant difference between ECG traces are detected by sampling the signal at rates between 250 and 500 Hz, but significant reduction in peak amplitude values and inaccurate interval measurements are obtained at 125 samples/s [4].

However, QoS in BSNs may not be fully described using only those parameters, because of its context-aware nature. For example, at application level, QoS may be regarded as guaranteeing the right number of sensors for monitoring the vital signals in accordance with the patients´ emergency state.

The available energy in the BSN is another very important parameter to take into account. In fact, if energy is carelessly consumed, the BSN may rapidly become completely useless due to lack of power. To prevent such failure, energy should be carefully saved using different approaches. For example, if the patient is in normal state then the sampling rate of sensors can be reduced to save power, or if the battery charge becomes low then its energy should be reserved to the more vital tasks of the patient. That is to say, the monitoring activity should adapt in accordance with the patient clinical state for energy saving. To save further energy, communication protocols should be simple, and data should be aggregated, eventually compressed, and transmitted in full-loaded packets, since computing demands much less energy than transmission. Attention must also be paid to delay, as it tends to increase linearly with the packet length.

Additionally, for efficiency reasons a large packet length may be chosen for non-critical situations. But as soon as an emergency occurs, the packet size can be reduced to meet the low delay QoS requirement, and signals considered irrelevant to this emergency episode are sampled at a lower rate, or not sampled at all.

Moreover, in emergency situations the computation power may be lowered to a minimum as all data must be forwarded, in opposition to the regular operation where, to save energy, the cardio-respiratory rhythm can be computed on-board before sending it. Or else, an ECG signal can be processed in the sensor itself to extract its relevant features. In this way, only information about an event is transmitted (e.g., QRS features and the corresponding timestamp of R-peak), hence reducing the traffic load and saving energy.

A BSN does not transmit only measurement data packets. Other packets may be present, such as those carrying control or alerting data. In this way, it is suggested that a high priority level should be assigned to data packets carrying alarming notification and measurements, and acknowledgement of correctly received packets; a medium priority level should be assigned to scheduled transmissions of data packets, and primary control packets (e.g. sensor configuration); and a low priority level should be given to periodic polling of nodes for network integrity check, and secondary control packets (e.g. link) [4].

Despite the number of e-health systems already developed [1], only a few encompass QoS support [7]. The QoS support and deployment provided in each project is diverse. Notwithstanding all the diversity, e-emergency systems should always support QoS in order to provide a pervasive, valuable and totally reliable assistance to any patient. This is the main goal in the case-study presented next.

4 Case-Study

We are implementing an experimental testbed to deploy QoS solutions based on a real clinical scenario. The scenario under study is based on a hospital room containing six beds with one patient per bed. Each patient is monitored by a personal BSN, and one base-station (BS) collects the vital signals of all patients. The signals being monitored are temperature (T), oximetry (OXI), arterial pressure (ART), respiration rate (RR), and ECG data, as shown in Figure 1. Each signal is collected by a dedicated wireless sensor. Patient´s vital signals are analyzed and/or correlated at BS. Our goal is to develop a solution which guarantees that every signal is delivered to the BS with the appropriated QoS, as specified next.

According to IEEE 1073 group, a wireless ECG electrode should generate 4 kbps of data, and the latency introduced by framing the data samples and the transmission delay should be below 500 ms. Since ECG signals are the most demanding in terms of QoS, we take this value as the maximum delay that any vital signal should have. Continuous healthcare monitoring normally uses a three leads ECG device, composed of two active electrodes plus one of reference. Since research is being done to eliminate the reference electrode, we assume that each active electrode is implemented by a wireless sensor (ECG0, ECG1). So, according to Table 1, each BSN produces a maximum aggregated rate of 10.424 kbps, hence resulting that the maximum total traffic inside the hospital room is 62.544 kbps. Besides guaranteeing this minimum goodput and latency below 500 ms, the e-health system must also guarantee low packet losses to every vital signal, low energy consumption and balanced energy drainage in every BSN.

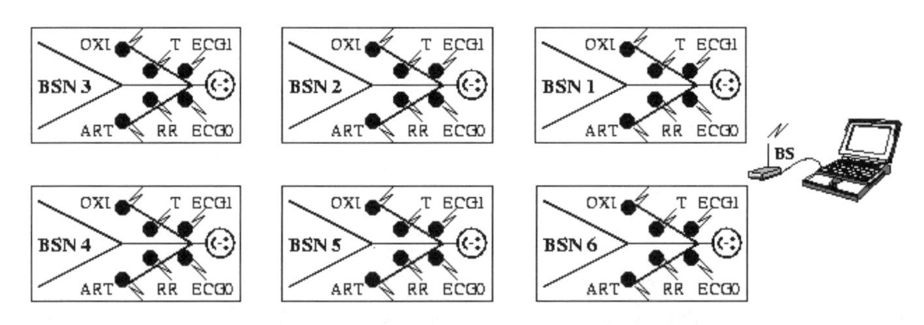

Fig. 1. Hospital room with a patient being monitored in each bed

In this case-study we have considered Bluetooth and ZigBee technologies. However, Bluetooth is unsuitable, since the protocol specifications allow a maximum of 7 active slaves (i.e. sensors) to be controlled by one master (i.e. BS).

ZigBee is a short range, low power, and low data rate standard for wireless sensor networks that supports a maximum rate of 250 kbps in the 2.4 GHz band. Therefore, a ZigBee WSN is able to handle the whole traffic generated inside the hospital room without congesting. Nevertheless, other factors which may affect QoS significantly have to be considered, such as the wireless channel access and transmissions errors. It is shown next why ZigBee is unsuitable for this case-study.

If traffic is to be sent within the same ZigBee PAN and short addresses are used, then a payload of 928 bits per packet is available to the applications. Ideally, all packets should be sent full-loaded in order to minimize the overhead for energy saving. Thus, to achieve the minimum required rate of 4 kbps, one packet carrying ECG data must be generated every 0.232 s. In addition, according to Table 1, each BSN should generate one full-loaded temperature data packet every 38.66 s, one oximetry data packet every 1.28 s, one arterial pressure data packet every 0.64 s, and one respiration data packet every 3.86 s. Since delivery delay must be below 500 ms, it is clear that only ECG data packets may be transmitted full-loaded.

Simulation results have shown that ZigBee is not adequate for several sensors to transmit ECG signals to a BS with full efficiency [8]. Either in acknowledged or in unacknowledged mode, the efficiency starts to drop when three or more ECG devices operate in the same RF channel. This is because ZigBee relies on CSMA-CA, a contention-based MAC protocol that is vulnerable to collisions. In acknowledged mode, lost packets may be retransmitted, but there is a maximum number (5) of retries allowed by ZigBee before declaring channel access failure. ZigBee seems to be more adequate for BSNs that do not have large amounts of data to transfer, only several small data packets per hour, like implanted medical sensors [9].

Another approach is using the beacon-enabled PAN mode described at IEEE 802.15.4 standard [10]. The BS sends regularly beacons which bound the superframes. These structures are divided into 16 identical slots. Any device wishing to communicate during the contention access period (CAP) shall compete with other devices using the slotted CSMA-CA. For low-latency applications or applications requiring specific bandwidth, the coordinator may dedicate portions of the active superframe to that application. These portions are called guaranteed time slots (GTSs). The GTSs form the contention-free period (CFP). The PAN coordinator may allocate up to seven of these GTSs, and a GTS may occupy more than one slot period. No transmissions within the CFP shall use a CSMA-CA mechanism to access the channel. The GTSs should be allocated dynamically to sensors accordingly with the QoS needs of the BSN. This TDMA-based transmission technique using slots is presently not available within any ZigBee profile.

Each full-loaded packet needs 4.256 ms to be completely transmitted at 250 kbps. Assuming that every packet is transmitted in individual slots, the beacon transmission interval must be above 4.256*16=68.096 ms. This means that the beacon order (BO) should be 3, implying a beacon interval of 122.88 ms (BI=3840*2BO bits, $0 \leq BO \leq 14$). Note that the next BO (4) implies a beacon interval of 245.76 ms, hence making impossible to transmit a packet every 232 ms. There is a significant waste of bandwidth because the packet never occupies the whole slot duration. In this case (BO=3), only 55.4% of the slot period is used to send a full-loaded packet.

Since twelve ECG sensors are in the hospital room, and ignoring the maximum number (7) of GTSs imposed by the standard, only four slots would be available for the other sensors to send data. A CAP having four slots and 55.4% of slot period utilization may transmit at most 34625 bps. Comparing this value with 2424*6=14544 bps of data sent by all the remaining sensors in the room, it is expected a large number of collisions during the CAP, hence degrading seriously the QoS of the system.

In order to find an alternative to ZigBee standard, the asynchronous superpoll paradigm was considered. In this model, the BS sends beacons to each BSN following a round-robin pattern, specifying which data should be received from its sensors. The BS only polls the BSN, and not each one of its sensors. For example, suppose a BS send one beacon carrying this information: {dst 3, ecg0 2, ecg1 1, art 1, oxi 1, sleep 150}. All sensors belonging to BSN 3 receive the beacon and read that information. Then, sensor ECG0 sends immediately two consecutive packets, next ECG1 sensor sends one packet, arterial pressure sensor sends one packet and, finally, the oximetry sensor sends one more packet. All these packets are sent consecutively. After transmitting, each sensor sleeps for 150 ms. After receiving all packets, the BS sends a beacon with a new set of requirements to the next BSN to get the data of its sensors. With this schema, the channel is guaranteed to be free whenever a sensor is about to transmit, and each slot is fully utilized in terms of data capacity. Since packets length and slot duration may change, each sensor knows the right time to send data by counting the number of transmitted packets after the beacon reception. As soon as a beacon addressed to its BSN is received, a sensor must always be in listening state until it is scheduled to transmit, a phenomenon known as overhearing. Such situation leads to a wasteful and unbalanced energy drainage in the BSN.

Since the superpoll model is energetically inefficient, the LPRT beacon-based protocol was considered instead [11]. Besides presenting the advantages offered by the superpoll model in terms of slot utilization efficiency, LPRT may lead to an efficient WSN in terms of energy consumption. In this protocol, the superframe is divided in a fixed number (1024 by default) of mini-slots, and starts with the transmission, by the base-station, of the respective beacon frame, which is followed by the Contention Period (CP). During the CP any station can transmit non-real time traffic using the CSMA/CA algorithm. The CFP comes after the CP.

Transmissions and retransmissions during the CFP are determined by the base-station using resource grant (RG) information carried by the beacon frame. Beacons also carry acknowledgment feedback and eventually other information, such as, the clock time of the BS for synchronizing the network regarding timing measurements. All data frames are acknowledged. A retransmission procedure helps to increase the reliability of the protocol. Due to the real-time constraints, only one retransmission attempt is scheduled in case of transmission failure during the CFP. Data packets are sent in contiguous, fixed, mini-slots of the CFP, according to the RGs. Small guard periods must be used to avoid the superposition of adjacent transmissions. So, after receiving a beacon, each sensor is able to calculate the time it may sleep before and after transmitting. Hence, energy saving is achieved because sensors only have to switch on the transceiver to receive beacons and data, or to transmit data. Since CFP uses a TDMA-based schema to access the channel, low-latency is provided for transmissions in single-hop networks. The hidden node problem is absent during the CFP, although it may occur during the CP. If there are other similar e-health WSNs causing interferences over each other in the near rooms, then distinct operating channels should be selected for each WSN (16 channels are available at 2.4 GHz band). In LPRT, the slots are efficient and dynamically used, unlike the slotted IEEE 802.15.4, where the low-level of granularity of the time-slots leads necessarily to poor bandwidth efficiency.

In order to study the suitability of LPRT protocol for this case-study, let us consider that the beacon interval is 232 ms (this value should be below 250 ms to prevent retransmitted packets to have a delay above 500 ms). Then, each superframe contains at most, per BSN, two packets carrying each one 928 bits of ECG data, one packet with 5 bits of temperature data, one packet with 167 bits of oximetry data, one packet with 334 bits of arterial pressure data, and one packet with 55 bits of respiration rate data. Thereafter, each BSN sends altogether 2417 bits of data per superframe. Since the headers and trailer of each packet require 136 bits, the overhead for transmitting all data is significant: 33.8%. This inefficiency is caused by the different signal sampling rates and by the delivery delay constraint (< 500 ms). Since packets transmission should respect the LIFS period specified at IEEE 802.15.4 (0.64 ms) to guarantee that the MAC sub-layer of the BS is able to process all incoming packets, each BSN takes 16.8 ms in the superframe to send its data. So, the CFP should occupy 100.8 ms to accommodate the whole traffic produced by the six BSNs, leaving 131.2 ms for the CP. This time is enough to allow the association of a new BSN arrived at the room, making the system scalable to a certain degree. The association should be allowed by the BS only if the allocation of resources to the new BSN does not compromise the overall QoS of the system already established. By this analysis, we predict that LPRT protocol is adequate to be used in this case-study.

We have considered that all sensors are monitoring the patients at the highest sampling rates. However, such scenario should only occur in critical clinical episodes. In non-critical clinical situations, the sensors should monitor the patient at lower sampling rates in order to save battery energy and channel bandwidth. Such policy would also improve the scalability of the system. Therefore, self-reconfiguring each BSN in accordance with the patients´ clinical state is an important paradigm to follow. We believe that the LPRT protocol associated with the self-reconfiguration of the BSNs could produce very interesting results in terms of QoS and scalability.

5 Conclusions

e-Emergency systems should be totally reliable and efficient in order to provide a pervasive and valuable assistance to any patient with risk abnormalities. Therefore wireless e-emergency networks need to support QoS at distinct protocol levels, as they clearly demand for reliability, guaranteed bandwidth, and low delays, due to their real-time nature. Energy consumption should also be minimized to extend the lifetime of the BSN.

We have concluded that both Bluetooth and IEEE 802.15.4/ZigBee standards are unable to satisfy the QoS requirements of the presented case-study. The superpoll paradigm was also rejected because it is not energetically efficient. Instead, by analyzing the LPRT, a protocol for low-power, real-time WSNs, we have concluded that it is suitable to fulfill such QoS requirements. The use of LPRT leads to an e-emergency system with low power consumption, low latency, reliability, flexibility, and high-throughput efficiency. The traffic generated by the sensors can be transmitted free of collisions and the system may be scalable to a certain degree. Currently, we are implementing this model in a testbed for experimental analysis.

Acknowledgments. Óscar Gama is supported by FCT (SFRH/BD/34621/2007), Portugal.

References

1. Kyriacou, E., et al.: e-Health e-Emergency systems: current status and future directions. IEEE Antennas & Propagation Magazine 49(1) (February 2007)
2. Paksuniemi, et al.: Wireless sensor and data transmission needs and technologies for patient monitor. In: The operating room and intensive care unit, Proc. 27th IEEE EMBC, China (2005)
3. Arnon, et al.: A comparative study of wireless communication network configurations for medical applications. IEEE Wireless Communications 10(1) (2003)
4. Lamprinos, et al.: Communication protocol requirements of patient personal area networks for telemonitoring. Technology & Health Care 14(3), 171–187 (2006)
5. Iglesias, A., et al.: Performance study of real-time ECG transmission in wireless networks. In: ITAB 2006, Ioannina, Greece (2006)
6. Pinna, et al.: The accuracy of power-spectrum analysis of heart-rate variability from annotated RR list generated by Holter systems. Physiol. Meas. 15, 163–179 (1994)
7. Gama, O., et al.: Quality of Service Support in Wireless Sensor Networks for Emergency Healthcare Services. In: Proceedings of 30th IEEE EMBC, Vancouver, Canada (2008)
8. Chevrollier, N., et al.: On the use of wireless network technologies in healthcare environments. In: Proc. 5th IEEE AWSN workshop, Paris, France, pp. 147–152 (2005)
9. Timmons, N.F., et al.: Analysis of the performance of IEEE 802.15.4 for medical sensor body area networking. In: Proceedings of the IEEE SECON (October 2004)
10. IEEE Std 802.15.4-2003, Wireless Medium Access Control (MAC) and Physical Layer (PHY) Specifications for Low-Rate Wireless Personal Area Networks (October 2003)
11. Afonso, J.A., et al.: MAC Protocol for Low-Power Real-Time Wireless Sensing and Actuation. In: Proceedings of 13th IEEE ICECS (December 2006)

Sentient Displays in Support of Hospital Work

Daniela Segura[1], Jesus Favela[1], and Mónica Tentori[1,2]

[1] Departamento de Ciencias de la Computación, CICESE, Ensenada, B.C., km. 107 Carretera Tijuana-Ensenada
{favela,dsegura,mtentori}@cicese.mx
[2] Facultad de Ciencias, UABC, Ensenada, B.C., km. 103 Carretera Tijuana-Ensenada

Abstract. Sentient computing can provide AmI environments with devices capable of inferring and interpreting context, while ambient displays allow for natural and subtle interactions with the environment. In this paper we propose to combine sentient devices and ambient displays to augment everyday objects with the development of sentient displays. These devices are aware of their surroundings while providing continuous information in a peripheral and expressive manner. A particular working environment that can benefit from the use of sentient displays are hospitals. We present the design of two sentient displays that provide awareness of patient's status to hospital workers. The first display is a flower vase that notifies about patients' urine outputs. The second display is a motion statue that notifies about patients' movements. Finally, we discuss the implementation of the flower vase to show the feasibility of using sentient displays in a hospital.

Keywords: Sentient Computing, Patient monitoring, Ambient Displays, Ambient Intelligence, Pervasive healthcare.

1 Introduction

Ambient Intelligence (AmI) environments are ubiquitous in the sense that they enhances the physical environment with heterogeneous computational and wireless communication devices naturally integrated and, at the same time, invisibles to the user (Weiser, 1991; Shadbolt, 2003). Hence, AmI applications need intelligent capabilities to be adaptive to users and reactive to context in order to provide high quality services based on their preferences while allowing non-expert users to interact with them in a simple, effortless and seamless way.

Sentient computing is an approach that allows AmI to best interact with their physical environment by becoming aware of their surroundings and react upon them (López de Ipiña and Lai Lo, 2001). Awareness is achieved by means of a sensor infrastructure that helps to maintain a model of the world which is shared between users and applications (Hopper, 1999). Indeed, sentient artifacts have the ability to perceive the state of the surrounding environment, through the fusion and interpretation of information from diverse sensors (Addlesee et al., 2001). However, it is not sufficient to make AmI environments aware of users' context they must be able to communicate this information to users while becoming an interface to the environment.

This vision assumes that physical interaction between humans and devices will be less like current computer paradigm and more like the way humans interact with the

J.M. Corchado, D.I. Tapia, and J. Bravo (Eds.): UCAMI 2008, ASC 51, pp. 103–111, 2009.
springerlink.com

physical world. For instance, a mirror augmented with infrared sensors and an acrylic panel could detect human presence and act as a message board to display relevant information when a user faces the mirror. Hence, AmI environments could be augmented with such displays that unobtrusively convey information to users without requiring their full attention. Indeed, the notion of what constitutes a computer display is changing. No longer is a display confined to the typical CRT monitor with a single user paying focused attention while interacting with virtual objects on the screen (Lund and Wilberg, 2007). Rather, computer displays are found in such diverse forms as small screens in mobile phones to ambient displays that provide peripheral awareness.

In this paper, by binding the ideas of sentient computing and ambient displays we propose the concept of *sentient displays* to define a new and appropriate physical interaction experience with an AmI environment. Such sentient displays will be capable to adequately monitor users' context, promptly notify relevant events and provide us with continuous information in a subtle, peripheral and expressive manner without intruding on our focal activity.

A working environment that can benefit from using sentient displays are hospitals. Given the work load of hospital workers it is not rare for them to miss important events, such as the need to change a urine bag that has been filled (Moran et al., 2006). Consequently, hospital workers have been held liable for their failure to monitor and promptly respond to patients needs (Smith and Ziel, 1997). Sentient displays located throughout hospital premises could be used for a diverse number of hospital applications, such as notifying hospital workers of a crisis or providing continuous awareness of the health status of patients.

2 Sentient Displays: Augmenting Natural Objects with Ambient Displays and Sentient Technologies

Research in pervasive computing has included the development of ambient devices that can become part of the background while acting as a digital interface to ambient information. As stated by Mankoff: *"Ambient displays are aesthetically pleasing displays of information which sit on the periphery of a user's attention. They generally support the monitor of information and have the ambitious goal of presenting information without distracting or burdening the user"* (Mankoff et al., 2003). For instance, the artist Natalie Jermijenko at Xerox Parc augmented a string with a motor and spin to convey the traffic's status to a user –the Dangling String (Weiser and Brown, 1995). The device rotates at a speed that depends on the amount of traffic in the highway captured through analog sensors. During periods of intense traffic, the string's movements are slightly audible as well. Thus, ambient displays are, unlike ordinary computer displays, designed not to distract people from their tasks at hand, but to be subtle reminders that can be occasionally noticed. In addition to presenting information, the displays also frequently contribute to the aesthetics of the locale where they are deployed (Lund and Wilberg, 2007). For instance, as part of the AmbientRoom project, several displays using light, sound or motion have been developed to augment a user's office (Ishii et al., 1998). Undeniably, ambient displays need computing devices capable of perceiving our surroundings by seeing or hearing the entities in the environment, what these entities are doing and where they are. To this aim, research in sentient computing has been focused in the development of sensors that attached to today's computing and communication technology are capable of perceiving a range of contextual information such as location, traffic status, user's presence and so on. The most popular sentient devices are indoor location systems (Werb and Lanzl, 1998; Addlesee et al., 2001).

While ambient displays provide a different notion of what constitutes an interface to the AmI environment sentient technologies allow such displays to be reactive and perceptive to dynamic changes in the environment. In this paper we propose the concept of *sentient displays* combining the ideas of ambient displays and sentient computing. Sentient displays are our everyday artifacts augmented with digital services capable of perceiving information from the environment and then using this information to extend the capabilities of such artifact. We envision a sentient artifact as an object that encapsulates the information perceived and then provides users with a new form of interaction with the environment. This interaction could be either by offering continuous awareness in subtle and peripheral awareness or by allowing users to change the status of such object in order to affect an AmI environment.

3 Opportunities for the Deployment of Sentient Displays in Hospitals: A Case Study

For nine months, we conducted a field study in a public hospital's internal-medicine unit, observing the practices of the hospital staff, who attended to patients with chronic or terminal diseases (Moran et al., 2006). Such patients are often immobile and incapable of performing the activities of daily living (ADL) by themselves. The study was conducted to understand: (1) the type of patients' information being monitored by hospital workers and (2) the way and the problems faced by hospital workers when monitoring patients.

3.1 Understanding the Activities Monitored by Hospital Workers

Hospital workers are responsible for providing integral and specialized care for patients. As part of the integral care, nurses monitor the activities of daily living (ADL) conducted by patients, such as, if a patient has taken his medicine, if he has walked or eaten. As a part of specialized care, nurses monitor the behavioral patterns in the activities that put at risk patients' health (e.g., pneumonia or a stroke), such as, if a patient is agitated, if a patient is bleeding or if the patient has respiratory insufficiency. These behavioral patterns associated to risk activities (RA) are monitored through the vital signs. In the following lines, we discuss common problems faced by hospital workers when monitoring patients.

Highly mobile hospital workers spend more than 50% of their time on-the-move, making it difficult for them to be aware of the status of the patients they are responsible for (Moran et al., 2006). A nurse commented: "*sometimes, I have patients that are placed in different areas of the hospital and if I am looking for a patient I am locked up in his room and I do not realize what is happening with my other patients, I am totally disconnected from those patients and in one or two minutes a thousand things could happen because I am not there I am over here*". In particular, issues related to hospital workers being on the move include maintaining awareness of their patients' status and being easily accessible when an emergency occurs.

3.2 Identifying the Issues Faced by Nurses When Monitoring Patients

To illustrate the problems faced by hospital workers when monitoring patients we present two real-scenarios that were observed in the hospital during the study.

Scenario 1: Monitoring patients' urine outputs

Nurse Refugio is informed, at the beginning of her working shift, that the attending physician has changed Pedro's medication to include cyclosporine. Pedro is a 56 years old man, who has a chronic renal failure and just had a renal transplant. So, to monitor Pedro's reaction to the new kidney and to the medicine being administered to him, Refugio needs to supervise the frequency and quantity of Pedro's urine. Therefore, she needs to promptly know when Pedro's urine bag is almost full or has reached a threshold to: (1) avoid spills by promptly changing the urine bag, (2) maintain his liquid balance timely updated and (3) quickly determine a crisis in his health status.

Scenario 2: Monitoring patients' movements

Jorge is a 60 years old man who just had a neurological surgery. He is partially immobile so he must perform physical exercises on his bed to improve his motors skills and blood circulation. Since the surgery, Jorge has been very anxious and has been having epilepsy strokes with abrupt movements resulting in falls from the bed – especially during the night. Consequently, Carmen, the nurse in charge of Jorge, wants to monitor Jorge's movements to find out: (1) if Jorge has performed his exercises, (2) if he is abruptly moving resulting in a possible epilepsy stroke and (3) if he has fallen from bed.

4 Designing Sentient Displays to Monitor Hospital Patients

The data from our study helped us discover opportunities for the development of sentient displays in support of hospital work. Particularly, we envisioned the implementation of two sentient displays aimed at creating an ambient connection between patients and nurses. The first display is a FlowerBlink that takes into account contextual information to notify nurses whenever a patient has urinated, as well as, the state of his urine bag. The second display uses data from an accelerometer wore by a patient to drive the swing of a pendulum in the form of a statue –and the more a patient moves the faster the pendulum moves. In the following section we describe the motivation, the sensing technologies required, and the design of both ambient displays.

4.1 The FlowerBlink

The FlowerBlink is a sentient display that notifies nurses the patients' urine outputs and the status of their urine bag (Figure 1). The flower is a wooden box containing twenty four artificial flowers: twelve emergency flowers with stems and twelve situation stemless flowers. The flowers are composed of a two-layered felt that enclose pistils covered with insulating tape. In each pistil a red or yellow led is embedded. The *emergency flowers* have stems with an embedded yellow light in their pistils (Figure 1a). All emergency flowers blink whenever an event or an emergency occur with a urine bag wore by a patient –if a urine bag is full. In contrast, *situation flowers* are flowers without stems that have a red light embedded in their pistils. This situation flowers are arranged in a matrix to represent patients' location in the unit. Each column in the matrix represents a room while a row represents a patient's bed (Figure 1b) –each room has three beds for patients. This arrangement allows nurses to quickly discover which patients' bag is about to spill. Situation flowers turn on whenever a nurse approaches the FlowerBlink or if the emergency flowers are blinking. While emergency flowers are blinking a situation flower turns on, indicating to a nurse the location of the patient related to that event. When a nurse personalizes the information shown by the FlowerBlink, by scanning her RFID tag, situation flowers convey patient's urine outputs instead of location.

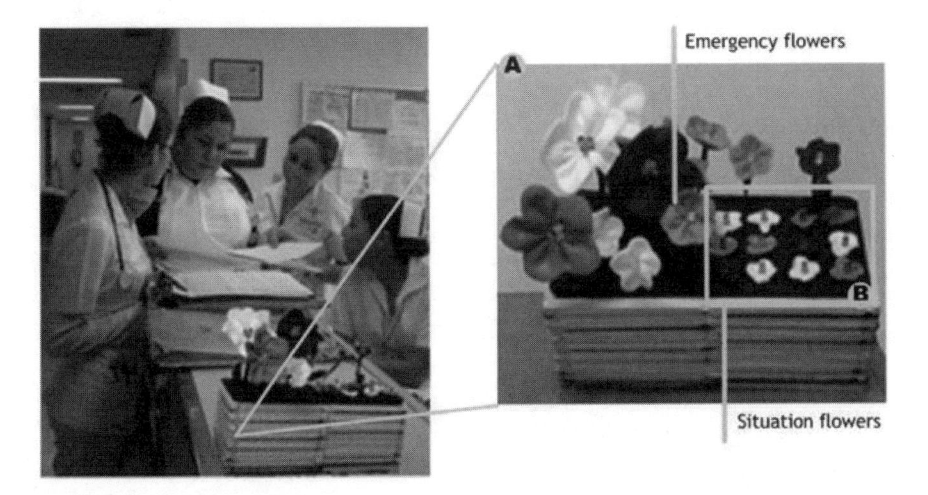

Fig. 1. The FlowerBlink placed in the nurse pavilion. (a) The flowers that notify of emergency events (b) The flowers that personalized their color based on the nurse's presence.

Going back to our scenario: Refugio uses the FlowerBlink to supervise the frequency and quantity of her patients' urine outputs. While Refugio is in the nurse pavilion, the lights in the emergency flowers of the FlowerBlink start to blink (Figure 1a), indicating her that the bag of a patient in the unit is almost full. Refugio approaches the FlowerBlink and realizes that the light that represents Pedro in the situation flowers has turned on (Figure 1b). She moves to the warehouse and gathers the medical equipment needed to change Pedro's urine bag. Then, she updates his liquid balance. A couple of hours later, while Refugio is discussing the evolution of a patient with the physician the emergency flowers and Pedro's situation flower start to blink. Refugio approaches the FlowerBlink and scans her RFID card to find out more about Pedro's urine outputs. Refugio realizes that Pedro has urinated four times in a period of two hours. She discusses this with Dr. Perez, who then decides to change Pedro's medication to avoid damage the new kidney.

4.2 The Motion Statue

The Motion Statue is a sentient display that notifies nurses of patients' movements. The Motion Statue is a decorative pendulum made with metal that represents one patient registered in the unit (Figure 2). To drive the pendulum, it uses two sentient artifacts: the CrowdPuller and the RhythmDetector. The CrowdPuller is a motor that moves back and forth a magnet attached to it (Figure 2a). The CrowdPuller is embedded in the base of the statue and while the magnet is moving it attracts the extremities of the statue. The RhytmDetector is an accelerometer attached to the wrist of a patient that sends information to a control unit (Figure 2b). The control unit is embedded in the base of the statue and drives the speed of the CrowdPuller's motor. Thus, the pendulum swings whenever a patient is moving and during abrupt patient movements it's faintly audible as well.

Fig. 2. A nurse consulting the motion that represents a patient under her charge

Going back to our scenario: Carmen uses the motion statue to represent Jorge and places it in the warehouse. Later, while Carmen is preparing medicines, Jorge's motion statue silently moves with a uniform rhythm indicating to her that Jorge might be performing his motion exercises. Carmen notices this and updates Jorge's nurse chart indicating that his exercises of the day are completed. Throughout the day, Jorge's motion statue is increasingly moving indicating to her that Jorge is becoming anxious. An hour later, she faintly hears something and realizes that Jorge's motion statue is moving rapidly. Carmen walks to Jorge's room where she notices that the patient is about to have an epilepsy stroke and that he is almost falling from bed. Carmen holds Jorge and gives him a medication to calm him down avoiding the epilepsy stroke.

5 The FlowerBlink: Implementing a Sentient Display

The FlowerBlink consists of two sentient artifacts and one ambient display (Figure 3). The first sentient artifact is the WeightScale that is attached to the urine bag and measures its weight (Figure 3a). The other sentient artifact is the PrescenceDetector that is a card carried out by a nurse that detects her identity when she scans the card in the FlowerBlink (Figure 3b). The ambient display is the flower vase with a set of flowers that display the status of the urine bag of the patient (Figure 3c). We embedded in the box a communication interface with 8 analog inputs, 8 digital inputs, and 8 digital outputs. We used the digital outputs that switch between 30VDC and 2 Amps to directly control the sentient artifacts. The output acts as a switch to ground voltages, and it is protected from transient voltages typical when switching inductive devices -relays, solenoids, motors. The outputs can be used to directly control devices requiring substantial power such as incandescent lights, high power LEDs, relays, solenoids and motors, making it possible to control the flower light. When the base station receives the information, it identifies the sensor that sent it, thereby identifying the location of the patient, and then turning on the red light of the corresponding flower. At the same time the flowers with stems begin to blink, trying to get the attention of nurses. We use the phidgets toolkit (Greenberg and Fitchett, 2001) to implement the FlowerBlink.

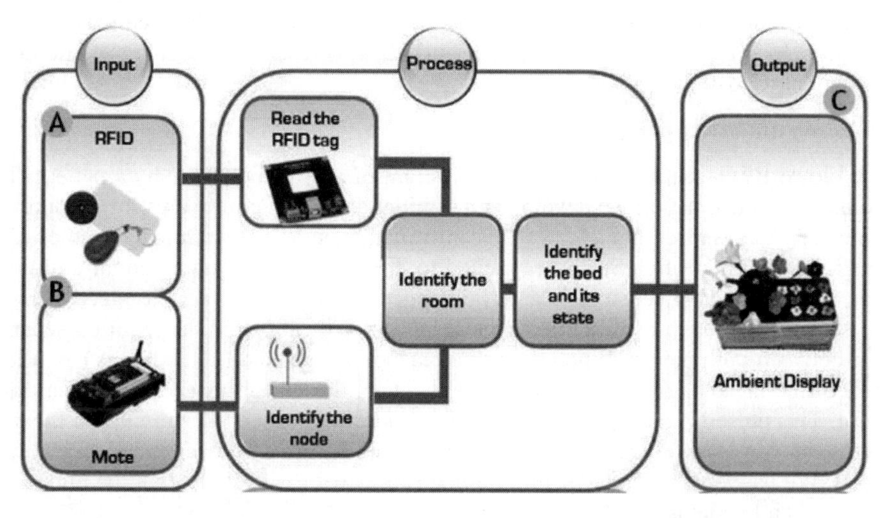

Fig. 3. The FlowerBlink components (a) The weightScale and the prescenceDetecter; (b) The FlowerBlink as an ambient display

Thus, three main services are provided through these components. The first service provides perceptible and silent awareness of patients' urine outputs. For example, a nurse can personalize the information shown by the FlowerBlink by using her PrescenceDetector card. A second service allows the seamless interaction with the sentient display. As the scenario shows, a nurse, by noticing that a flower with stems is blinking and consulting the stemless flowers, can infer which patient bag must be changed. Finally, the third service allows the unobtrusive information sensing to avoid interference with digitalized artifacts attached or wore by patients. In the following lines we described how the components work to support the services just described.

Providing perceptible and silent awareness

The idea of this service is to adequately manage how the information will be presented to the user. An ambient display that changes too fast can distract the user whiles a display that changes too slowly can pass unnoticed (Johan et al., 2000). The FlowerBlink uses the communication interface embedded in the flower vase to increasingly change the intensity with which a flower blinks. In addition, using the PresecenceDetector a nurse can personalize the information shown by the display avoiding to cramp the display while unrestricting its usefulness by presenting just enough information to her –how much a patient has urinated.

Enabling a simple, effortless and seamless interaction

The idea of seamless interaction consists in allowing users to have an implicit and natural interaction with the display. Users do not have previous experience interacting with ambient displays; hence, novel metaphors must be used to make them intuitive. The FlowerBlink, by using different sizes in the flowers stems, allows nurses to naturally discover the information shown by the display. In addition, by embedding a light in each flower of the vase we are extending the capabilities of an artifact without altering the traditional means of interaction with it. This will result in a reduction of the cognitive load by learning how displays work thus increasing the amount of attention on content (Gross, 2003).

Enabling unobtrusive information sensing

The FlowerBlink requires to monitor the weight of a urine bag wore by a patient. To monitor this information we developed a sentient artifact that measures the amount of urine in a bag and communicates this information wirelessly through a mote –the weightScale. This weightScale is attached to a urine bag wore by a patient allowing thus an unnoticeable sensing. The weightScale is made of two acrylic pieces which are separated through a spring and a push button. We calibrated the required separation between both pieces. When the urine reaches a threshold (i.e., when it has filled 80% of the urine bag) the button is pressed. Then, the sensor generates an electronic pulse. This pulse is read by the mote that is responsible for the transmission of this information wirelessly. When the bag is replaced the button goes back to its normal position. We use motes to avoid saturating the rooms with wires that could be obtrusive to nurses and patients.

6 Conclusions and Future Work

In this paper, we discuss the concept of *sentient displays* in support of hospital work. We show that sentient displays are capable of becoming aware of users' context and then present continuous and expressive information in a subtle and unobtrusive way. This type of awareness will allow hospital workers to promptly identify patient' needs, save time and avoid errors.

The ambient displays presented here are just preliminary designs that give a hint of the potential of this technology in healthcare. We plan to conduct an in situ evaluation of the displays developed to assess their impact within the hospital. We plan to explore a new setting where this type of technology could be useful –in particular, in nursing homes. Workers at nursing homes specialized in the care of elders with cognitive disabilities face working conditions that are similar to those in hospitals. Such workers also use common strategies to monitor patients' status. This monitoring is done manually, making it time consuming and error prone. This is another healthcare scenario where sentient displays can prove useful.

Acknowledgments. We thank the personnel at IMSS General Hospital in Ensenada, Baja California, Mexico. This work was partially funded by CONACYT through scholarship provided to Daniela Segura and Monica Tentori.

References

1. Addlesee, M., Curwen, R., Hodges, S., Newman, J., Steggles, P., Ward, A., Hopper, A.: Implementing a Sentient Computing System. IEE Computer 34(8), 50–56 (2001)
2. Greenberg, S., Fitchett, C.: Phidgets: easy development of physical interfaces through physical widgets. In: 14th annual ACM symposium on User interface software and technology, Orlando, Florida (2001)
3. Gross, T.: Ambient Interfaces: Design Challenges and Recommendation. In: International Conference on Human-Computer Interaction, Crete, Greece. Lawrence Erlbaum, Hillsdale (2003)
4. Hopper, A.: The Cliord Paterson Lecture: Sentient Computing. Philosophical Transactions of the Royal Society of London 358(1773), 2349–2358 (1999)
5. Ishii, H., Wisneski, C., Brave, S., Dahley, A., Gorbet, M., Ullmer, B., Yarin, P.: AmbientROOM: integrating ambient media with architectural space. In: Conference on Human Factors in Computing Systems (CHI), Los Angeles, US. ACM Press, New York (1998)
6. Johan, R., Skog, T., Hallnäs, L.: Informative Art: Using Amplified Artworks as Information Displays. In: Designing augmented reality environments, Elsinore, Denmark. ACM Press, New York (2000)
7. López de Ipiña, D., Lai Lo, S.: Sentient Computing for Everyone. In: The Third International Working Conference on New Developments in Distributed Applications and Interoperable Systems, Krakóow, Poland, Deventer, The Netherlands (2001)
8. Lund, A., Wilberg, M.: Ambient displays beyond conventions. In: British HCI Group Annual Conference (2007)
9. Mankoff, J., Dey, A.K., Hsieh, G., Kientz, J., Lederer, S., Ames, M.: Heuristic evaluation of ambient displays. In: CHI, Lauderdale, Florida, USA (2003)
10. Moran, E.B., Tentori, M., González, V.M., Martinez-Garcia, A.I., Favela, J.: Mobility in Hospital Work: Towards a Pervasive Computing Hospital Environment. International Journal of Electronic Healthcare 3(1), 72–89 (2006)
11. Shadbolt, N.: Ambient Intelligence. IEEE Intelligent Systems 18(2), 2–3 (2003)
12. Smith, K.S., Ziel, S.E.: Nurses duty to monitor patients and inform physicians. AORN Journal 1(2), 235–238 (1997)
13. Weiser, M.: The Computer for the 21st Century. Sci. American 265(3), 94–104 (1991)
14. Weiser, M., Brown, J.S.: Designing calm technology. PowerGrid 1(1), 10 (1995)
15. Werb, J., Lanzl, C.: Designing a positioning system for finding things and people indoors. IEEE Spectrum 35(9), 71–78 (1998)

University Smart Poster: Study of NFC Technology Applications for University Ambient

Irene Luque Ruiz and Miguel Ángel Gómez-Nieto

Department of Computing and Numerical Analysis, University of Córdoba,
Campus de Rabanales, E-14071, Córdoba, Spain
iluque@uco.es, mangel@uco.es

Abstract. In this paper we present our current work for the application of the Near Field Communication technology in University scenarios. NFC applications fall into the ambient and content-aware systems where users interact with the surrounding objects interchanging information between them. We present the use of Smart Posters devoted to a University Centre in which TAGs transmit information to mobile phones when the device touch the tag. We support the use of the NFC technology for more complex interchanging information and ticketing applications.

Keywords: browsing paradigm, ambient and context-aware computing, NFC, mobile phones.

1 Introduction

Ubiquitous Computing paradigm [1] believes that the technology itself doesn't matter, what really matters is its relationship with us, so this paradigm considers the natural human environment and allows the computers themselves to vanish into the background. In other words, they are interested in "invisible" computers that would allow us to focus on life beyond computational devices [2].

Ubiquitous Computing takes into account the social perspectives of technology, and different paradigms that have emerged in the last decades, such as: Pervasive Computing, Ambient Computing and Context-Aware Computing between others.

Hence, in physical browsing, users can access to services or information about an object by physically selecting the object itself. The enabling technology for physical browsing are tags, RFID-based that contain information and the devices used for interacting with the environment, that have interface capabilities, such as display, sound and keypad; read and write memory; processing power and communication capabilities (i.e. a mobile phone or a PDA).

1.1 NFC and Ubiquitous Computing

NFC (Near Field Communication) technology is nowadays an emerging technology, and simply consists in mobile phones (ubiquitous computing devices) [3] incorporating RFID characteristics [4], thus, allowing the interaction with the environment's objects around us.

NFC is a standard-based, short-range wireless connectivity technology that lets different devices communicate when they are in close proximity. It is based on RFID

J.M. Corchado, D.I. Tapia, and J. Bravo (Eds.): UCAMI 2008, ASC 51, pp. 112–116, 2009.
springerlink.com

(Radio Frequency Identification) technology. NFC enables simple and safe two-way interactions among electronic devices, allowing consumers to perform contactless transactions, access digital content and connect devices with a single "touch" [5].

Nevertheless, NFC has also the ability to write information onto the RFID-chip, therefore bidirectional communication can be established between a mobile phone and a NFC reader and, therefore, it is possible to develop complex applications as these oriented to: authentication, peer-to peer, ticketing or payment.

In this way, NFC could change human to environment interaction. Users only present their mobile phone in front of objects equipped with RFID tags and would get different kind of services offered by the objects (tags), giving place to the Internet of the Things [6]. The Smart Poster [5] is a concept close related with the Internet of the Things and it defines how a phone number, SMS or URL can be stored in an NFC tag or transport them between devices and also makes possible to initiate a phone call, send a SMS or go to an URL by reading a tag with a NFC phone, generating an environment where the surrounding objects are transformed into a source of information.

2 Smart Posters in the University Scenarios

The university environment is especially appropriate to the development of the NFC technology. Bologna process [7] promotes that thousands of users from the same or different universities/countries are moving in an open and controlled environment, users provided, the vast majority of them, with a mobile device.

The goal of our work is the development of Smart Posters that interact with the users in this university environment, providing information, allowing the establishment of a communication between the user and the environment and offering services from the surrounding objects simply touching the objects (tags) with the mobile devices.

2.1 Scenario Design

In the development of this work we have used mobile phones Nokia 6131 NFC [8], Omnikey USB CardMan 5321 readers [9] and Mifare 1K contactless cards for tags. As programming environment Eclipse 3.2 [10] has been used, and SDK for the Nokia 6131 NFC [8], which includes API contactless communication contacts JSR-257. A Java application has been developed for the development of the tags.

The scenario chosen to carry out our work has been a University Smart Poster devoted to a Faculty or High School. This Smart Poster aims to provide information to the user in the campus, just when the user with the phone touches a selected object from the Poster. The work has consisted of the following activities: (a) Selecting what information will be provided by the Smart Poster, (b) Selection of objects (tags), (c) Smart Poster design and (d) Study of the object-user interaction and poster services.

Figure 1 shows the prototype of the Smart Poster designed, with the associated tags. As is shown, the Poster contains objects associated with four different groups:

- *Organizational structure*. Each of these objects maintains information about: people and locations, phone numbers, e-mail, Web addresses, etc.

Fig. 1. Smart Poster corresponding to a Centre of University of Córdoba

- *Degrees that are taught at the University* These tags keep information about the characteristics of qualifications, subjects, training, courses, etc.
- *Departments related with the University.* These tags maintain information as the previously commented to the center (people, phone numbers, web addresses, etc.).
- *General information about the center*, as its location, map to access within the campus, etc.

Tags present in the Smart Poster described above establish a one-way communication with the mobile device (tag-to-mobile). So, when the mobile device is close to the tag, the mobile collects the information held in the tag. Later, the user determines how to use the collected information (i.e. store contacts, use the phone number, activate a Web address, etc.). The information is stored in tags using the standard-NFC Forum. This standard recognizes different types of records to be stored following the NDEF specification [5]. An NDEF message contains one or more NDEF records, each one carrying a payload of an arbitrary type. Records can be linked together to support larger payloads.

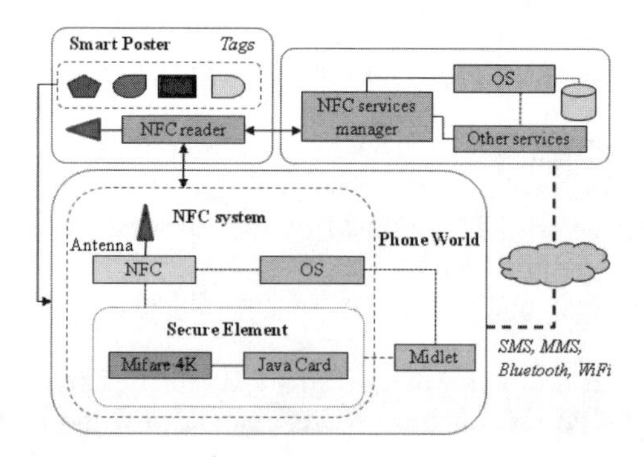

Fig. 2. A bidirectional interaction scenario where a midlet is in charge of interpreting the tags using the NFC system of the phone

However, the mobile device is only able to recognize certain types of records (only well-known records) and not all fields for some services (i.e. all possible fields of the VCard standard). This requires the development of a midlet that interacts through the reader and the NFC services manager, i.e., to get a ticket for accessing to the centre secretary (see Fig. 1).

In the scenario shown in Fig. 2, a midlet stored in the mobile device is executed by the user. The midlet incorporates the functionality needed to read different types of records and information stored in the tags of the Smart Poster. In addition, the midlet incorporates functionality to allow the user, when approaching the mobile to the reader placed in the Smart Poster, to send requests that are collected by the NFC services manager, processed and later released in different ways:

- Sending information through the NFC reader, for example: a text, an URL, an image, and so on.
- Enabling a service automatically on the mobile device, for example Bluetooth and communicating between them.
- Storing information in the secure element of mobile device, for example, a ticket or the turn to access the secretary of the centre.
- Establishing a peer-to-peer communication. For this scenario we currently are studying the use of Nexperts technology [11].

3 Discussion

Near Field Communication (NFC) is nowadays an emerging technology and ideal for the development of Internet of the Things because it enables mobile phones to contactlessly interact with the surrounding objects of the environment.

University scenarios are clear examples of ubiquitous environments. Current European University directives promote students and teachers moving among universities/countries to participate and interact with the university environment as if they were in their own centre. Hence, it is necessary to pay attention to the development of an active and intelligent environment devoted to promote this movement of people.

We are working in the use of the NFC technology for building these active environments. Our research is oriented to the development of Smart Posters, authentication and authorization systems and more complex systems where ticketing and payment are involved.

Acknowledgments. This work was supported by the Ministry of Industry, Tourism and Trade (Project: PROFIT FIT-330100-2007-76).

References

1. Weiser, M.: The Computer for the 21st Century, pp. 94–10. Scientific American (1991)
2. Aarts, E., Marazano, S. (eds.): The New Everyday: Visions of Ambient Intelligence. 010 Publishing, Rotterdam, The Netherlands (2003)

3. Yan, L., Zhang, Y., Yan, L.T., Ninq, H. (eds.): The Internet of Things: From RFID to the Next-Generation Pervasive Networked Systems (Wireless Networks and Mobile Communications). Taylor & Francis Group, USA (2008)
4. Ward, D.M.: The Complete RFID Handbook: A Manual and DVD for Assessing, Implementing, and Managing Radio Frequency Identification Technologies in Libraries. Neal Schuman Publishers, New York (2008)
5. NFC Forum. NFC Forum Data Exchange Format Specification, http://www.nfc-forum.org/specs/
6. Rukzio, E., Leichtenstern, K., Callaghan, V., Holleis, P., Schmidt, A., Chin, J.: An experimental comparison of physical mobile interaction techniques: touching, pointing and scanning. In: Dourish, P., Friday, A. (eds.) UbiComp 2006. LNCS, vol. 4206, pp. 87–104. Springer, Heidelberg (2006)
7. Bologna Process and European Higher Education Area, http://www.ond.vlaanderen.be/hogeronderwijs/bologna/
8. Nokia Corp., http://www.nokia.com/
9. Omnikey GmbH, http://www.omnikey.com/
10. Eclipse, an open development platform, http://www.eclipse.com/
11. Nexperts. NFC Business Platform, http://www.nexperts.com

Touching Services: The Tag-NFC Structure

Gabriel Chavira[1], Salvador W. Nava[1], Ramón Hervás[2], Vladimir Villarreal[3], José Bravo[2], Julio C. Barrientos[1], and Marcos Azuara[1]

[1] Faculty of Engineering, Autonomous University of Tamaulipas, México
{gchavira,snava,jcbarris,m_mazuara}@uat.edu.mx
[2] MAmI Research Group, Castilla La Mancha University, Spain
{ramon.hlucas,jose.bravo}@uclm.es
[3] Technological University of Panamá, Panamá
vladimir.villarreal@utp.ac.pa

Abstract. In an Ambient Intelligence Environment (AmIE), sensing technologies are crucial. Near field communication (NFC) technology is a step further in achieving the ideal vision. We propose the implementation of an AmIE with the single use of NFC technology (AmIE-NFC), which was developed in an environment with already existing computer infrastructure. In this environment, services from devices are controlled or requested simply by touching. We put a tag-NFC on each element or device. The tag-NFC contains the necessary information to connect an AmIE-NFC system. This work explains the tag-NFC structure to implement an AmIE-NFC.

Keywords: Ambient Intelligence, NFC, Touching Interaction, Location-based Services.

1 Introduction

Ambient intelligence environment (AmIE) is where "humans will be surrounded by intelligent interfaces supported by computing and networking technology that is embedded in everyday objects, which should be aware of the specific characteristics of the human and there should also be unobtrusive - interaction" [1].

However, in order for this vision to be realized, it is necessary to handle the information about all the elements in the environment, this is, context aware information. Dey defines context as "any information that can be used to characterize the situation of an entity. An entity is a person, place, or object that is considered relevant to the interaction between a user and an application, including the user and application themselves" [2] whereas a computer system is a context aware system if it "uses context to provide relevant information and/or services to the user, where relevancy depends on the user's task" [2].

An AmIE is a context aware system and needs to obtain information about all of the entities from the environment. In this process the sensing technologies are crucial, although embedding sensing technologies in the environment is a matter of debate in terms of cost, the users' ability to adapt to them, their widespread use, their embedment in daily life devices, etc.

The near field communication (NFC) technology overcomes some of the problems mentioned, as it can be easily embedded in a cell phone, whose use is inherently

J.M. Corchado, D.I. Tapia, and J. Bravo (Eds.): UCAMI 2008, ASC 51, pp. 117–124, 2009.
springerlink.com

familiar to people. You want something to communicate – touch it (no configuration required). Transactions are inherently secure (because the transmission range is so short) and it can be used easily in conjunction with other protocols to select devices and automate connection set-up [3].

At the moment, as of May 2008, the cell phone's cost is the greatest obstacle (tags are inexpensive). However, Nokia has announced a new cell phone, available by the end of the third quarter 2008, with a price that is significantly lower than those currently in the market.

In our previous work, we looked at adapting the radiofrequency identification technology (RFID) so as to perceive inputs (localization and identification) in an implicit way, without requiring user effort [4] [5]. We detected limitations in some contexts or situations. In another phase of our research work, we used an RFID-NFC combination [6] to implicitly capture the user's presence in the working area of the intelligent environment. The user utilized an NFC-enabled cell phone that could request and accept certain services through touching interaction.

This work explains the tag-NFC structure for implementing an AmIE with the single use of NFC technology (AmIE-NFC) to get context information. In the next section we explain the basic concepts of NFC technology and tag types. Then, we show a scenario where an NFC-enabled cell phone is used to obtain services very easily. Finally, section 6 makes some concluding remarks.

2 NFC Technology

In 2002, given the increasing number of new devices appearing in the market and their need to interconnect with one another, Sony and Philips developed the Near Field Communications (NFC) technology. NFC is a short-range wireless connectivity technology that combines radiofrequency identification (RFID) technology and interconnection technologies, using magnetic field induction to enable communication between electronic devices in close proximity. The NFC forum was formed in 2004, "to advance the use of NFC technology by developing specifications, ensuring interoperability among devices and services, and educating the market; and promote the use in consumer electronics, mobile devices, and PCs " [7]. In May of 2008 the forum had 149 members.

NFC operates in the unlicensed high frequency band of 13.56 MHz, with a bandwidth of almost 2 MHz, supporting data transmission speeds of 106, 212, and up to 424 kbits/s within a range of 10 cm on average (since the transmission range is so short, NFC-enabled transactions are inherently secure). The system is compatible with ISO 14443, but incompatible with the EPC global standards [8]. The first NFC specifications were published by ECMA International in the open standard 340 "NFC Interface and Protocol". One year later, they were adopted by ISO/IEC with number 18092.

NFC communication is established between two devices; namely the initiator and the target. What is generally known as the initiator, as implied by the name, initiates and controls the information exchange. This element is called reader in RFID terminology. On the other hand, the target is the device that responds to the initiator's request. The target is called a tag in RFID terms. The initiator and target used in our environment proposal are shown in Figure 1.

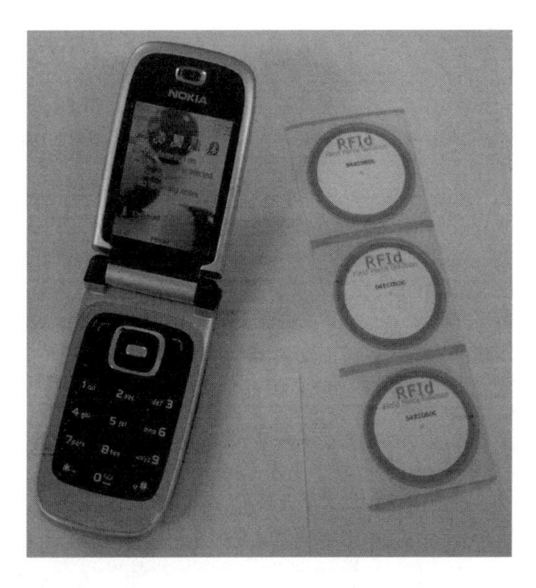

Fig. 1. NFC technology devices: initiator (cell phone) and target (tag)

2.1 NFC-Tag Types

The NFC Forum mandates four tag types to be operable with NFC devices [7]. Types 1 and 2 are based on ISO 14443-A, the international standard for contactless smartcards that supports MIFARE. This standard has 96 bytes availability in memory, expandable to 2 Kbytes, which can be read and re-written, although users can configure it to become read-only. Data transmission speed is 106 kbits/s.

Type 3 is based on the Japanese Industrial Standard (JIS) X 6319-4, also known as FeliCa. Its memory limit is 1 MByte and its speed may be either 212 kbit/s or 424 kbit/s. This memory can be pre-configured to be either read and re-writable, or read-only.

Finally, type 4 is based on ISO14443A and B. Its memory availability is variable, up to 32 KBytes per service while data transmission speeds reach up to 424 kbit/s. This type of memory can be pre-configured to be either read and re-writable, or read-only.

The NFC forum has defined a common data format called NFC Data Exchange Format (NDEF) specification, which contains technical specifications for Record Type Definitions (RTDs) and three specific RTDs: TEXT, URI, and Smart Poster. NDEF is based on MIME and is, therefore, very similar in concept.

An NDEF message contains one or more NDEF records. Each record can hold a different type of object. The first record defines the context and records the amount of the entire message.

3 An AmIE-NFC Scenario

John arrives at the building door, where his office and other workspaces (laboratory, other members' offices, and meeting room) of his research group are located. With his cell phone he touches the tag at the side of the main door of the building and the cell phone reminds him that he has an important comment for George who is already working at his desk. For this reason John decides to go to the laboratory (where George is). At the instant John touched the tag, all the members of the research group, who are working in a computer, receive a message indicating that John has entered the building.

In a corridor John can observe (on a public display) a summary of the research group's current work, such as deadlines of the congresses in which they will participate, the last versions of the papers being written, the identity and location of each person working in the building, etc.

When John arrives at the door of the laboratory he can observe who is in inside by looking at a little display. He can also see the degree of advance of the different activities (along with notes on projects, programs, articles, etc.) that the members of the group are developing. Before entering he touches the tag of the next door. Already within the laboratory he can observe a reminder of all "notes to comment on", on a public display of the laboratory, which has been stored in his cell phone. Meanwhile, all users who have "notes to comment on" to John, can see a reminder indicating that John entered the laboratory on their computers.

While John talks to George, John places his cell phone near the tag on the display of George´s computer to show a file. After commenting on it, they decide to show it to everyone in the laboratory. To this end, John touches the public display with his cell phone.

Before John leaves the laboratory, George decides to send him a paper for checking, but, due to its large size, it does not fit in the cell phone's memory. He therefore, decides to send the file to John so that it can be checked from any computer in the AmI environment.

When John leaves the laboratory, he runs the exit service in his cell phone to warn the AmI environment that he is coming out of the laboratory. When John arrives at his office and touches the tag in the door, his cell phone shows the list of people who came to see him while he was out, as well as the messages left for him.

4 The AmIE-NFC Framework

Our proposal for implementing an ambient intelligence environment with NFC technology (AmIE-NFC) has been developed in a setting with an already-existing computer infrastructure (e.g. office, research lab). The setting was to have computers, LANs, printers, servers, public displays, and very likely, but not indispensably; Internet connections.

The previous environment will be complemented with two NFC-enabled elements: the cell phone and the tag. Users were provided with one cell-NFC and a tag-NFC was placed on each:

- Device. Those electronic or computer devices that give a service to the user (e.g.: Display, printer, computer).
- Element. Allow us to obtain data from the environment, although the user does not perceive or receive any services (e.g. the door although it does not have electronic locks).

We will use NFC technology to request a service with a somewhat explicit task that generates savings in terms of effort via touching interaction (i.e. if you need taking the control or requesting a service from a device, a simple touch is enough). This "touch" may take place among the NFC devices in two situations; a cell-NFC touches a tag-NFC or a cell-NFC touches another cell-NFC. All elements or devices of our environment have tag-NFC (printers, computers, displays, desks, and doors).

Figure 2 illustrates the information flow and relationships among the main elements of our proposal, including NFC technology, computer infrastructure (LAN/Internet, Bluetooth connection, databases), system and context management, intelligent devices, and services devices.

When users touch a tagged device with their cell-NFC, the application installed executes automatically by reading the service and the information necessary to connect a Bluetooth server from the tag-NFC. The cell-NFC will connect to the Bluetooth server by sending a requested service, an ID tag-NFC and the necessary information. The Bluetooth server connects to the services server through a LAN or Internet connection (they can be Web services). The answer generated will be sent directly to the cell-NFC or to the devices, regarding this type of service, that are required by the AmIE-NFC infrastructure, which we call "services in the AmI area" (Figure 3). The figure describes a framework to use NFC in an ambient intelligence environment and explains a tag structure proposal.

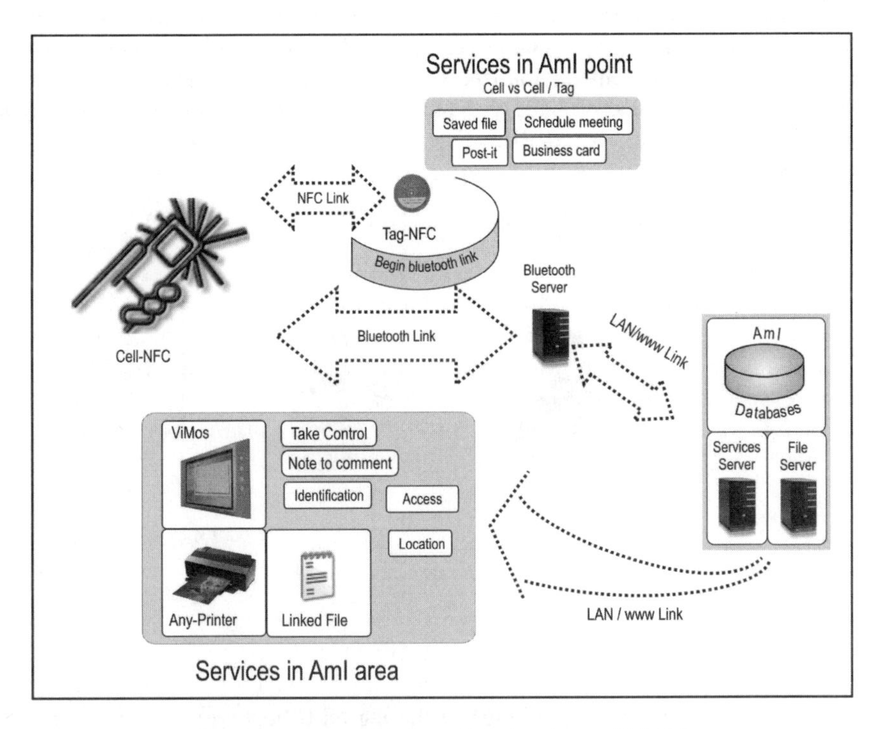

Fig. 2. Devices and information flow in AmIE-NFC

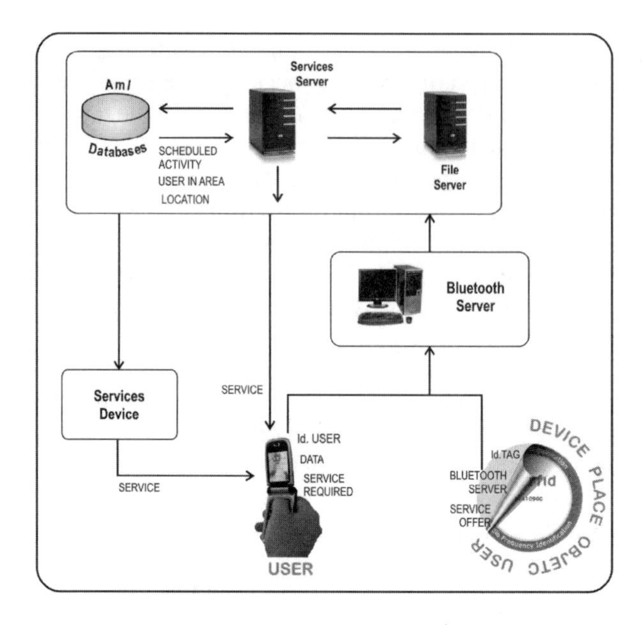

Fig. 3. Data flow in services in the AmI area

Sometimes users may need to exchange information with other users (i.e. business card). This is possible just by bringing near their cell-NFC. In this case, the service is satisfied merely by their interaction (the necessary information saved in cell-NFC's memory), without the need of AmIE-NFC infrastructure. The information interchange is realized from cell-NFC to cell-NFC. We call these services "services in the AmI point" (Figure 4).

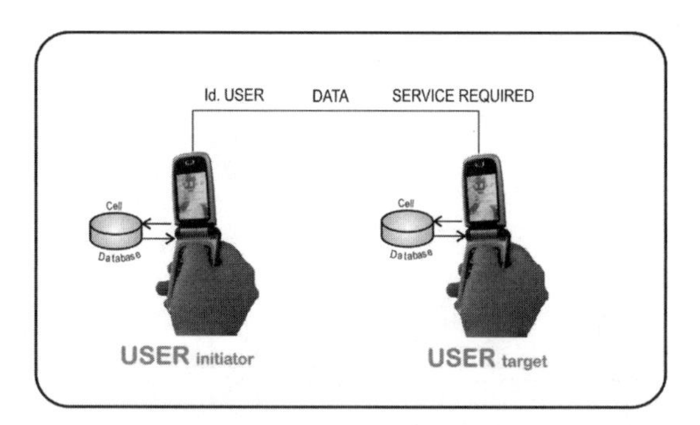

Fig. 4. Data flow in the services in the AmI point

5 Tag-NFC Structure

In our proposal we will take advantage of the tag-NFC being automatically read by cell-NFC, when touching. The corresponding application installed is executed automatically. The tag-NFC can be used in two situations:

- One service-one tag. If a user perceives a single service when touching the tag-NFC (e.g. the tag only opens the door).
- Several services-one tag. The tag-NFC contains several services (e.g. the tag not only opens the door, but also may contain notes for specific users or for all users who touch it, a file for the office's owner). In this case, data distribution is fundamental for optimal advantage of tag-NFC.

When a tag-NFC is associated with one service, this service is executed by default. If it contains several services, some of these are executed by default and others are based on the context (it can be a service for everyone, for a group or for someone in particular). In addition to the services, the tag-NFC must include a tag ID, a path of Bluetooth services (several if one or more are off-line) and information about the tag-NFC content. We define a structure for the tag-NFC that comprises the following elements (Figure 5):

- Tag ID. A default serial number that will be associated with the device and its location in the AmI.
- Tag Structure. This defines the fields that come next.
- Bluetooth Servers. They contain at least one MAC address of a Bluetooth server, although they will regularly be three. These servers can be permanent in an area or, in the areas where there are no computers; there will be laptops that have just arrived in the area.
- Area services. They require the infrastructure of the AmI since they are needed for consulting or modifying a database.
- Point services. They are services that do not require the infrastructure of the AmI, and are directly satisfied by other NFC devices (e.g. exchanging business cards, leaving or reading post-its).
- Temporary user profile. They adapt the user's devices configuration to activities developed in the area. When the user "leaves" the area, the devices return to their previous configuration.

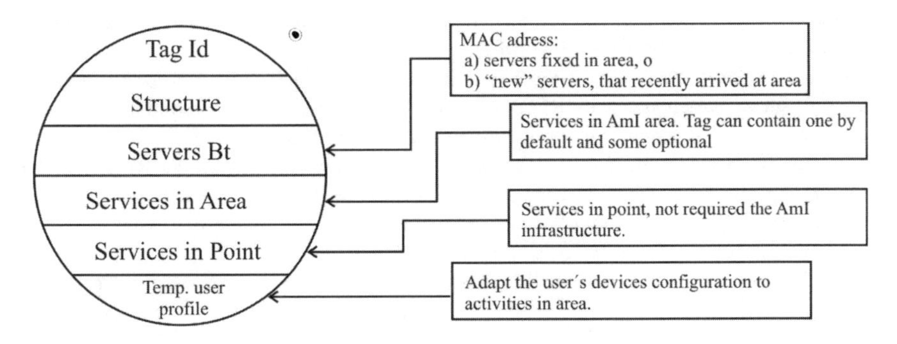

Fig. 5. Tag-NFC structure

6 Conclusions

NFC technology has a promising future and can contribute to the implementation of an AmIE-NFC to great advantage. The embedment of NFC technology in cell phones will facilitate its widespread use by taking advantage of what is inherently familiar to people.

We have developed the first phase of an AmIE-NFC at the MAmI research group; the AmIReG (Ambient Intelligent Research Group). In the daily use of the system we have observed that putting several services in tags provides advantages but cannot be used in all situations.

Users gladly accept all the services if they save efforts but there are those who are obstinate in accepting the tags where there is no perception of savings. In AmIReG, when we put a tag in the door so that users may get into or leave the system; for users this represents an over-exertion and sometimes they forgot to do it. This perception of over-effort disappears when we install an NFC-enabled electronic door lock.

Sometimes the user requires several services in the same place and does not like to move to each one of the devices to obtain them (for example, when this person is at his writing-desk and does not want to go to the printer or public display). In the quest to find an answer in these situations, we are developing a SerBo (service board); where we put several tags to access services from one place.

Although in this work we explain a tag structure, we believe that it is possible to improve it and we continue to exploring, aiming to obtain the best combination in tag structure for all possible situations.

References

1. Information Society Technologies Advisory Group ISTag, Ambient Intelligence: from vision to reality, European Commission, p. 31 (2003)
2. Dey, A.K.: Understanding and Using Context. Personal and Ubiquitous Computing Journal 5(1), 4–7 (2001)
3. ECMA_International, Near Field Communication -white paper- (2004)
4. Bravo, J., Hervás, R., Chavira, G.: Ubiquitous computing at classroom: An approach through identification process. Journal of Universal Computer. Special Issue on Computer and Educations. 11(9), 1494–1504 (2005)
5. Bravo, J., Chavira, G., Hervás, R., Nava, S., Sanz, J.: Visualization Services in a Conference Context: An approach by RFID Technology. Special issue of Ubiquitous Computing and Ambient Intelligence. Journal of Universal Computer Science (2006)
6. Chavira, G., Nava, S.W., Hervás, R., Bravo, J., Sánchez, C.: Combining RFID and NFC Technologies in an AmI Conference Scenario. In: Eighth Mexican International Conference on Current Trends in Computer Science (ENC 2007), Morelia, Michoacán, México (2007)
7. `http://www.nfc-forum.org` (cited April 28, 2008)
8. Want, R.: An Introduction to RFID technology. IEEE Pervasive Computing 5(1), 25–33 (2006)

Interaction by Contact for Supporting Alzheimer Sufferers

José Bravo, Carmen Fuentes, Ramón Hervás, Gregorio Casero, Rocío Gallego, and Marcos Vergara

MAmI Research Lab
Castilla-La Mancha University
Paseo de la Universidad 4, 13071, Ciudad Real, Spain
Jose.Bravo@uclm.es

Abstract. Alzheimer disease patients need holistic attention by caregivers, due to the fact that they present with disorders of memory and orientation. However, communication on the state of each patient is sometimes deficient. In this work we present a program to complement and support the daily activities of an Alzheimer's centre. This is achieved by adapting Near Field Communication Technology. By means of touching tags with mobile phones. Day centre caregivers can manage patient information through such pieces of equipment (tags) placed in the patients' environment, i.e. on patients, places, and devices used by the patients. With this simple interaction, some problems involved in the daily routine of a day centre can be solved, since basic information about the patient, orientation, door security, therapy and other support activities, is relayed directly to the caregivers. Information about incidents is appropriately gathered and received by caregivers, physicians and relatives. A complement to therapy, to be used at home, is also presented here.

Keywords: Ambient Intelligence, Ubiquitous Computing, NFC, Context Awareness.

1 Introduction

The Ubiquitous Computing paradigm proposes the idea of placing computational devices around us, in an effort to make the traditional computer disappear [1]. This idea, from Mark Weiser in Xerox Lab, now complemented by Ambient Intelligent vision [2], proposes the adaptation of sensorial capabilities as inputs to the system, which means that users may avoid using computers when carrying out some tasks in their traditional daily activities. In essence, the idea is that the user should focus on the task and that technology should disappear. The fact is, however, that many people do not use computers in this way in their daily activities. It has to be recognised that, despite computers have clear advantages, they are not well integrated into care environments and Weiser's vision is not being played out successfully in these scenarios [3]. To ensure that this happens, it is necessary to develop applications for different devices which permit a natural interaction. This involves embedding these into everyday objects and activities. Applying these ideas to healthcare scenarios and to aid people with dementia, there are numerous initiatives that propose technologies to assist and support Alzheimer sufferers. These include products to help in problems with speaking,

J.M. Corchado, D.I. Tapia, and J. Bravo (Eds.): UCAMI 2008, ASC 51, pp. 125–133, 2009.
springerlink.com

hearing, eyesight, memory, cognition, etc. However, choosing the correct device for each patient is still difficult. While one person responds appropriately to a monitoring system, another may prefer a simple sound to refresh his or her memory. So, there are devices that can be adapted to the individual patient. In this sense we know that sensors have to be unobtrusive and wireless in such a way that, when placed around us in the environment, they make it possible to obtain data about temperature, the presence of gas, whether someone has fallen or is missing from their room and so on. Actuators allowing automatic switching on of lights, intrusion detection alarms, etc. have to be considered. Others devices like clocks, calendars or locator devices reinforce memory and give appropriate aid to patients with the disease within their own homes. Finally, mobility problems, location, reminders of the schedule patient routine, etc., are also important.

In this work we present a technological adaptability which is not specifically for people with Alzheimer but rather for the carers of those who have this disease. In giving that care, natural interaction should be the best way. However, after trying to obtain service from the environment with an identification technology (RFID) and not achieving suitable results, our proposal is based on a combination of RFID and the mobile phone: the NFC. It allows not only identification, but processing, storage and communication capabilities with a simple touch between mobile phone and tags.

Below, we give an introduction to Alzheimer's disease and the technology applied, along with the corresponding ontology. We then put into practice the idea of treatment of information based on a simple interaction, transforming it at three levels: care-assistants, relatives and physicians. A complement for supporting therapy is also studied. Finally, conclusions and future work are set out.

2 The Context: An Alzheimer Day Centre

Alzheimer's disease is a neurodegenerative illness where the patient displays cognitive impairment and behavioural disorders. Typically, it is characterized by a progressive loss of memory and of other mental abilities, as nerve cells (neurons) die, and different parts of the brain atrophy. The disease usually continues for 10-12 years, although this may vary considerably between patients.

Alzheimer's patients may be admitted to a day centre to give respite to their relatives and to improve their own quality of life. Surveillance and patient care is the main objective of this kind of centre, contributing to the patients' ongoing improvement in independent social functioning and combining rehabilitation and therapeutic activities.

These centres are organized as a set of rooms for rehabilitation, therapy, crafts, audio-visual displays and a dining room. In these, a great variety of activities (physical, cognitive, social, etc.) are devised according to the patients' needs and disease level. To achieve this, a professional team of physicians, physiotherapists, assistants, psychologist, etc., is needed and a large supply of auxiliary equipment is required for interventions.

Keeping this in mind, the project's target is to make the job of the day centre staff easier. The idea is to achieve this aim by the automatic generation of medical reports, and the rapid identification of adverse incidents which can then be controlled by caregivers and relatives. To accomplish this task, a proposal based on technological adaptability has been conceived, specifically, that easy-to-use devices such as NFC mobiles, RFID tags, surveillance cameras, etc. will be placed inside the day centre. A fundamental part of this work lies in the identification of the individual patient. This will be achieved by sticking a RFID tag on each patient, who will carry it with them day and night. This tag will store important information for use by relatives, as well as by the day centre staff. Each group will use the information in a different way. For instance, day centre staff can use it to adapt to the changes in the patients' state of health or activities.

2.1 The Technology: NFC

In order to create applications for supporting daily activities in an assisted environment, it is necessary to adapt sensorial capabilities. Our experience, which bears mind Weiser's vision, is through identification technologies [4] [5].

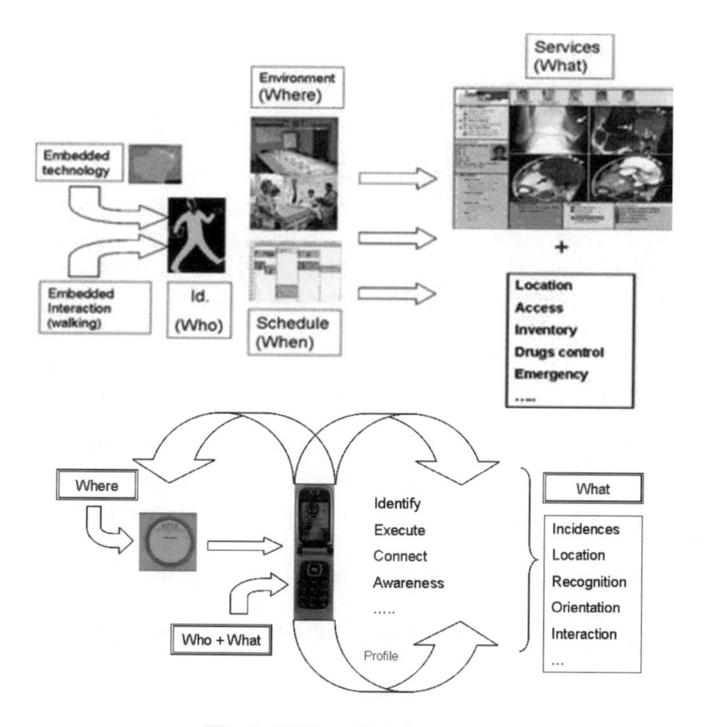

Fig. 1. RFID and NFC models

On the one hand RFID technology offers an easy method for interacting with the environment. Only wearing tags, people can obtain services relevant to the context in which they are living. But some disadvantages have to be considered. Firstly, even a single interaction supposes the patient having to pass near to an antenna, and the fixed nature of such antennae and reader carries associated infrastructure costs. On the other hand, the NFC technology changes the model (see fig. 1). While in the first case,

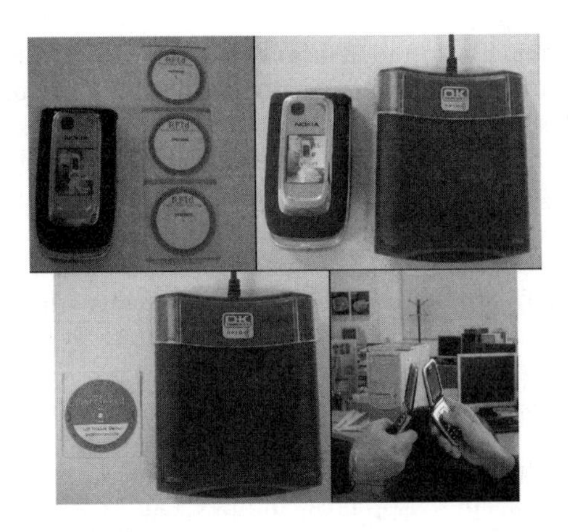

Fig. 2. NFC devices and operation modes

readers and antennas are fixed, in the second one, they are mobile. In addition, in the first case, tags are mobile, in the second one, they are mobile or fixed [6].

NFC is a short range technology using a frequency of 13.56 Mhz. The system consists of two elements: the Initiator, which controls the information exchange (called reader in RFID); and the Target, which responds to the promptings of the initiator (called tag in RFID). In addition, two operation modes are available: Active and Passive. In the Active one, both devices generate their own field of radio frequency to transmit data (peer to peer). In the Passive one, only one of these devices generates the radiofrequency field, while the other is used to load modulation for data transfers. Figure 2 shows the three types of NFC devices: mobile phone, tag and reader. As well as that, the four modes of operation are presented.

It is important to mention that, by just bringing a mobile phone near a tag, the information contained in the tag can run the corresponding mobile phone application automatically. This is due to the PushRegistry of J2ME and the JSR 257 API of NFC. To achieve this, one needs to enter the name of the applications to be executed into the tag. The following lines show two examples for therapy and rehabilitation activities.

MIDlet-Push-1:
ndef:mime?name=application/Therapy,Presentation.VisualMIDlet_INCIDENCES,*

MIDlet-Push-2:
ndef:mime?name=application/Rehabilitation,Presentation.VisualMIDlet_INCIDENCES,*

2.2 Ontology

Now that the day centre and the technology have been described, the next step is to illustrate the context ontology. A. Dey defines context as "*any information that can be used to characterize the situation of an entity. An entity is a person, place, or object that is considered relevant to the interaction between a user and an application, including the user and application themselves*" [7].

Dogac proposes the creation of context ontology to achieve a better scenario understanding in Ambient Intelligence [8]. At this point we define the most important aspects of our scenario as follows:

User: The user is an active entity requiring and consuming services that interact implicitly with an intelligent environment which is aware of the user's identity, profile, schedule, location, preferences, social situation and activity. This entity is composed of:

- *Dependant person.* User with a dependency grading who requires various types of special care. These will be those favoured by the services available.
- *Assistant.* User who helps in particular tasks related to patient care.
- *Physician.* Medical professional entrusted with the monitoring of the dependant person's state of health.
- *Nurse.* User with enough professional training to take specialized care of the dependant person.
- *Relative.* User who belongs to family or caring circle of the dependant person.

Services: When we talk about services, we are referring to all the activities that the system might offer the user to satisfy their needs. These services include:

- *Information management.* This means processing the context information, filtering it so it can be visualized and understood for later decision making. Most of the users would benefit from this service. In addition, other services such as identification of user, access, display control and interaction between user and carers are managed.
- *Monitoring.* To understand what is happening in the environment, it's necessary to know the origin of the information using sensors. The user, as well as the environment, thus needs to be monitored.
- *Decision making.* Automatic processes mean that action can be taken based on the information available at the exact moment an incident happens.

Environment: In a day centre, the entities that form the environment are those physical entities that could be found in the next list:

- *Door.* In a place like a day centre, it's necessary to control doors to avoid the possibility of patients escaping (or intruders entering).
- *Outside area.* Places outside the building, i.e. the yard or the swimming pool.
- *Room.* The day centre is composed of a group of rooms. Each of them would have certain attributes to differentiate it from each others.

Device: Mechanisms capable of modifying the context by the explicit or implicit interaction of the user or to collect information about the context or users. Some of the devices considered are listed below:

- *Tag.* Is a RFID tag that contains information about the place, object or person it is attached to. Depending on the kind of tag (on a place, object or person) the information stored in it and its structure will vary.
- *NFC Mobile Phone.* Device capable of reading near RFID tags. It is a special type of mobile phone that contains an NFC reader and allows information exchange between two mobile phones, as well as reading and writing the information stored in RFID tags.
- *Sensor.* Bluetooth-enabled device recording the dependent person's state of health.
- *Visualization system.* System used to show the user the information he/she needs.

- *Door actuator*. Its function is related to the *Door control* service. This service will send an opening signal to the device, so that it executes the order.
- *Server*. Server computer that stores and provides the information for each user of the system.

Four ontologies will be obtained from these four entities, allowing them to be joined in context ontology.

According to Noy and Musen [9], an ontology fusion is to generate ontology from the initial ones, and an alignment consists of establishing different kinds of links between two or more ontologies, while keeping the original ones.

Nowadays, there are three methodologies (FCA-Merge, Prompt and Onions) for ontology fusion. After analyzing them, the decision made was not to follow any of them, because they are based on criteria irrelevant to our context. For example, they use name concordance in different ontologies to link these entities. Our fusion proposal was made from the knowledge of the day centre's context, looking for the relationship between entities belonging to different ontologies, in such a way that we could obtain definite context knowledge.

Our context ontology resulted from the fusion of four ontologies. The *User* (physician, dependent person, assistant, etc.) is in the *Day centre*, being able to use the active Devices and, at the same time, to benefit from Services offered.

When the fusion of the four ontologies was made with less abstraction, it resulted in a too large a diagram. That's why only the most significant part of the diagram is shown in this document (Fig. 3).

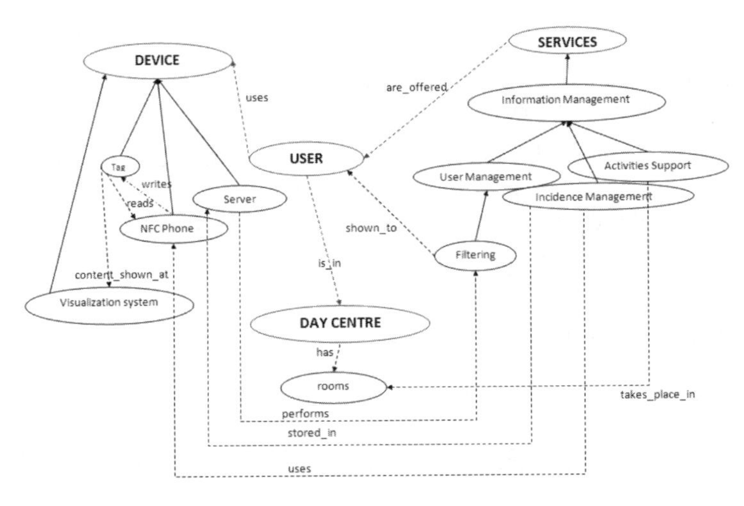

Fig. 3. Ontological fusion

This figure aims to show what information is known about the incidents which occur. When an incident happens (*Incidence Management*) at the day centre, it is written or read from or to the mobile phone (*NFC Phone*) and stored in RFID tags (*Tag*). This stored information could be shown in the different *Visualization systems* or stored in the *Server*, which will filter the incident in such a way as to show just the information necessary for the *User*, as will be detailed in the next section.

3 Managing Information

One of our system's main services is the management of patient incidents, in order to make the difficult job of the caregivers easier.

At present, when a caregiver notices a patient incident, he or she needs to stop doing their job, leave their patients with another caregiver, and write the incident in a notebook manually. There are many clinics and centres which follow this protocol, causing numerous problems. These include the interruption of day to day work, lack of formalism and the impossibility of searching for the information and generating accurate reports for physicians at a later date.

We need to address this in a way that will help caregivers with the simplest possible interaction, without making their job even harder. Reaching this objective is discussed in the next three stages:

The first step to accomplish our task successfully is to formalize the type of incident a patient may have. It is not a good idea for each member of staff to record the same incident in a different way. So, a long list of all possible incidents and issues has been devised. In this, we have separated each area of the day centre into a separate category (see Figure 4).

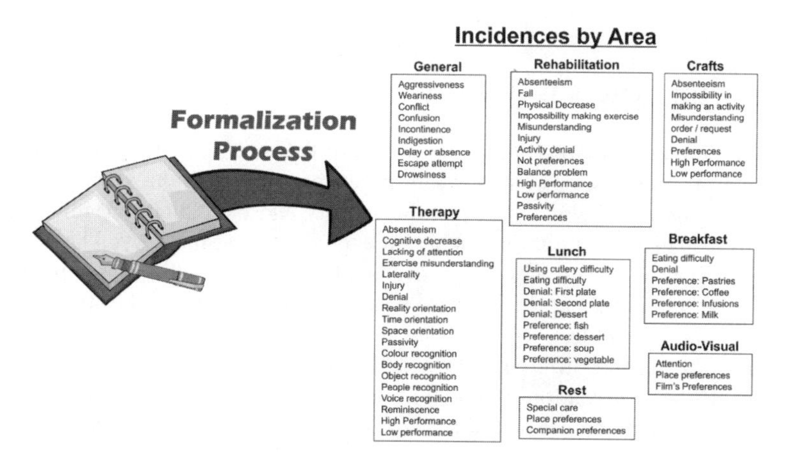

Fig. 4. Formalization process

In addition, each centre could add its own incidents through a common server interface or even a website. This is a concept that we have called: meta-incidents.

With the incidents already formalized, we are now able to offer a form to the user so that he/she can add the incidents to the system in a consistent way. This should be as simple as possible and we have been able to use NFC technology to carry out this task. In combination with a mobile phone, NFC technology allows a user to "insert" the incident on the patient himself/herself, via the NFC tag. This can then be accessed easily by others.

Not every incident should necessarily be displayed in the same way to every member of the staff. Physicians would want to know how incidents relate to the patient's health, while assistants are more interested in the behavioural implications. What is more, there may be incidents that are not relevant at all unless they happen several times over a limited period of time.

Overall, a special format for the information is needed which will demonstrate each person's incidents and activity without overwhelming the carer with irrelevant data. We would recommend that they take some action after several incidents or take particular note if an incident has been repeated several times in a week, paying attention to the patient's record. Using this record, we also could automatically generate a full medical report for the physician. Figure 5 summarizes these ideas.

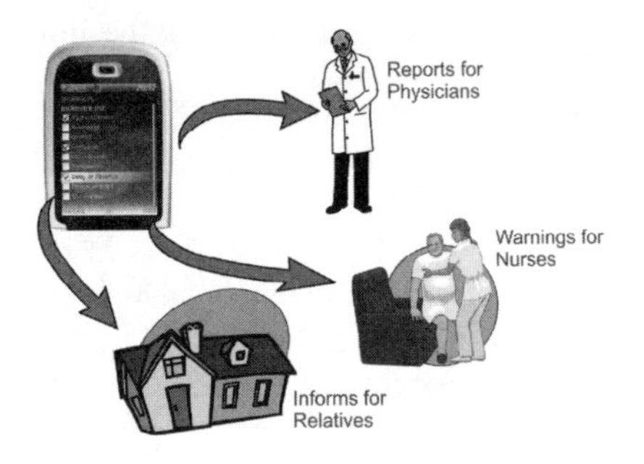

Fig. 5. Personalized Information

4 Supporting Therapy Activity

In order to demonstrate some activities we have focused on therapy as an example. For this we propose a simple interaction to get information about each patient from the server. In the case below, with just the activity and the patient tag involved, it is possible to visualize on a public display an exercise to test memory. The exercise is contained within the server, which then sends it to a display thanks to a wireless video sender. Those taking part in the exercise wear some tags which makes the interaction shown in figure 6 possible.

This device can offer different data: video, voice, text, etc. Also, the corresponding exercise structure is given, only, touching tags. All the data about each patient is accessible to relatives who are introduced to the appropriate centre web site. For that,

Fig. 6. Complement to Therapy

some advice is offered in order to find the best type of memory information. This complement to therapy can be managed at home. For that we have implemented some applications through TDT. These have been developed with MHP language and are automatically generated from information of the day centre server. So, relatives wishing to aid patients only have to interact with a television set, using the remote control.

5 Conclusions and Future Work

In this work we have demonstrated the technological adaptability of NFC as a complement to the Alzheimer day centre activities for caregivers. Keeping in mind the initiatives on Ambient Intelligence, we have focused on the ease of interactions which use touching tags. With NFC, numerous services can be managed. In addition, different activities can be complemented through the visualization of information reinforcing patients' memories. The way the information is treated promotes adaptability for assistant, relatives and physicians. In this paper, an optimization offering data about patient incidents is proposed. In future work we intend to consider complement therapy games, door control, surveillance cameras, etc. Also, an exhaustive study of incident statistics could be useful data for those researching Alzheimers disease. Finally, a visual complement to day centre therapies have been a implemented by a TDT system and offered at home.

Acknowledgments. This work has been financed by the FIT-350300-2007-84 ALIADO project of the Industry Ministry.

References

1. Weiser, M.: The Computer for the 21st Century. Scientific American 265(3), 94–104 (1991)
2. ISTAG, Scenarios for Ambient Intelligence in 2010 (February 2001), `http://www.cordis.lu/ist/istag.htm`
3. Want, R.: People First, Computers Second. IEEE Pervasive Computing 6(1), 4–6 (2007)
4. Bravo, J., Hervás, R., Chavira, G., Nava, S.: Modeling Contexts by RFID-Sensor Fusion. In: 3rd Workshop on Context Modeling and Reasoning (CoMoRea 2006), Pisa, Italy (2006)
5. Bravo, J., et al.: Visualization Services in a Conference Context: An approach by RFID Technology. Special issue of Ubiquitous Computing and Ambient Intelligence. Journal of Universal Computer Science 12(3), 270–283 (2006)
6. Bravo, J., Hervás, R., Chavira, G., Nava, S., Villarreal, V.: From Implicit to Touching Interaction: RFID and NFC Approaches. In: Human System Interaction Conference (HIS 2008), Krakow, Poland (2008)
7. Dey, A.: Understanding and Using Context. Personal and Ubiquitous Computing 5(1), 4–7 (2001)
8. Dogac, A., Laleci, G., Kabak, Y.: Context Frameworks for Ambient Intelligence. In: eChallenges 2003, Bologna, Italy (October 2003)
9. Fridman, N., Musen, M.: An Algorithm for Merging and Aligning Ontologies: Automation and Tool Support, Stanford (1999)

Secure Integration of RFID Technology in Personal Documentation for Seamless Identity Validation

Pablo Najera, Francisco Moyano, and Javier Lopez

Computer Science Department
University of Malaga,
E.T.S.I. Ingeniería Informática, 29071, Málaga, Spain
{najera,moyano,jlm}@lcc.uma.es

Abstract. Seamless human identification and authentication in the information system is a fundamental step towards the transparent interaction between the user and its context proposed in ambient intelligence. In this context, the IDENTICA project is aimed to the design and implementation of a distributed authentication platform based on biometrics (i.e. voice and facial image) and personal documentation. In this paper, we present our work in this project focused on the secure integration of RFID technology in personal documentation in order to provide seamless identity validation. Our actual work status, first results and future directions are described in detail.

Keywords: Biometry, identity verification, privacy, RFID, security.

1 Introduction

As computing gets pervasive, new scenarios arise where automatic identity validation is required in order to allow access or grant pertinent privileges to a user without human intervention. Taking advantage of emerging technologies (e.g. smart cards, voice recognition and RFID tags), the IDENTICA project will provide a distributed authentication platform based on biometrics and personal documentation in order to make easier human identification. Embedding RFID technology in personal documentation requires adequate security mechanisms that provide access control and ensure confidentiality and integrity of data stored in a tag: this point is where UMA is interested in.

2 Focus

Provided biometry-based identity recognition techniques (i.e. facial image, iris and voice recognition), as well as optical document interpretation techniques such as OCR (Optical Character Recognition) and ICR (Intelligent Character Recognition) from other research groups, our work focuses on the integration of RFID technology in personal documentation including the secure storage and access of biometrics and personal data from RFID tags to the authentication platform.

In our research on how RFID technology can be securely incorporated in the area of human authentication preserving data privacy, several matters must be kept in

J.M. Corchado, D.I. Tapia, and J. Bravo (Eds.): UCAMI 2008, ASC 51, pp. 134–138, 2009.
springerlink.com

mind, such as eavesdropping and man-in-the-middle attacks, non-authorized access to a tag leading to data leakage or corruption, as well as counterfeited tags. In order to prevent potential attacks, several steps need to be taken.

First of all, it is necessary to make a state of the art study in order to determine which security mechanisms (e.g., authentication protocols, symmetric and asymmetric encryption algorithms, key generation algorithms or secure ID-disclosure protocols) are used nowadays in RFID technology and documents.

From this point on, we must proceed to the design and implement of the modules that will establish the link between tag's memory and the authentication platform. In order to enhance data privacy, we must design secure tag identification mechanisms (preventing ID disclosure and later bearer tracking), mutual authentication and key generation protocols with the purpose of providing access control, preventing tag counterfeiting and securing the communication channel.

As a project contribution, it has been proposed to design an infrastructure to manage the keys required for access the tag and establish a secure communication channel. Until now, it has not been developed a robust, reliable key management infrastructure oriented to RFID-based systems. Due to this lack, actual applications tend to apply the same key for all the tags or rely on weak key generation protocols (i.e. based on predictable values) degrading the whole system security.

In addition, it has been proposed to apply the biometric features used in the project to develop strong key generation protocols. Thanks to this biometric information, it could be possible to establish that a reader can only access data inside a tag if the person (who it is going to be authenticated) is present. Moreover, it could be possible to generate access keys to RFID tags from biometric measures of the bearer of the document, and so in this context, there would be no need of accessing a key management infrastructure. Biometric features that will be discussed include facial, iris and fingerprint recognition.

3 Tasks Performed and Future Work

With the purpose of defining secure mechanisms to avoid tag ID disclosure to unauthorized third parties and provide mutual authentication as well as encrypted communications between RFID tags and readers, we required to perform a through study on the RFID-related security mechanisms proposed in the literature.

As a result, we have identified several protocols and techniques based on different onboard RFID tag capabilities to fulfill the previous security requirements.

A wide range of out-of-tag techniques have been proposed in order to prevent unauthorized readers from identifying a tag. These schemes range from the Faraday cage (where a tag is confined in a physical shield to block its output) to active jamming techniques (as in the blocker tag [3] or soft blocking [4] solutions, where a device broadcasts a signal to prevent readings). These techniques do not force specific tag features, so enabling the use of security-naked RFID tags, at the cost of actively involving the user in ensuring his privacy.

In the analysis of solutions for non-cryptographic tags, we found several proposals based on tag pseudonyms (such as tags with rewritable memory [5,7] or sets of pseudonyms [6]) where the tag stores one or several alternative IDs that a legitimate reader will be able to link to its real ID, thus avoiding identification from evil readers, but not tracking attacks until pseudonyms sets are refreshed. More exotic solutions for non-cryptographic tags include the error-prone approach of tags with antenna energy analysis [8], where the tag estimates reader to tag distance based on signal/noise ratio and restricts data disclosure to far-off readers.

In the area of security solutions oriented to cryptographic tags, we have identified a wide range of proposed solutions that differ in the on-board circuitery required (thus tag cost and suitable applications) as well as security level provided.

Among these schemes, we would like to highlight a few different or interesting ideas shared by sets of solutions. In the simpler approaches, hash-lock schemes [9] only require a tag to integrate hash functions in order to implement access control schemes. As a reader demonstrates the knowledge of a shared secret, the tag can be unlock to read its contents and relocked to avoid future data disclosure.

Reencryption functions [11] allows a tag to provide a metaID that has been encrypted an undefined number of times, but the reader only requires to decrypt it once to recover the real ID. This scheme prevents ID disclosure and tracking protection, at the cost of implementing asymmetric cryptographic circuitry onboard.

Another interesting approach pointed out in this research is based on PUF (Physical Unclonable Function) functions [10] that provide a unique fingerprint thanks to the unpredictable behavior of logic gate and wire delays of each RFID tag. Thanks to this signature, tag authentication schemes turn out to be unclonable.

Last on our highlights of the security mechanisms for RFID tags research, asymmetric cryptography that seemed to be out of scope for the highly constrained RFID technology, it is turning out to be feasible thanks to elliptic curve cryptography [12] that requires only a fraction of the number of logic gates providing a similar level of security than traditional asymmetric cryptography.

Even if our study found hundreds of papers on RFID security, most of literature seems to focus on the higher constrained technology branch (i.e. EPC Class 1 Gen1&2 tags) oriented to tagging products in the supply chain and beyond raising important privacy threats. Due to cost restrictions, these are the less capable tags providing an inadequate security level for personal documentation.

Cryptographic mechanisms for ePassports cover a comprehensive set of countermeasures: Passive Authentication (i.e. data on tag signed by the issuer); Basic Access Control that prevents skimming and eavesdropping by authenticating the reader and encrypting messages; and Active Authentication where the tag is validated proving the possession of a private key. But key generation for BAC is based on basic owner and document data, thus enabling several attacks [1].

Regarding RFID technology branches, semi-active and active tags where discarded due to size and cost reasons. In passive RFID technology, UHF (i.e. EPC Class 1 Gen1&2/ISO 18000-6), LF (i.e. ISO 18000-2, ISO 11784/11785) and part of HF (i.e. ISO 15693) RFID branches were designed for substantially different applications and do not provide adequate cryptographic features [2]. So we concluded that passive HF tags based on ISO 14443 are the most suitable RFID tags for our purposes due to computational resources and onboard security facilities.

Further work on actual RFID tags conforming ISO 14443 revealed candidate choices including SmartMX P5C family from NXP, the slightly slower SLE66CLX from Infineon and the expected RF360 from Texas Instruments. The SmartMX tag was selected due to higher performance and actual availability. The SmartMX P5CT072 features a 72kB EEPROM and a cryptographic coprocessor that supports RSA, ECC, 3DES and AES functions.

In order to embed the security mechanisms on a tag, a complete implementation of tag operating system seemed required. The resulting ROM mask would be provided to NXP in order to manufacture the final ICs. To achieve this goal, Keil's PK51 developing environment and SmartMX DBox testing box were selected. Similar alternatives like Ultra-SmartMX from Ashling did not allow squeezing the full range of IC features according to providers. However, a different approach based on NXP tags preloaded with the JCOP operating system have eliminated these hardware requirements and simplified the development process.

Further work will define and test new security schemes defined in the project. In particular, robust key generation algorithms for authentication and encryption derived from owner's biometrics, instead of basic personal and document data as used in actual ePassports, are expected.

A different approach will work on an adequate infrastructure for RFID key management in the proposed scenario that will handle pre-established high entropy keys without requiring on-reader robust key generation algorithms.

4 Conclusions

Seamless identity validation through RFID technology in personal documentation requires adequate security mechanisms. We have shown the first steps and results of our work where a biased effort towards the most constrained RFID technology branches, not suitable for personal documentation, has been detected in actual proposed solutions and specific mechanisms for electronic documentation have turned up to present weaknesses. Our future directions will focus on an adequate RFID key management infrastructure and key generation schemes based on biometric data that will potentially solve the deficiencies in actual solutions.

Acknowledgments. This work has been partially supported by the Spanish Ministry of Industry through the IDENTICA project (FIT-360503-2007-3). The first author has been funded by the Spanish Ministry of Science and Education through the National F.P.U. Program.

References

1. Avoine, G., Kalach, K., Quisquater, J.: ePassport: Securing International Contacts with Contactless Chips. In: Tsudik, G. (ed.) Financial Cryptography and Data Security - FC 2008, IFCA. Springer, Cozumel (2008)
2. Phillips, T., Karygiannis, T., Kuhn, R.: Security standards for the RFID market. Security & Privacy Magazine 3(6), 85–89 (2005)
3. Juels, A., Rivest, R., Szydlo, M.: The blocker tag: Selective blocking of RFID tags for consumer privacy. In: Proceedings of the 10th ACM Conference on Computer and Communications Security, October 27-30, 2003, pp. 103–111 (2003)

4. Juels, A., Brainard, J.: Soft Blocking: Flexible Blocker Tags on the Cheap. In: Workshop on Privacy in the Electronic Society, pp. 1–7. ACM Press, New York (2004)
5. Kinoshita, S., Hoshino, F., Komuro, T., Fujimura, A., Ohkubo, M.: Low-cost RFID privacy protection écheme. IPS Journal 45(8), 2007–2021 (2004) (in Japanese)
6. Juels, A.: Minimalist Cryptography for RFID Tags. In: Blundo, C., Cimato, S. (eds.) 4th Conf. Security in Comm. Networks, pp. 149–164. Springer, Heidelberg (2004)
7. Golle, P., et al.: Universal Re-encryption for Mixnets. In: Okamoto, T. (ed.) Proc. RSA Conference Cryptographer's Track, pp. 163–178. Springer, Heidelberg (2004)
8. Fishkin, K.P., Roy, S.: Enhancing RFID Privacy via Antenna Energy Analysis. IRS-TR-03-012, Intel Research Seattle (2003)
9. Weis, S.: Security and Privacy in Radio-Frecuency Identification Devices, Masters Thesis (May 2003)
10. Tuyls, P., Batina, L.: RFID-Tags for Anti-Counterfeiting. In: Pointcheval, D. (ed.) CT-RSA 2006. LNCS, vol. 3860, pp. 115–131. Springer, Heidelberg (2006)
11. Kinoshita, S., Ohkubo, M., Hoshino, F., Morohashi, G., Shionoiri, O., Kanai, A.: Privacy Enhanced Active RFID Tag. In: Proceedings of ECHISE 2005 (May 2005)
12. Martínez, S., Valls, M., Roig, C., Giné, F., Miret, J.M.: An elliptic curve and zero knowledge based forward secure RFID Protocol. In: RFIDSec 2007 (July 2007)

Semantic Model for Facial Emotion to Improve the Human Computer Interaction in AmI

Isaac Lera, Diana Arellano, Javier Varona, Carlos Juiz, and Ramon Puigjaner

Universitat de les Illes Balears. crt. Valldemossa km 7.5, Illes Balears 07122, Spain
{isaac.lera,diana.arellano,cjuiz}@uib.es

Summary. To facilitate the Human Computer Interaction will be needed to create a friendly interface in AmI. Thus, in this work we propose a semantic knowledge model to represent facial emotions of a virtual character. These emotions depend on the personality, preferences, goals of the character and the environment events. The virtual character's emotions are the nonverbal communication interface between the system and the user. Furthermore, we develop an algorithm to deduce the resultant emotion using previous information, and we represent it with a set of activated facial animation parameters.

Keywords: Facial Emotion, Human Computer Interaction, Semantic Model.

1 Introduction

Ambient Intelligence (AmI) researchers claim for system awareness which is embedded in all types of objects to attend human necessities in a transparent and unobtrusive way. AmI discipline involves Human-Computer Interaction (HCI) field [1], where researchers work in the creation of graphical interfaces, interaction devices, voice synthesizers, movement capture devices, among other fields. AmI researchers try to obtain an interaction more "human-like" although this task is rather difficult due to the complexity of the human communication. The way to communicate the message (linguistic message, non-linguistic conversational signal, emotion, person identification), can take place (facial expression, head movement, tone of voice, etc.), where it take place and the actions of the speech listener [2] are factors that difficult the computational analysis of the human language.

In this work, we propose a knowledge semantic model to represent facial emotions of a virtual character. The virtual character's face "humanizes" the system with the aim of facilitate the feedback communication. Furthermore, we represent personality traits and preferences which clarify their character so that we can emphasize and understand better its message. Our model involves two HCI fields: the context identification through events and the affective feedback among the system and the users. Facial expression will change according to character's personality, preferences, goals, and the events of the environment.

J.M. Corchado, D.I. Tapia, and J. Bravo (Eds.): UCAMI 2008, ASC 51, pp. 139–148, 2009.
springerlink.com

2 Related Work

In the field of emotions and personality one work is the one of Egges *et al.* [3] which present a generic personality and emotion simulator where mood can have any number of dimensions. The relationship between emotions, personality, and moods is done using matrices with empirical values that relate these elements according to their influence on each other. Kshirsagar *et al.* [4] presents a layered approach for modeling personality, moods and emotions. They adopt the Five Factor Model (FFM) of personality, and Bayesian Belief Network for the implementation of the model.

In the semantic representation, García-Rojas *et al.* [5] proposed an ontology in order to support the modeling of emotional facial animation in virtual humans using the standard MPEG-4. The structure of the ontology specifies a facial expression defined by an archetypal or intermediate emotion profile that utilizes psychological models of emotions. The ontology allows storing, indexing, and retrieving prerecorded synthetic facial animations that express a given emotion. In [6], García-Rojas *et al.* presented an ontology that lays out the knowledge of previous work on body animation expressions within MPEG-4 framework. The animations are classified into a model of emotion and some parameters are presented in order to enhance the face expressivity.

In the area of HCI within AmI discipline, Pantic [2] introduces the importance of identify the emotional state in AmI environments for an affective computing. Also, she proposes a method for automatic recognition of facial expressions though the position of facial points.

3 Semantic Data Model

We propose a generic knowledge model which represents elements as the ambient and events that surround a character, his personality and preferences. We also design an algorithm that permits to obtain the emotions of a character using the elements mentioned before. We improve context representation using ontologies because they permit to define new knowledge and easily reuse it. Also the applicability of inference rules allows to improve the outcoming of emotions, adaptability and evolution of the character in front of new events. Finally, ontologies are able to develop an automatic process using artificial intelligence techniques. Ontology repositories, which are external data sets, provide new specific context knowledge for every situation.

The figure 1 represents the base-model; the upper layers show the future knowledge modules for adaptation in AmI or other environments. Ontologies define the model requirements such as the personality, actions, emotions, objects, characters, animals, goals, environment, etc. The upper layer is used to access the ontology data. Moreover, this layer will be composed of query engines, an inference engine, and a database manager. AmI Module interprets all events in temporal order and fixes the sequence of credible facial expressions. However, in our case we consider the event as independents facts. The events are classified in three types: indirects, directs, and inferreds. Indirect events are obtained

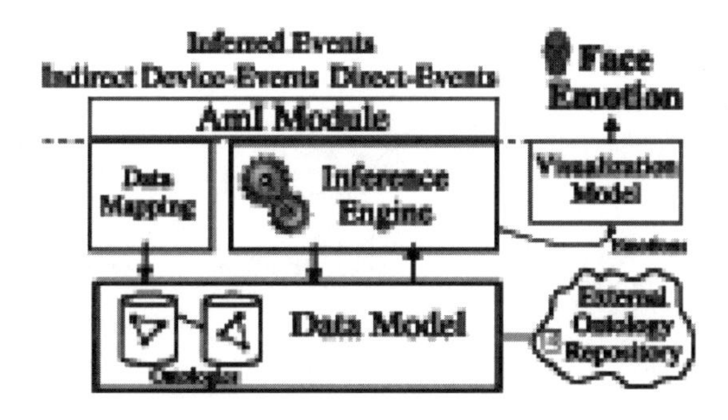

Fig. 1. Architecture based on two layers data information and inference engine

by the context interpretation using external sensors. Direct events are user's orders, i.e. "turn on the TV", there are not any kind of interpretation in that sentence. The inferred events are done by superior entities for specific adaptations as: management of hospital patients, teaching, or security systems. In this work we have focused on inferior architecture's layers: context representation, character's personality and preferences, and an algorithm to obtain character's facial emotions.

3.1 Event Ontology: Environment Description

The interaction depends on the context that provides new information, and it affects the user's behaviour. Events trigger the facial reaction. This ontology defines concepts that describe the events presented in generic domain, which cause a change in the affective state of the character. The model is based on action description, more precisely on four questions: *Who* (represents persons and animals present in the action), *Where* (represents the place and its description), *What* (represents the action, contains a verb and the complements), and *When* (is a period of time).

In this ontology, we use concepts that have been already defined on external ontologies, such as: person, animal, pet, material object, location, and time interval. They are obtained from ontologies with a different purpose from ours. Importing external knowledge, more complex structures without reach to understand specific domains can be represented. Thus, we save code and time development.

3.2 Action Categorization Ontology

Scenarios where several action produce different emotions are plausible. To solve that, we classify them as group of emotions. So that, there are actions emotionally similar although they are not identical. This classification is represented in the action categorization ontology.

Every action has a unique genre. This link represents a "standard" vision of this action. For example, most people think action "to kill a person" has a

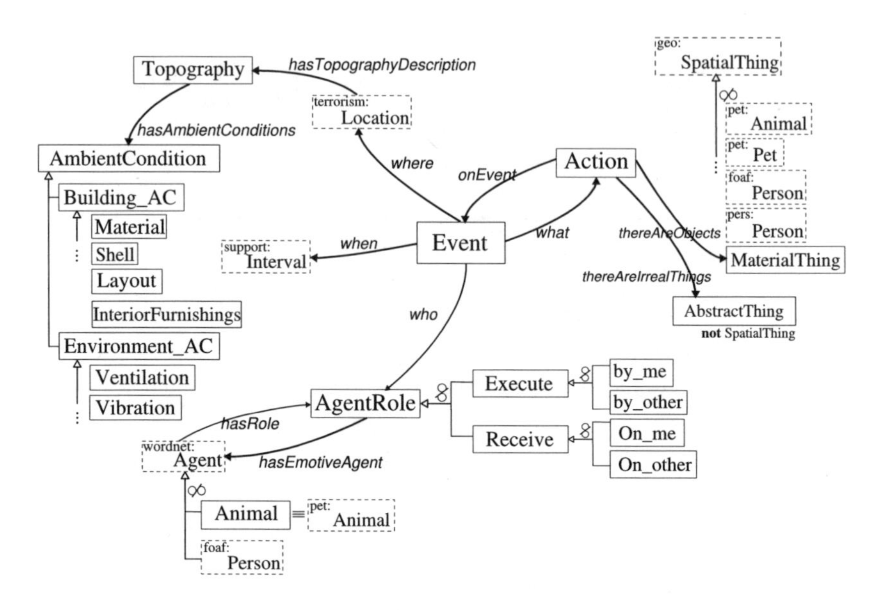

Fig. 2. Simplified Event Ontology Diagram

negative connotation, thus, it belongs to negative genres: *horror*, *war*, or *thriller*. We identify a set of genres such as: *action*, *comedy*, *family*, *history*, *mystery*, *thriller*, among others. The greater the number of genres, the better the quality of interpretation of the emotion categorization. However, this number has an operational threshold that is going to be explained in next section. An perfect solution would be a binary relationship between one genre and one action.

3.3 Personality-Emotion Ontology

The core of the ontology is composed by the character, the personality, the emotional categorization, and the emotions. Starting from the top left of the fig. 3, a *character* has *goals* and *preferences*. A *goal* is considered as the occurrence of an event with the highest desirability degree, for instance, "keep a warm temperature at home". *Preferences* can be divided in two types: *sporadic* based on motivational states (thirsty, hungry, etc.), or *not sporadic* which can be: an agent (person or animal), an action, a material object, or preferences from the environment. All preferences are categorized in a class named *PreferenceEmotiveScale*, in concordance with the seven item Likert scale.

In our ontology the scale is given by the following values: *7-StronglyGood*, *6-VeryGood*, *5-Good*, *4-Moderate/Indifferent*, *3-Bad*, *2-VeryBad*, *1-StronglyBad = Phobia*. Therefore, everything that could like, dislike, or make the character feel indifferent, is grouped in this class.

Another aspect of the character is his *personality*. In this ontology the personality model we have used is the one proposed by Rousseau, which main goal is to classify personality traits in a structured way, identifying their impact on the

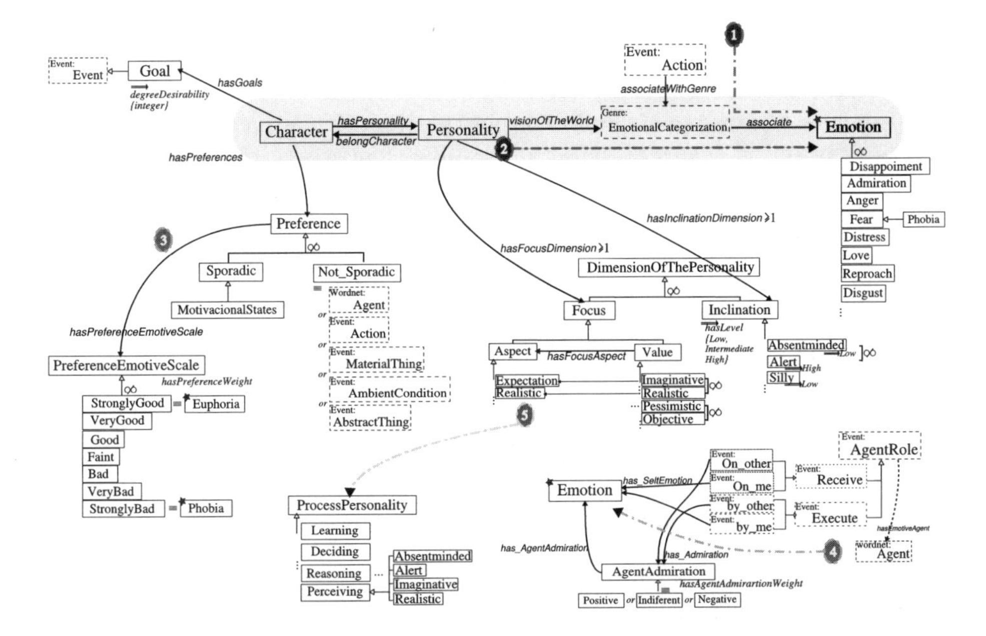

Fig. 3. Simplified Personality-Emotion Ontology Diagram

behavior of the character, moods and relationships. The classification is based on the different processes that the agent can perform: perceiving, reasoning, learning, deciding, acting, interacting, revealing, and feeling emotions. Each process is considered in two levels that form the class *DimensionOfThePersonality*: *Focus* and *Inclination*. They are defined as the aspect an agent focuses on while performing the process, and the tendency the agent has to perform the process [7]. In our ontology, *personality* is linked with the classes *Focus* and *Inclination* through the relations *hasFocusDimension* and *hasInclinationDimension*, respectively. Depending on the values that *focus* and *inclination* have assigned, we can derive a process represented by the class *ProcessPersonality*. The reason to choose the described model is its completeness and the possibility that it offers to derive other personality models from it, as is the case of the Five Factor Model (FFM) [8].

The personality of the character will determine his *visionOfTheWorld* which can be categorized in genres, grouped in the class *EmotionalCategorization*. The genres we have considered were described in the previous section. The combination of the perception according to the personality and the emotional categorization of an event result in a set of emotions, which are the ones proposed in the OCC model [4], except *happy-for* and *fear-confirmed*.

In our model, the level of appreciation of one agent for another is considered in the class *AgentAdmiration*. It can have three levels, or weights: *Positive*, *Indifferent* or *Negative*. An event can also be considered from the point of view of the agent who performs the actions or from the agent who receives the action. If the agent is the one who performs the action (*me*) the set of possible emotions is: *relief, remorse, pity, pride, satisfaction,* or *shame*; and the emotions for the agent that receives the action (*On_other*) are: *pride, pity, surprise,* or *gloating,* depending on the degree of *admiration* for the other. If the agent is the one who receives the action (*On_me*) the set of possible emotions is: *relief, gratification, pride, shame, surprise,* or *disgust*; and the emotions for the agent that performs the action (*by_other*) are: *disappointment, joy, love, reproach, gratitude, resentment, surprise, disappointment,* or *gloating,* depending on the degree of *admiration* for the other.

4 How to Use This Model to Obtain Emotions

The inference engine layer contains the algorithm to obtain emotions from the environment and the character. The developed generic guideline to obtain a set of final emotions is not a closed sequence of steps. The process estimates emotions and their intensities during event generation including external factors. Intensity is the weight of each emotion. The set of higher intensity emotions will be the ones we represent. We distinguish three phases:

(I) In the first phase, we call *base emotions* to the ones obtained from actions (figure 3 tags: ① ②) and the ones obtained in case there are preferences with extreme intensities of the scale (*stronglygood* and *stronglybad*) (*Preference Extreme Emotive Scale, PEE*; fig. 3: ③). Every action has its own cultural interpretation, for example, people culturally associate the action *kissing* to emotions like: happiness and love (fig. 3: ①). However, each person has his own interpretation, its different vision of the world due to its education and other external factors. Those persons can associate emotions as shame or reproach to *kiss* action (fig. 3: ②). We obtain the base emotion combining both interpretations. Extreme values of the preference scale represent the euphoria or anxiety disorder, suffered by people due to strong and persistent feelings of fear or happiness.

(II) In the second phase, we control intensities of the base emotions, which depend on environment actions. Character preferences about objects, persons, or animals, increase or decrease positive or negative emotions respectively (fig. 3: ③).

(III) In the last phase, there are three steps but all of them generate additional emotions with their intensities that add new emotions or increase the

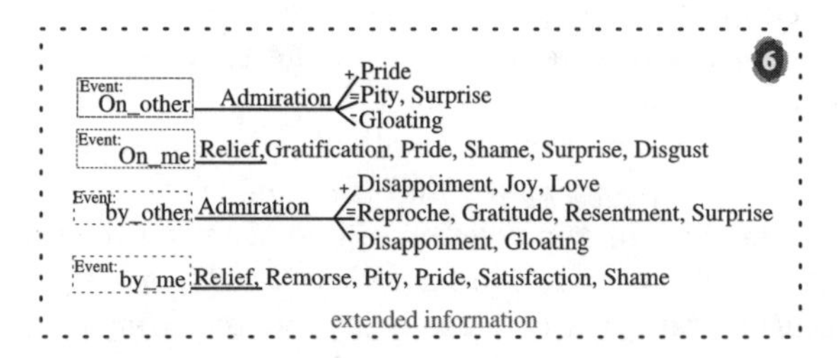

Fig. 4. Extending information about person admiration

intensity of the existent ones. In the first step, the algorithm consider the emotions caused by other people who take part in the action (fig. 3: ④, fig. 4). In the second step is checked if the event is similar to the goal and if so, new emotions with intensities corresponding to the goal's desirability degree are triggered. In the last step, first order logic rules treat particular cases that depend on specific personality traits or a character role. Rules give flexibility and improve process accuracy.

5 Visualization of the Emotional States

We develop an algorithm that works with the facial parameters specified in the MPEG-4 standard, and mixes the universal, also called basic, emotions: *joy*, *happiness*, *sadness*, *anger*, *fear*, *surprise*, to obtain intermediate emotions: *hate*, *disappointment*, *love*, among others. To visualize facial expressions, we have considered each emotion E_i associated with a set of activated facial animation parameters (FAPs), named P_i. Each P_i has a set of values (range of variation) $X_{i,j}$, such as $P_i = \{X_{i,j}\}$, where j is the number of the FAP ($j = \{1, ..., 64\}$). $X_{i,j}$ has the form $[minValue, maxValue]$. For intermediate emotions, the ranges $X_{i,j}$ were generated as a sub-range of a basic emotion, or a combination of two basic emotions, depending on their relation according to the Whissell model [9].

6 A Case Study

We present a group of events that can occur in AmI. We suggest a first event where children are playing with fire. In this case the character has two personalities: one more aggressive and other one more sentimental. In the second event the user is grateful to the system for its support. Finally, in the third event the system set the temperature, illumination, and music of a room.

In the next phase we define three possible genres for each event: *Dangerous Situation*, *Congratulation*, and *Setting Room* respectively. The genre definition depends on each situation. For instance, other event as "House thief"

Table 1. Join Personality-Categorization and Event-Categorization Tables for Aggressive and Sentimental Personality

Traits		DangerousSit.	SettingHouse	Traits(P2)	DangerousSit.
Perceiving	Absentminded	Surprise	–	Alert	Anger, Disgust, Gloating
	Imaginative	Sadness, Fear	–	Realistic	Anger
Reasoning	Rational	Hope, Surprise	Joy	Rational	Disappointment, Surprise
	Objective	Sadness	Satisfaction	Pessimistic	Sadness
Learning	Curious	Pity	–	Incurious	–
	Open-minded	–	–	Intolerant	–
Deciding	Insecure	Fear	–	Self-Confident	–
	Thoughtful	Sadness	Pride	Impulsive	Anger,Hate
Acting	Active	Disgust	–	Zealous	Hate
	Diligent	Disappointment	–	Perfectionist	Annoyed, Disappointment
Interacting	Extroverted	Disgust	Joy	Extroverted	Disgust, Reproach, Resentment
	Neutral	–	–	Hostile	Reproach, Resentment
Revealing	Open	–	Gratification	Open	–
	Honest	Disgust, Sadness	–	Honest	Disgust, Reproach
F.Emotions	Sensitive	Sadness	Joy, Liking	Emotionless	Disgust
	Unselfish	–	Joy	Selfish	–

Table 2. Algorithm Sketch applies in three AmI events

	Events					
	Ev1(**P1**): PlayWithFire		Ev1(**P2**): PlayWithFire		Ev4: SettingHouse	
Step	Case	E(+I)	Case	E(+I)	Case	E(+I)
(1) **PEE**	ϕ	–	ϕ	–	ϕ	–
(1) **(1∩2)**	*Danger.*	Surprise(2), Sadness(5), Fear(2), Hope, Pity, Disgust(3), Disappointment	*Danger.*	Anger(3), Disgust(4), Gloating, Disappointment(2), Surprise, Sadness, Hate(2), Annoyed, Resentment(2), Reproach(3)	*Sett.*	Joy(4), Satisfaction, Pride, Gratification, Liking
(2) **Pref.**	–	–	–	–	*AmbientCond.:*	(+1PE) Temperature, (+1PE) Illumination, (+2PE) Music
(3) **Rule**	R1	Reproach(**2**)	R1	Reproach(**2**)	ϕ	–
(3) **Agent**	⑥	Reproach, Disappointment, Resentment, Surprise	⑥	Reproach, Disappointment, Resentment, Surprise	ϕ	–
(3) **Goal**	ϕ	–	ϕ	–	ϕ	–
	Final Emotions					
	Sadness(**5**) Disgust(**3**) Surprise(**3**) 		Disgust(**4**) Reproach(**6**) Anger(**3**) 		Joy(8) Satisfaction(**4**) Liking(**4**) 	

could be a dangerous situation. The character's preferences are: admiration for a group of children with Faint scale (WordNet:Agent); admiration for the user with VeryGood scale (WordNet:Agent) and he "likes" a temperature around 23°C (event:ambientCondition), illumination at 40% (event:ambientCondition) and chill-out music (event:AbstractThing).

According to character's personality, table 1 shows personality traits for each genre. Only event *Dangerous Situation* is used with aggressive and sentimental personalities.

Table 2 represents the simplified process to obtain emotion. First row contains the triggered event and the first column, the algorithm phases. The two first columns are the same event: children play with fire for both personalities. Our character does not have any extreme preferences (PEE). Afterwards, the algorithm produces a set of emotions belonging to (1∩2), which corresponds to character's personality and action genre (fig. 3 tags: ① ②). Following, step 2 increases or decreases intensity emotions caused by character preferences (fig. 3: ③). In the fourth event, the character's references about temperature, music, and illumination increase the weight of the emotion according to its scale, in the fourth event. In step 3, we define rule R1 as: *If* $Children \wedge \text{"}causethesituation\text{"} \wedge DangerousSituation \mapsto Reproach(2)$. Rule R1 is triggered in the first and second event; being considered emotions elicited by person admiration (fig. 4) and goal's character. Finally, we choose the three emotions with higher intensities to be represented.

7 Conclusion and Future Work

Human Computer Interaction is an essential discipline to improve the interaction between AmI system and user. In this work we focus on the facial representation of virtual character for system request feedback. Human Face transmits non-verbal communicative cues and is able to get on with the users. Thus, we presented a knowledge model using ontologies to define facial emotions of virtual agents. These emotions depend on the personality, preferences, and goals of the character and the events that take place in the context.

We dealed with the "emotional" part of the system so that in our model we represent personality traits and preferences about: agents (persons and animals), physic and abstract things (music, perfume, car,...), ambient conditions; goals and the emotions. Also, the environment is represented using events which are described as an answer to four questions: Who, When, What, and Where. We have proposed a model so it can be easily adapted to any situation. Furthermore, we developed an algorithm to obtain a final set of emotions and respectively intensities. This set of emotions is represented with our algorithm based on MPG4. We show the right work of the purpose with a case study of an AmI.

To sum up, this idea is under development. We are working on two applications to rapidly characterise the personality and the events. In addition we are centering our efforts working in a model of AI to handle the event history and character's behaviour generation. Additionally, we are researching its applicability in other fields as automatic characters in a virtual worlds, virtual teachings, guide of museum, and computer games.

Acknowledgment

This work is supported by the projects *TIN2007-60440* and *TIN2007-67896* from Spanish Ministry of Education and Science.

References

1. Te'eni, D., Carey, J.M., Zhang, P.: Human-Computer Interaction: Developing Effective Organizational Information Systems. John Wiley & Sons, Chichester (2005)
2. Pantic, M.: Face for ambient interface. In: Cai, Y., Abascal, J. (eds.) Ambient Intelligence in Everyday Life. LNCS (LNAI), vol. 3864, pp. 32–66. Springer, Heidelberg (2006)

3. Egges, A., Kshirsagar, S., Magnenat-Thalmann, N.: Generic personality and emotion simulation for conversational agents. Comput. Animat. Virtual Worlds 15(1), 1–13 (2004)
4. Kshirsagar, S., Magnenat-Thalmann, N.: A multilayer personality model. In: Proceedings of the 2nd international symposium on Smart graphics, vol. 24, pp. 107–115 (2002)
5. Garcia-Rojas, A., Vexo, F., Thalmann, D., Raouzaiou, A., Karpouzis, K., Kollias, S., Moccozet, L., Magnenat-Thalmann, N.: Emotional face expression profiles supported by virtual human ontology, 17(3-4), 259–269 (2006); VRLab. - EPFL, CH-1015 Lausanne, Switzerland
6. García-Rojas, A., Vexo, F., Thalmann, D., Raouzaiou, A., Karpouzis, K., Kollias, S.: Emotional body expression parameters in virtual human ontology. In: Proceedings of 1st Int. Workshop on Shapes and Semantics, pp. 63–70 (2006)
7. Rousseau, D.: Personality in computer characters (1996)
8. McCrae, R., John, O.: An introduction to the fivefactor model and its applications. special issue: The fivefactor model: Issues and applications. Journal of Personality 60, 175–215 (1992)
9. Whissell, C.: The dictionary of affect in language. Academic Press, London (1989)

Bridging the Gap between Services and Context in Ubiquitous Computing Environments Using an Effect- and Condition-Based Model

Aitor Urbieta[1], Ekain Azketa[1], Inma Gomez[1], Jorge Parra[1], and Nestor Arana[2]

[1] Software Technologies Area, Ikerlan Technological Research Centre, 20500 Mondragon, Spain
{aurbieta,eazketa,inma.gomez,jparra}@ikerlan.es
[2] Computing and Electronics Department, Mondragon Unibertsitatea, 20500 Mondragon, Spain
narana@eps.mondragon.edu

Summary. In the vision of ubiquitous computing, environments are populated by smart devices which adapt depending on the user context in order to meet user needs and look after them. Nevertheless, current devices describe their functionalities without taking into account context changes derived from service execution, making difficult the automation of the environmental behaviour. Thus, this work tries to cover the gap between services and context, and to integrate them by defining an ontological model for semantic representation based on effects and conditions.

Keywords: Semantic Services, Context, Effects, Conditions, Ubiquitous Computing, Context-awareness.

1 Introduction

A ubiquitous environment establishes a mechanism to provide users with all the functionality of the devices, components and local and distributed software applications in a flexible, integrated and almost transparent way for the end-user [19]. All this implies that the computational devices are integrated, embedded and distributed in the physical media and surroundings around us with the goal of offering user-centered functions. These environments must be observed as the medium that allows users to complete a set of tasks by using the available services. So environments' devices must be aware of the existence of other ones and must be able to establish coalitions among them with limited human intervention if any at all [7].

In the ubiquitous computing paradigm, physical environments are described as full of computational resources that behave in a non-intrusive way with the user, facilitating user interaction and access to resources and information everywhere and everytime. One of the main features of ubiquitous applications is the capability to adapt to a dynamic context [2], thus, they must be aware of the changes that occur in the environment continuously, such as the location and state of users and devices. This vision tries to get away from the paradigm of the

J.M. Corchado, D.I. Tapia, and J. Bravo (Eds.): UCAMI 2008, ASC 51, pp. 149–158, 2009.
springerlink.com

user in front of a PC as the only interaction mechanism with software components. Also tries to promote the use of mobile and embedded devices that makes these highly open and distributed, which need ad-hoc deployment and execution, that is to say, situation customized deployments, integrating available software and hardware resources at a suitable place and time. This is only possible if devices are regarded as autonomous and independent resources in the system. Thus, the SOA (Service Oriented Architecture) paradigm is very suitable for this kind of environment since, in this architectural model, the applications and devices that support them are abstracted like loosely-coupled services that can be integrated into bigger systems.

Web Services [3] allow communication between different platforms and operating systems. They only share syntactic information without any machine-understandable semantics. Thus, dynamic service composition and discovery are difficult tasks in traditional service infrastructures and human intervention is often required. The Semantic Web Services [8] paradigm tries to keep the intervention of end-users to the minimum, automating wherever possible the discovery, integration, invocation and interoperability of Web Services. Bearing in mind that the Semantic Web [1] vision, by means of ontologies, allows the semantic description of any domain of knowledge, adding domain concepts to Web Services produces the Semantic Web Services. These allow processes to express and describe which activities can be performed in such a way that it is possible to discover services in a particular environment and establish a collaboration among them to develop high level tasks with higher added value in a dynamic way [8].

Although we have the necessary infrastructure to achieve the goal proposed by Weiser in [19], the reality is that we are a long way from doing so. The main fact that explains this paradox is that the majority of the devices are isolated. Furthermore the features they offer are described in a syntactic or semantic way, but exclusively based on inputs and outputs, so the functionalities offered by services cannot be adequately consumed by users or other devices that populate the environment. That is why it is necessary to have mechanisms that describe services with more information related to the service behaviour. A way to achieve this is by using effects (produced in the context after the correct execution of the service) and conditions (necessary for the correct execution of the service).

This paper is structured as follows: In Section 2, the state of the art about effects and conditions representation approaches is described. This is followed by Section 3, where we present our ontological model for effects- and conditions-based service representation. After that, in Section 4, an example is described comparing different service description approaches. Finally in Section 5, conclusions and a set of new areas of research are outlined.

2 Related Work

Several research efforts have been developed by the semantic services and context modelling communities for effects and conditions representation.

2.1 Effects in Semantic Service Representation

In order to consume services it is essential to know their descriptions. This information (IOPEs = Inputs, Outputs, Preconditions and Effects) can be divided into two aspects:

- The first aspect is related to information transformation, represented by inputs and outputs (signature-based description [10]), which can be described in two ways: The simplest description is based on providing only the syntactic signature of services and delegating to the service requester the task of understanding its functionality. A more advanced technique consists of adding semantics to signature-based description. By this way, all environment elements use the same concepts to express their descriptions, where not only operational information but also functional is explained.
- The second aspect, specified by effects and preconditions (specification-based description [22]) is related to behaviour and changes in the context (representation of the world) that the service execution produces.

Nowadays, there are several languages which allow describing service effects and conditions in a semantic way. OWL-S (Ontology Web Language for Services), WSMO (Web Service Modelling Ontology), SWSO (Semantic Web Services Ontology), WSDL-S (Web Service Semantics) and SAWSDL (Semantic Annotation for WSDL) are some of them. On a previous work [16], where we show a detailed analysis of some of their characteristics, we conclude that these languages offer the possibility of defining effects and conditions in services (at Operation, Capability or Atomic Process level), but they only mention the level of the ontology where they are defined and rule-based languages which can be used for it. That is to say, **they neither detail how to do it nor provide an ontological model for the representation of effects and conditions of the services**.

2.2 Effects in Context Modelling Approaches

In the next lines the main context modelling approaches for ubiquitous environments are grouped analyzing whether the proposed model supports effects and condition representation.

- In the first group we can find approaches such as COMANTO (COntext MAnagement oNTOlogy) [15] which describes an ontological model for representing most common environmental concepts. Although they represent services, they do not support neither effects nor preconditions.
- CoDaMoS (Context-Driven Adaptation of Mobile Services) [11] and AmIGO (Ambient intelligence for the networked home environment) [12] approaches are members of the second group. In this group, the mechanism used for service representation is OWL-S, but they only use inputs and outputs.
- In the last group, we can find similar approaches such as SOUPA (Standard Ontology for Ubiquitous and Pervasive Applications) [4] and CONON (CONtext ONtology) [18] which do not integrate contextual information with

services descriptions. However, in the case of CONON they use ECA (Event-Condition-Action) to define the environmental behaviour, and in the case of SOUPA, the paradigm used for defining agents behaviour is BDI (Belief-Desire-Intention). This paradigms could be employed for services effects and preconditions representation in some cases.

After this analysis we can appreciate that in **none of the cases is proposed a complete model for effects and conditions representation, neither their relation with context information**.

2.3 Bridging the Gap between Services and Context

As it has been stated in previous sections the semantic services and context model have been considered as non related aspects by the community. Nevertheless, we think that they must be considered as complementary fields. Nowadays, there are two main approaches that try to integrate these two fields in a unique model:

- Yau and Liu [20, 21] present an ontological model for postcondition (referring to effects) and precondition representation taken the SAW (Situation-Awareness) paradigm as base for that. The approach integrates OWL-S with SAW ontology creating a new ontology that supports the representation of different kinds of situations such as preconditions and postconditions. One of the main characteristic of the approach is the possibility to add temporal information to postconditions and preconditions by means of an special kind of situation (Temporal Situation).
- Young-hee et al. [6, 5] propose an ontological model that supports only service effects representation. In this approach the effects are described by means of three factors: the contextual information affected by the effect (i.e: Brightness), the contextual value (i.e: High) and the affected target (i.e: Eye). The affected target can be human (based on human five senses) or other kind of target.

These approaches give an advance in the integration of semantic services with context by means of effects and conditions. Our approach goes far away since it **supports service postconditions and multiple effects for a single service Capability** (minimum invocable unit). Furthermore, our model **supports representation of effects and conditions with lower or higher grade of abstraction and expressivity** for its later use in service matchmaking and composition processes.

3 Ontological Model for Effects- and Conditions-Based Service Representation

The aim of the present research work is to develop an ontology for effects and conditions representation and in this way to cover the existing lack between current semantic service description languages and context modeling approaches. Our model (see Fig. 1) is described at the following sections.

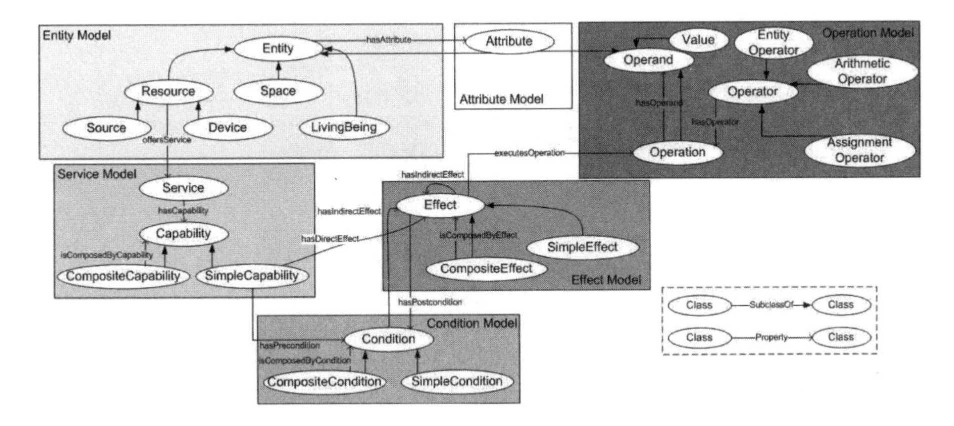

Fig. 1. Effects and Conditions ontological model

3.1 Entity Model

A user notices the effects of a service as changes in the state of the world surrounding him. The Entity class groups and represents all the entities forming that world. The entities can be Living beings, Spaces or Resources. Living being subclass is divided into three groups: Person, Animal and Vegetal. Similarly, Resource subclass can be a Source or a Device. All of them have characteristics and properties, such as temperature, color, owner, etc. which are represented in Attribute class of Attribute Model and joined with Entity class by means of *hasAttribute* relation. Besides, Resource class has *offersService* relation with Service class which belongs to Service Model. The entities to be represented are so different that it is necessary to organize and categorize them according to relations of subsumption, meronymy and instantiation, as described by Singh and Huhns in [13].

As has been widely demonstrated in context modelling related works and analyzed by Strang and Linnhoff-Popien in [14], ontology oriented approaches are very suitable to model context information such as spaces, people, objects, etc (examples described in Section 2.2). The representation of the different entities of the environment falls beyond the scope of the present work, because the current context modelling ontologies might be used, making appropriate changes in order to facilitate effect representation.

3.2 Attribute Model

The world is formed by entities, and the state of the world or the context at a specific moment in time is represented by the set of values that the attributes of those entities have in that particular time. The attributes are information that can be represented by simple data types (integer, string, etc.), as well as by other entities. For example: the SerialNumber attribute of a mobile phone Device Entity can be an integer dataType value, whereas its Owner attribute can be an individual of Person class, which is a subclass of LivingBeing class.

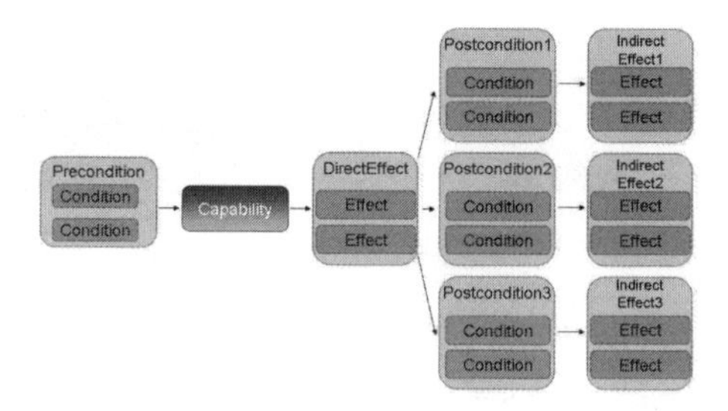

Fig. 2. Capability structure

3.3 Service Model

Resources of Entity Model offer different Services which have Capabilities (see Fig. 2). This connection is specified by *hasCapability* relation. A Capability can be SimpleCapability or CompositeCapability, which is composed by some other capabilities by means of *isComposedByCapability* relation. SimpleCapability has two important relations: the *hasPrecondition* one relates a simple capability with preconditions (Condition class of Condition Model) and *hasDirectEffect* relates it with direct effects (Effect class of Effect Model).

3.4 Effect Model

The effects are considered as changes in the values of the entities attributes, that is to say, changes in the state of the world. The Effect class is related to Operation class by means of *executesOperation* relation. The operations represent the actions which make these changes in the values of the entity attributes, and consequently, in the context. We distinguish two kind of effects:

- **Direct Effect:** This kind of effect is directly related with the entity or device that offers the service. This effect type supports the representation of changes related with the properties of the services entity and its relations. The Direct Effect is described by means of *hasDirectEffect* relation with SimpleCapability class and also has a relation with the PostCondition, by means of *hasPostCondition* relation with Condition class. Moreover, it supports *hasIndirectEffect* relation with Effect class, for the definition of non-conditional indirect effects.
- **Indirect Effect:** This effect type is used to describe the effects derived from the direct effect, that is to say, the indirect changes produced by the direct effect in the other entities of the environment. The indirect effect is described by means of *hasIndirectEffect* relation with Condition class representing the PostCondition.

3.5 Condition Model

The model supports the definition of two kind of conditions related with their situation in the service description:

- **Precondition:** Represents any condition that must be fulfilled before the service execution (see Fig. 3).
- **Postcondition:** Represents any condition that is associated with a direct effect and determines the possible indirect effect (see Fig. 3).

A condition can be a SimpleCondition or a CompositeCondition which is a Condition composed by others. The representation of service conditions is beyond the scope of this research, because as OWL-S, WSDL-S and SAWSDL mention, it is possible to use rule-based languages like: SWRL (Semantic Web Rule Language), DRS (Declarative RDF System), KIF (Knowledge Interchange Format), etc. to represent them in a correct way.

3.6 Operation Model

The Operation class groups the actions which change the values of the attributes. The operations are composed by an unique operator (Operator class) and some operands (Operand class), which can be constants, other operations or attribute values (Attribute class) of entities. Different kinds of operators exist, being the most significant ones grouped into AssignmentOperator, ArithmeticOperator and EntityOperator classes. The assignment operator (=) is used to assign a concrete value to an attribute. The arithmetic operators (+, -, etc.) are valid to increment, decrement, etc. values. Finally, the entity operators (create, delete, etc.) are useful to create and delete individuals of entities in the model.

4 Example

Let us take as an example a room with a dimmed lamp switched off and an automated blind closed. A requester (user, service or device) wants to increase the light level in the room. Depending on the service description approach used, the following is determined:

(a) Using **signature-based syntactic service descriptions**, service discovery mechanism reveals that both services describe how to be invoked (operational properties) but not what they do (functional properties). Therefore, the requester must **have prior knowledge** of what lamps and blinds do, and must use **some kind of reasoning** to decide which of them is better suited to the final objective of increasing the light level.
(b) Using **signature-based semantic descriptions**, both of them announce their operational and functional properties expressed by using semantic concepts of a shared ontology. As the requester knows that ontology, he knows these concepts, so **he does not need to rely on prior knowledge** to identify what these services do. However, **he has to reason** about which service fulfills the need of increasing the light level.
(c) Combining **signature-based and specification-based semantic descriptions**, both services announce the operational and functional characteristics and the effects they produce in the environment. Every description

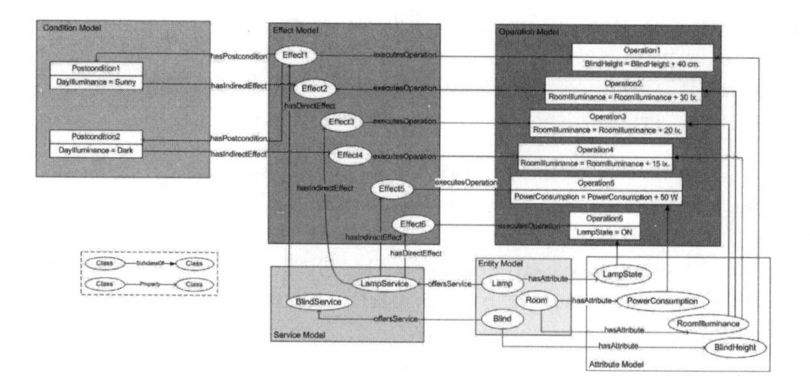

Fig. 3. Instances representing services effects

is related to some semantic concepts of a shared ontology whose instances are shown in Figure 3.

The BlindService service produces three effects: Effect1 executes Operation1 (raises the blind -BlindHeight- 40 cms, direct effect), Effect2 executes Operation2 (increases the room illuminance -RoomIlluminance- 30 luxes on sunny days, indirect effect), and Effect4 executes Operation4 (increases the room illuminance 15 luxes on dark days, indirect effect). Effect2 and Effect4 are indirect effects related with postconditions and represented using effects and conditions because they depend on the luminosity of the day. The LampService service also produces three effects: Effect3 executes Operation3 (increases the room illuminance 20 luxes, indirect effect), Effect5 executes Operation5 (increases the room's electric consumption 50 watts, indirect effect), and Effect6 executes Operation6 (assigns the ON value to the lamp state -LampState-, direct effect).

The effects of the lamp and the blind are related to their attributes, but they also alter attributes of other entities of the environment (case of the room). It is also possible that an attribute may be affected by several different effects of distinct services. For example, the RoomIlluminance is modified by Effect2 and Effect4 of BlindService and Effect3 of LampService as well. Imaging that the user wants to raise the room illuminance (the RoomIlluminance attribute of the Room entity), but he wants the energy consumption (PowerConsumption attribute) to be as low as possible. The automatic discovery mechanism reveals the existence of LampService and BlindService which produce the effects desired by the user (to increase RoomIlluminance). With this information, the selection system,**without prior knowledge or reasoning of the user**, chooses the BlindService because, as the day is sunny, it offers more illuminance with a negligible energy cost. Thus, the discovery system is the one that has the knowledge and makes the reasoning and not the user.

Thanks to effects- and conditions-based approach it is easier to automate the service selection and integration process, due to the fact that it is oriented to users or devices needs, describing **which the service effects are** instead of **what is necessary to invoke them**.

5 Conclusions and Future Work

In this paper an ontological model has been presented for effects- and conditions-based service representation. This model facilitates service invocation by users, as well as the discovery and integration of them. This makes it possible to search, match, integrate and invoke services based on the effects that occur after the invocation and not only based on the inputs and outputs, as it is done nowadays. So we believe that the next logical step of this research work is to extend the current model, adding time related information to effects and conditions and consider the uncertain context information. After that, the next step will be the definition of a method for services matchmaking based on effects and conditions and continue with techniques of services integration. In this way, we will manage to complement and improve existing methods of ubiquitous service discovery [9] and integration [17] exclusively based on inputs and outputs.

References

1. Berners-Lee, T., Hendler, J., Lassila, O.: The semantic web (2001)
2. Bottaro, A., Bourcier, J., Escoffier, C., Lalanda, P.: Autonomic context-aware service composition. In: 2nd IEEE International Conference on Pervasive Services (2007)
3. Champion, M., Ferris, C., Newcomer, E., Orchard, D.: Web services architecture. W3C Tehnical Report (2002)
4. Chen, H., Finin, T., Joshi, A.: The SOUPA Ontology for Pervasive Computing. In: Ontologies for Agents: Theory and Experiences. Whitestein series in software agent technologies edition. Springer, Heidelberg (2005)
5. Kim, G.H., Kim, D.H., Hoang, X.T., Lee, Y.H.: Group-aware service discovery using effect ontology for conflict resolution in ubiquitous environment. In: ICACT 2008 (2008)
6. Kim, G.H., Kim, D.H., Hoang, X.T., Lee, Y.H., Lee, G.S.: Semantic service discovery using effect ontology for appliance service in ubiquitous environment. In: ICUT 2008 (2008)
7. Lassila, O., Adler, M.: Semantic gadgets: Ubiquitous computing meets the semantic web. In: Spinning the Semantic Web (2003)
8. McIlraith, S.A., Zeng, T.C., Zeng, H.: Semantic web services. IEEE Intelligent Systems and Their Applications 16(2), 46–53 (2001)
9. Ben Mokhtar, S., Preuveneers, D., Georgantas, N., Issarny, V., Berbers, Y.: Easy: Efficient semantic service discovery in pervasive computing environments with qos and context support. Journal Of System and Software (2007)
10. Paolucci, M., Kawamura, T., Payne, T.R., Sycara, K.P.: Semantic matching of web services capabilities. In: Horrocks, I., Hendler, J. (eds.) ISWC 2002. LNCS, vol. 2342, pp. 333–347. Springer, Heidelberg (2002)

11. Preuveneers, D., Van den Bergh, J., Wagelaar, D., Georges, A., Rigole, P., Clerckx, T., Berbers, Y., Coninx, K., Jonckers, V., De Bosschere, K.: Towards an extensible context ontology for ambient intelligence. In: Markopoulos, P., Eggen, B., Aarts, E., Crowley, J.L. (eds.) EUSAI 2004. LNCS, vol. 3295, pp. 148–159. Springer, Heidelberg (2004)
12. IST Amigo Project. Amigo d3.2 amigo middleware core: Prototype implementation and documentation. Technical Report IST-2004-004182
13. Singh, M.P., Huhns, M.N.: Service-Oriented Computing: Semantics, Processes, Agents. John Wiley and Sons, Chichester (2005)
14. Strang, T., Linnhoff-Popien, C.: A context modeling survey (September 2004)
15. Strimpakou, M.A., Roussaki, I.G., Anagnostou, M.E.: A context ontology for pervasive service provision. In: AINA 2006: Proceedings of the 20th International Conference on Advanced Information Networking and Applications (AINA 2006), vol. 2, pp. 775–779 (2006)
16. Urbieta, A., Azketa, E., Gomez, I., Parra, J., Arana, N.: Analysis of effects- and preconditions-based service representation in ubiquitous computing environments. In: IEEE 2nd International Conference on Semantic Computing (2008)
17. Urbieta, A., Barrutieta, G., Parra, J., Uribarren, A.: A survey of dynamic service composition approaches for ambient systems. In: Ambi-Sys - First Workshop on Software Organisation and MonITToring of Ambient Systems. ICST (2008)
18. Wang, X.H., Zhang, D.Q., Gu, T., Pung, H.K.: Ontology based context modeling and reasoning using owl. In: PerCom Workshops, pp. 18–22 (2004)
19. Weiser, M.: The computer for the 21st century (1991)
20. Yau, S.S., Liu, J.: Hierarchical situation modeling and reasoning for pervasive computing. In: SEUS-WCCIA 2006, pp. 5–10 (2006)
21. Yau, S.S., Liu, J.: Incorporating situation awareness in service specifications. In: ISORC 2006: IEEE International Symposium on Object and Component-Oriented Real-Time Distributed Computing, pp. 287–294 (2006)
22. Zaremski, A.M., Wing, J.M.: Specification matching of software components. In: SIGSOFT 1995: Proceedings of the 3rd ACM SIGSOFT symposium on Foundations of software engineering, pp. 6–17 (1995)

Modeling the Context-Awareness Service in an Aspect-Oriented Middleware for AmI

Lidia Fuentes and Nadia Gámez

Departamento de Lenguajes y Ciencias de la Comunicación, Grupo GISUM
Universidad de Málaga
Boulevard Louis Pasteur, Málaga, Spain
lff@lcc.uma.es, nadia@lcc.uma.es

Abstract. Ambient Intelligence applications are gradually becoming integrated into our environment and will be deployed in a wide range of small devices and appliances. Middleware platforms will be useful to express the different requirements of these devices in terms of their commonalities and variabilities. These ambient intelligence applications should be highly reconfigurable in order to deal with changes in the environment. Consequently, one of the services that the middleware has to provide is precisely this managing of the environment i.e., the context. Context-awareness and context adaptation are crosscutting concerns that cannot be well-encapsulated in single modules using traditional technologies, which hinders application design maintenance and reusability.

In order to overcome these problems we use of Aspect-Oriented Software Development for modeling the context-aware service. There are several classifications of the different kinds of context at application level. We specify these context types at middleware level and model them using aspects. Finally, we compare the aspect-oriented version of context-awareness versus the non aspect-oriented one.

Keywords: Ambient Intelligence, Context-Awareness, Aspect-Oriented, Middleware.

1 Introduction

The development of Ambient Intelligence (AmI) applications has given rise to new challenges that have still not yet been fully resolved. This kind of application must support: (1) embedded software for a wide range of small devices with different capacities and resources, (2) context-awareness (3) customer tailored-configuration (4) adaptability and reconfigurability in order to deal with context changes and (5) maintainability, because the hardware and software technologies in AmI are changing and evolving continuously. We propose a middleware platform that deals with all these issues, while at the same time avoiding the necessity to develop them from scratch for each application. Furthermore, we consider that a software product line approach would be very useful to express the different requirements of devices in terms of commonalities and variabilities of a family of middleware platforms. Normally, a middleware platform provides all the common services most used by distributed applications. In AmI applications however, the resource constraint is an important limitation, so only a specific middleware platform configuration that is compatible with the device characteristics must be installed. Furthermore, the middleware platform for AmI must to be reconfigurable to deal with the context changes.

J.M. Corchado, D.I. Tapia, and J. Bravo (Eds.): UCAMI 2008, ASC 51, pp. 159–167, 2009.
springerlink.com

Context-awareness is a recurrent requirement in AmI and in pervasive computing. In our approach therefore, one of the main services that the middleware has to provide is context management. Context is any information that can be used to characterize the situation of an entity. An entity is a person, place or object that can be considered relevant to the interaction between a user and an application, including the user and applications themselves [1]. Depending on the application domain, these contexts may differ greatly and change quickly. Thus, ambient intelligence applications for such domains should be highly reconfigurable and adaptable to deal with a wide number of different contexts. Context-aware requirements imply a number of crosscutting concerns that are not related to the core logic of the middleware functionality. This means that using traditional technologies, such crosscutting concerns, cannot be well-modularized in single modules due to the tyranny of the dominant decomposition [2]. Consequently, they appear scattered and tangled across different design modules, hindering the correct design, maintenance, evolution as well as the independent reuse of its modules [3].

Aspect-Oriented techniques [4] allow the encapsulation of crosscutting concerns, such as context-awareness, into well-localized modules, overcoming problems arising from scattered and tangled representations [5, 6].

In our approach, the different middleware services, such the context-awareness service, are provided as a set of user-configurable aspects. Implementing these middleware services by means of aspects would also contribute to the modularization, reconfigurability and adaptability of the middleware. Regarding the context management, we can manage the context in two parts. Firstly, the context acquisition collects all data relative to the context and subsequently, the context adaptation makes the appropriate changes depending on the context. The modularization of the context acquisition has been properly treated. Some frameworks like the Context Toolkit [7] allow this modularizarition,, but the context adaptation has still not been well-encapsulated in single modules using traditional technologies.

This problem of encapsulating context adaptation can also be found in [8] but this time at the application level. To solve it, the context has been divided into smaller, more manageable modules, each of which models all structural and behavioural context-awareness code related to a specific context type. In this approach, a modular design is proposed for these context types using AOSD. In our case this kind of context is treated at middleware level in order to provide a well encapsulated context-awareness service for the middleware platform.

Following this introduction, this paper is structured as follows. In Section 2 we explain how the context can be subdivided into small categories at middleware level. In Section 3 we give some background on AOSD and the architecture of the aspect-oriented middleware supporting the context-awareness service. In Section 4 our approach modelling the context-awareness service using aspects is presented and we compare our approach to non aspect-oriented ones. Finally, Section 5 comments on related work, and Section 6 outlines some conclusions and future work.

2 Context at Middleware Level

There are several classifications about context, such as [1][9][8]. In [8] the overall context space is divided into eight sub-categories: device, location, user, social, environmental, system, temporal, and application context. AOSD is used to modularize the context. We will follow this approach but at middleware level. Then, we will transform (Table 1) each category from application level to middleware level.

Table 1. Type of context mapping

Kind of Context	Application Level	Middleware Level
Device	User's computing devices	Device constraints
Location	Applications device location	Middleware device location
User	Knowledge about user (app.)	Knowledge about user (MW)
Social	Uuser interactions info	There is no correspondence
Environment	Relating to the physical world	Execution environment
System	Running infrastructure (app.)	Running infrastr. (app. & MW)
Temporal	All data related to time	All data related to time
Application	Core functionality	Application restriction

At application level the **device** context encapsulates information describing the user's computing device. At middleware level the device context refers to all the device constraints that the middleware has to be aware of, e.g. if the battery is less than a certain value the graphics in the application may need to be changed to text.

Location is an important element of context [1] that can serve as a source for higher-level contextual information, e.g., proximity to other entities. Regarding the middleware, the this context is the information about position that can be relevant to adapt the middleware itself or to notify the applications about the location.

User context captures all knowledge pertaining to the user known to an application. User preferences may provoke changes in the middleware context requiring it to be reconfigured.

For the applications, the **social** context is any information that influences the interactions of one user with the others. For the middleware, we do not find any correspondence for this kind of context, because the social context is intrinsically related with the applications and the users.

Environmental context relates to the physical world in which the application runs, e.g., temperature, noise and lighting conditions [1]. In the case of middleware, this kind of context contains the properties of the execution environment. For example, if one device is missing, the environment may change and some services may not be executed in the same way as before.

System context refers to all contexts related to the (technical) infrastructure in which an application runs [8]. This kind of context is exactly the same as those described in the environment context properties. Therefore, we can join these two context types at level application (environment and system) and make correspondence with the environment context in the middleware application.

The **temporal** context refers to all data related to time that affects the application behaviour. It is the same at middleware level, but affects the configuration of the middleware.

The **application** context relates to the core functionality of the application. However at middleware level, the application context refers to the application restrictions. These applications may vary and may require different services to those which were required before their variation. For example, where an application needs to be secure only at certain moments, then the optional security service will be active only during these moments.

3 Aspect-Oriented Middleware for the Context-Awareness Service

In this section, firstly we summarize the AOSD terminology that will be used throughout this paper. Then, we outline the architecture and functioning of our aspect-oriented middleware platform family approach, focusing especially on the context-awareness service.

3.1 Aspect-Oriented Software Development

We use the glossary defined in [10] to describe the AOSD terminology used in this paper. Separation of concerns is the in-depth study and realization of concerns in isolation for the sake of their own consistency (adapted from [11]). Each concern is an interest, which pertains to the system development, its operation or any other matters that are critical or otherwise important to one or more stakeholders. Crosscutting concerns are those that cannot be modularly represented within the selected decomposition. Consequently, their elements are scattered and tangled within elements of other concerns. In AOSD, a new unit for modularizing an otherwise crosscutting concern has been named aspect. In an aspect, the crosscutting behavior is modeled as advices, where an advice is an aspect element, which augments or constrains other concerns at join points matched by a pointcut expression. A join point is a point of interest in some artifact in the software lifecycle at which two or more concerns may be composed. A pointcut is a predicate that matches join points. Possible join points are determined by the join point model, which defines the kinds of join points available and how they are accessed and used. Once crosscutting concerns are encapsulated in aspects, they have to be composed with the other concerns in the system. The term used to refer to this kind of composition is weaving, this being defined as the integration of multiple modular artifacts into a coherent whole.

3.2 Architecture of the Aspect-Oriented Middleware

Context-awareness cuts across system and application layers. One way to solve this issue is by encapsulating the context-awareness in aspects. On the other hand a middleware oriented approach can effectively address such issues by providing development and runtime support and by striking a balance between awareness and application transparency [12,13].

Some reconfigurable and adaptable middlewares have already been developed. However the main functionality of middleware is not the management of context-awareness. We propose to build an Aspect-Oriented middleware for AmI where the context-aware and other crosscutting concerns are encapsulated using aspects. In this way, operations related with the context-awareness will be not spread across the middleware functionality. In our proposal the middleware will be composed of a microkernel, which will be responsible for the basic functionality, and other services that the middleware will provide. Some of these services will help to deal with context-awareness and will be modelled using aspects.

The architecture of our middleware is shown in Fig. 1. Like most of the traditional middleware platform architectures, we follow a layered approach. The application level is at the top and the middleware itself is immediately below. In the first sub-layer of the middleware are the application services like context-awareness, security, error-handling, and so on. These services use the base services in the next layer, such as communication, lookup and etcetera. Below these layers is the microkernel, which is comprised of the container, the factory, the context manager and the service manager. Due to space limitations, here we will focus on the context-manager, for other components of the microkernel see [14].

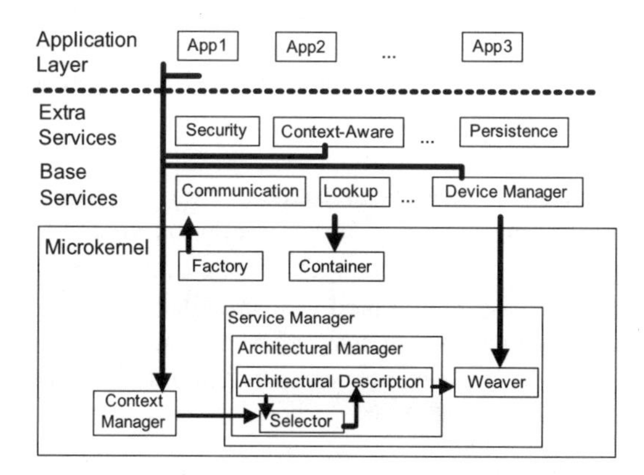

Fig. 1. Aspect-oriented middleware architecture

The microkernel, through the context manager, has to know the context properties before running an application. With this information and with the architectural description, the service manager runs the automatic selector facility in order to instantiate one particular middleware architecture of the middleware family. This is a static configuration of the middleware. Using the context properties, the factory (in the microkernel) will instantiate the specific selected services. In order to run the application, the service manager has the weaver that is in charge of the composition between the services, the microkernel and the application. If a change occurs in the context (user properties, device constraints and so on) the microkernel will use the selector again and a new middleware of the family will be instantiated. The context-awareness service that has access to the context manager is responsible for this instantiation. This is the dynamic reconfiguration of the middleware. In the next section, we detail the design of this context-awareness service.

4 Context-Awareness Service Using Aspects

Here we present our aspect-oriented approach for modelling the kinds of contexts for context acquisition and the adaptation of the middleware to the context changes. We detailed the device context type and due to space limitations have only briefly outlined the other kinds.

4.1 Device Context

As we said before, device context refers to all the device constraints that the middleware has to be aware of. To manage this context, in the context acquisition, the context-awareness service has to monitor the current capacity and status and the correct functioning of all the devices that are using the applications. On the other hand, in the context adaptation, one component of the microkernel called context manager has to reconfigure the middleware and adapt the applications to the new context device situation. This scenario is typical to apply the subject-observer pattern, with the context as subject and the part of the context-awareness service responsible to context device, is the observer. This pattern has been refactored using aspects [15], and it has been demonstrated that the aspect-oriented version is an improvement on the traditional solution regarding modularisation [16]. Therefore, we model this pattern in its aspect-oriented version.

In this version, the aspect-oriented microkernel weaver is the one which calls the update method of the observer to notify it that the observed entity has changed. More specifically, when the device context changes, if for example the energy of the battery is below a given value, the aspect-oriented weaver calls the advice responsible for the battery that contains the context-awareness service. In order to detect the device context changes we have to declare several pointcuts that intercept the following (different) situations:

1. The operating system detects an event or signal about errors in devices.
2. The monitoring service gives a specific value for a device or detects errors.
3. A middleware service sends a message to the microkernel informing of an error in a device.
4. An application sends a message to the microkernel informing of an error in a device.

In this way we have all the pointcuts that intercept the possible device context changes and are able to declare the advice (placed in the context-awareness service part responsible for each device) that must be called. This advice contains all the information with the issues that the middleware has to deal with in order to adapt itself to the new context. These issues have to be defined by the application in an XML document that the middleware loads at the beginning, at the same time as the architectural representation of the application (also in an XML document). An example of the content of that document is: if the energy of the battery is less than a particular value, the middleware can reconfigure itself and adapt the application in order to reduce the intensity of the screen light, not show some images, switch off the wifi or bluetooth (if they are not essential) or not use some unnecessary services. In order that the necessary changes are performed, the context-awareness informs the context manager. This manager runs the selector, which with the new device constraints, instantiates a new appropriate middleware.

4.2 Other Kinds of Contexts

Due to space limitation we cannot go into more detail here on the other context types, but in general terms they are similar to the device contexts. We try to use the subject-observer pattern in its aspect-oriented version in all the situations possible. Then we declare pointcuts for all the message or events that imply a changed context. For example, for user context changes, we can intercept the message that the application sends to the microkernel with the new user context or the event that the location service sends informing about a new location position.

In summary we will define all the pointcuts that can intercept context changes for every kind of context and the context-awareness service will have the advices that have the behaviour to perform when a context changes. All the join points that can be intercepted are specified in the join point model of our platform. Examples of these join points are:

1. Messages that one service sends to others
2. Messages that one service or the applications send to the microkernel
3. Events that some services can throw
4. Events that the microkernel can throw
5. Events or signals the operative system can throw

4.3 Comparison between AO and Non-AO Context-Awareness Concerns

If we compare the aspect-oriented version of the context-awareness to the non aspect-oriented version using a subject-observer pattern, in both cases we can see, as is demonstrated in [15], that the aspect-oriented version improves the modularity of the non-aspect oriented one.

In the case of the device context type, we have to declare pointcuts for at least four situations and associate an advice to each pointcut. All these pointcuts and advices are included in the context-awareness service and the other services and the application does not need to do anything to be aware of the context. In this case, it is the weaver who calls the update method for adapting to the context. However, if aspects are not used, the context-awareness service has to ask periodically for each situation or has to receive information about the devices. For example, it has to ask the monitoring service for the battery level, or if an application sends a message to the microkernel informing of an error in a device, the microkernel has to notify the context-awareness service. This implies much more tangled and scattered code than in the AO version, where this is solved by declaring some pointcuts that intercept the monitoring message and the message from the application to the microkernel.

5 Related Works

There are several middlewares that try to manage the context. For example, GREEN [17], CARISMA [18] and the work of Yau et al. [19] are middleware platforms for pervasive systems with special focus on context-awareness. GREEN [17] is a configurable and re-configurable publish-subscribe middleware for pervasive computing. CARISMA [18] is a mobile computing middleware which exploits the principle of reflection to enhance the construction of adaptive and context-aware mobile applications. In [19], a reconfigurable context-sensitive middleware for pervasive computing is presented. These platforms use different technologies for dealing with context-awareness, but none of them use aspect-orientation, as we do. An important advantage of using aspect-orientation is that not only context-awareness is made transparent to the application, but also other crosscutting concerns, such as persistence or error handling, can be made transparent using the same paradigm.

Other proposals which claim to provide an aspect-oriented middleware level [20], are really non-fully aspect-oriented platforms for developing aspect-oriented applications. Our proposals try to bring the aspect-oriented benefits to the middleware itself, in order to deal with heterogeneity of devices and communication technologies and the high dynamism of AmI environments. As in the approach presented in [8], there are so many others that try to design and implement the context management using aspects, but all of them treat the context at application level. However, we treat it at middleware level where the problems encountered are quite different.

6 Conclusions and Future Work

In this paper we have considered the problems relating to AmI applications and how a middleware approach can solve them. We have adopted a layered middlware structure, using a microkernel and several services in order to obtain a more reusable and adaptable middleware. Furthermore, we have proposed the use of an aspect oriented middleware that allows dynamic reconfiguration, because this is a very important issue for context-aware AmI applications.

Mainly, in this paper we have focussed on the design of the context-awareness service of the middleware. In order to achieve good modularization of this service we have used AOSD techniques. Aspect-Orientation has been demonstrated as a useful technique for improving the encapsulation of a wide number of crosscutting concerns, such as context-awareness [21]. Then, aspect-orientation allows better system modularisation, thereby improving system maintenance, evolution and reusability. The context-awareness service has been divided into several categories and each one has been designed using aspect-orientation separately.

As future work, we are working on the development of the rest of the services that the middleware will provide and on the internal structure of the microkernel.

Acknowledgements. This work has been supported by the Spanish Ministry for Science and Technology (MCYT) Project TIN2005-09405-C02-01, the European Commission Grant IST-2-004349-NOE AOSD-Europe and the European Comm. STREP Project AMPLE IST-033710.

References

1. Abowd, G.D., Dey, A.K., Brown, P.J., Davies, N., Smith, M., Steggles, P.: Towards a better understanding of context and context-awareness. In: Gellersen, H.-W. (ed.) HUC 1999. LNCS, vol. 1707. Springer, Heidelberg (1999)

2. Tarr, P., et al.: N degrees of separation: Multi-dimensional separation of concerns. In: AOSD, pp. 37–61. Addison-Wesley, Reading (2004)
3. Clarke, S.: Extending standard uml with model composition semantics. Science of Computer Programming 44(1), 71–100 (2002)
4. Filman, R.E., et al.: Aspect-Oriented Software Development. Addison-Wesley, Reading (2004)
5. Lopes, C.V., Bajracharya, S.: Assessing Aspect Modularizations Using Design Structure Matrix and Net Option Value. In: Rashid, A., Akşit, M. (eds.) Transactions on Aspect-Oriented Software Development I. LNCS, vol. 3880, pp. 1–35. Springer, Heidelberg (2006)
6. Greenwood, P., et al.: On the Impact of Aspectual Decompositions on Design Stability: An Empirical Study. In: Ernst, E. (ed.) ECOOP 2007. LNCS, vol. 4609, pp. 176–200. Springer, Heidelberg (2007)
7. Salber, D., Dey, A., Abowd, G.: The context toolkit: aiding the development of context-enabled applications. In: CHI 1999 (1999)
8. Munnelly, J., Fritsch, S., Clarke, S.: An Aspect-Oriented Approach to the Modularisation of Context. In: PerCom 2007 (2007)
9. Schmidt, M.B., Gellersen, H.-W.: There is more to context than location. Computers and Graphics 23(6), 893–901 (1999)
10. Berg, K.v.d., Conejero, J.M., Chitchyan, R.: Aosd ontology 1.0 public ontology of aspect-orientation. Technical Report AOSD-Europe Deliverable D9, AOSD-Europe-UT-01, Universiteit Twente (2005)
11. Dijkstra, E.W.: SelectedWritings on Computing: A Personal Perspective. Springer, Heidelberg (1982)
12. Berstein, P.: Middleware: A model for distributed services. Communications of the ACM 39(2), 86–97 (1996)
13. Geish, K.: Middleware challenges ahead. Computer 34(6), 24–31 (2001)
14. Fuentes, L., Gámez, N.: A Feature Model of an Aspect-Oriented Middleware Family for Pervasive Systems. In: NOMI 2008 (2008)
15. Hannemann, J., Kiczales, G.: Design Pattern Implementation in Java and AspectJ. In: Proc. of OOPSLA 2002, pp. 161–173 (2002)
16. Garcia, F., et al.: Modularizing design patterns with aspects: A quantitative study. In: Rashid, A., Akşit, M. (eds.) Transactions on Aspect-Oriented Software Development I. LNCS, vol. 3880, pp. 36–74. Springer, Heidelberg (2006)
17. Sivaharan, T., et al.: A configurable and re-configurable publish-subscribe middleware for pervasive computing. In: Proc. of Symposium on Distributed Objects and Applications (2005)
18. Capra, L., et al.: Carisma: Context-aware reflective middleware system for mobile applications. IEEE Transactions on Software Engineering (2003)
19. Yau, S., et al.: Reconfigurable context-sensitive middleware for pervasive computing. IEEE Pervasive Computing 1(3), 33–40 (2002)
20. Pinto, M., Fuentes, L., Troya, J.M.: A component and aspect dynamic platform. The Computer Journal 48(4), 401–420 (2005)
21. Tanter, E., et al.: Context-aware aspects. In: Löwe, W., Südholt, M. (eds.) SC 2006. LNCS, vol. 4089, pp. 227–242. Springer, Heidelberg (2006)

An Agent-Based Component for Identifying Elders' At-Home Risks through Ontologies

Marcela D. Rodríguez, Cecilia Curlango, and Juan P. García-Vázquez

Engineering School, Autonomous University of Baja California, Mexicali México
{marcerod,curlango,juanpablo_garcia}@uabc.mx

Abstract. As elders age, performing everyday tasks can be filled with risks that threaten their health and independence. We present an agent-based component that can be used to build AmI systems. This component uses the ELDeR ontology that we propose to model the context information inherent to the elder's daily activities and that enables inferring risks.

Keywords: ontology, ADL, agents, context-aware, elders.

1 Introduction

Aging is often accompanied by the loss of close companionship and many health problems that may often lead to a decrease in of quality of life. In order for elders to receive timely care, a current technological trend is to integrate the Ambient Intelligence (AmI) vision into the elders' home environment in order to support "ageing in place". In this approach, the home environment should provide facilities to enable them to autonomously carry out their daily activities by warning them or their caregivers of any risk they may face. For reaching this end, the AmI system should include: i) *Pervasive technology* to track the elders' behavior and interaction with household objects. Different approaches have been proposed for sensing the elder's actions and behavior in order to recognize activities of daily living [1]. ii) *Context-aware and intelligent systems* that monitor activities of daily living (ADLs) and recognize when an elder faces an abnormal situation or risk. We follow the approach of using agents as a tool for designing and implementing AmI systems [2]. Agents can provide the characteristics of intelligence and context-awareness required for inferring the elders' activities and predicting if they are facing a risk. iii) *Human-computer interaction* techniques to provide direct assistance to the older person, such as activity-appropriate prompts and reminders or to notify caregivers of the status of an older person living alone include [1].

AmI systems for supporting elder's independent lifestyle, need not only mechanisms for capturing the elder's context, but also for representing this context and disseminating it to other system agents that will process it, and take the appropriate action to alert the elder of the feasibility of facing a risk, or to notify members of his care-network of the situation. To address the above issues, in this paper we present an agent-based component that facilitates inferring elders' risks by using an ontology, which is a representational model of the elders' context information that is captured by pervasive technology.

J.M. Corchado, D.I. Tapia, and J. Bravo (Eds.): UCAMI 2008, ASC 51, pp. 168–172, 2009.
springerlink.com

2 Related Work

Several development platforms for AmI environments use ontologies for addressing different issues related with managing context information. CoBrA is an agent architecture for pervasive context-aware systems. It uses ontologies for supporting knowledge sharing and data fusion in an intelligent meeting-room domain. For reaching this end, CoBrA provides an ontology that describes places, agents and events and their associated properties [3]. The Gaia infrastructure for ubiquitous applications uses ontologies to verify the validity of first order predicates used to represent context information. [4]. I.L.S.A is an agent-based architecture for creating home-based systems for assisting elders. To our knowledge, I.L.S.A is the only project that proposes an ontology for this domain. The aim of this ontology is to provide a common vocabulary for ILSA-related concepts and their relationships, such as support for agent configuration, long-term storing of monitored results, client and environment states and status communication [5]. Our approach is to define an ontology that represents the context-information and rules that help to infer if an elder is facing a risk in his home. As we mentioned before, we propose to use agents for facilitating implementing systems that support the elders' independent living by providing an agent-based component that infers risks [2]. This agent-based component consists of an agent that assesses the elders' context and infers whether there is a risk by querying an ontology. The ontology represents an elder adult's context, the rules that determine when a risk occurs based on the context, and the agents that alert of the situation. We have called it *ELDeR* ontology, as it has the aim of Enabling Living inDependently of Risks.

3 ELDeR Ontology

The *ELDeR* Ontology provides a common language for agents that are involved in the context of elder health care. The contents of this ontology will be provided by the programmer and will have the structure shown in the diagram seen in figure 1. We find the *ADL* at the top level which is composed of at least one *Action*. For example, the ADL *handwashing* is composed of actions such as *openFaucet, takeSoap*. These *Actions* will be identified by the programmer according to the pervasive infrastructure used for sensing or inferring these actions. Each *Action* can take place at one or many *Locations,* and at a moment in *Time*. For example, *openFaucet* may take place in a *bathroom* or *kitchen* at a different times during the day. In addition, *Actions* may have *Features* that cause *Risks* which should be specified in a *Rule*. For instance *openFaucet* is associated with *waterTemperature,* and the *Rule* is that if it is *54 ° C or more there is a Risk of scalding*. *Rules* are also associated with *Location* and *Time* since if an *Action* does not take place in a specific *Location* and/or *Time* there might be a *Risk*. For instance, *takingMedication* must occur at the same time each day as recommended by a physician. *Risks* are linked with *Notification Agents* that notify of the elder's situation appropriately as defined by the programmer. The identity of these agents is also specified in the ontology.

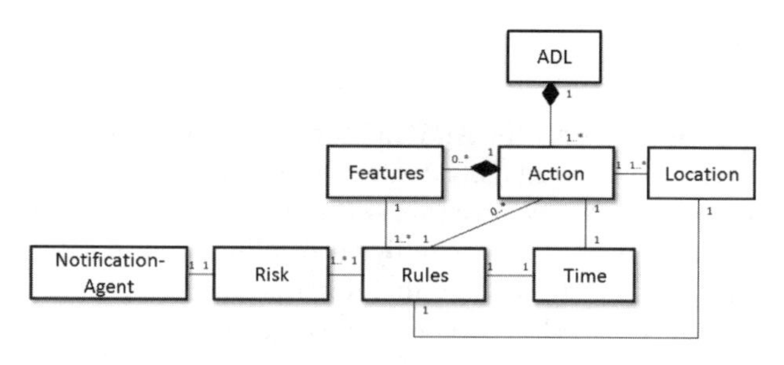

Fig. 1. ELDeR Ontology

4 Assessing Risks

We designed a component that contains an agent which is able to infer its context and the risks that are inherent to it by querying the ELDeR ontology. We called this agent *RiskAssessing Agent (RA-Ag).* This agent was designed based on type of agents proposed in the SALSA middleware, which provides a library of classes for implementing and handling the execution model of agents, which consists of the components for perceiving information, reasoning, and acting which may involve communicating with other agents [2]. According to the SALSA Agent life cycle, the *RA-Ag* begins by perceiving information from agents acting as proxies to components such as sensors, or software components with specialized algorithms that monitor or infer the elders' location and actions. The proxy agents communicate with the *RA-Ag* by sending the SALSA messages *sendPresence()* and *sendDataSensor().* In addition, it perceives the time of day. When the *RA-Ag* receives information about an *Action*, it also receives the *Features* that are associated with it. From the perceiving state the *RA-Ag* goes to the reasoning state. In this state, the agent uses the perceived information (*Location, Actions and Features*) to query the ELDeR Ontology to obtain the set of all *Rules* and *Risks* that may be taking place according to the perceived information. The *RA-Ag* then transforms the *Rules* and *Risks* into first order logic predicates which it evaluates with its inference engine. Evaluating the *Rules* with the perceived information yields the *Risks* that are involved as well the identity of the agent that should be apprised of the current situation. The *RA-Ag* transitions to the acting state in order to inform the elders' situation to the appropriate *Notifying Agent* which will take the appropriate measures for the *Risk* such as warning the elder or notifying to the elder's caregiver. A *sendNotification()* SALSA message would be sent which would contain the *Location, Action* and *Features* involved in the *Risk* so the *Notifying Agent* can alert the Elder or the Elder's caregiver depending on the gravity of the detected situation. *Notifying Agents* are SALSA agents which the programmer creates by identifying the appropriate notification mechanisms.

5 Usage Scenario

To demonstrate how it would be possible to build an AmI system that can identify risks using the component we propose and the SALSA framework of SALSA, we present the following scenario which is further illustrated in figure 2.

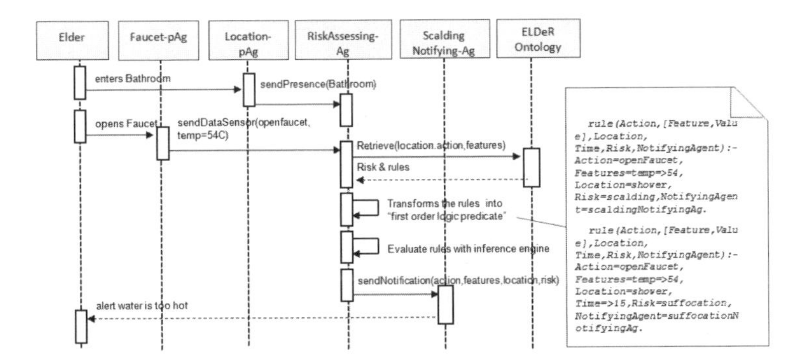

Fig. 2. Sequence diagram of usage scenario

Mr. Jones enters his bathroom to take a shower. The Location proxy agent *(Location-pAg)* informs the risk assessing agent *(RA-Ag)* of the Mr. Jones is in the bathroom by sending the *sendPresence(Bathroom)* message. When Mr. Jones opens the shower faucet, the *Faucet-pAg* informs the *RA-Ag* that the faucet is open and that the water temperature is 54°C with the *SendDataSensor()* message. Then, the *RA-Ag* retrieves from the ELDeR ontology the risks and rules. In this case, scalding and suffocation are the possible risks. The *RA-Ag* now transforms the rules and risks into the predicates presented in figure 2, and adds them to the prolog knowledge base. Next the RA-Ag uses the Prolog inference engine to evaluate the rules and determines that because the water temperature is 54°C, Mr. Jones runs the risk of scalding himself. Finally, the *RA-Ag* informs the *scaldingNotifying-Ag* by sending a *sendNotification()* message indicating the elder runs the risk of scalding in the shower with water that is too hot.

6 Conclusions and Future Work

We have presented the design of an agent-based component and the ontology ELDeR which defines the elders' context and the risks that can be inferred. According to our ontology, the context that should be monitored by the agent in order to predict risks when an older adult is performing an ADL are: the actions that make up the ADL, its features that determined a risk, and the location and time in which the ADL is being performed. We consider that the ELDeR ontology is general enough so that it can be used by any agent-based system that supports independent living of elder adults by predicting risks. We plan to implement several scenarios of use to validate our ontology, and to identify how it is useful not only for identifying risks but for generating the logs of elder's past activities that can later be analyzed in order to find abnormal behavior.

Acknowledgements. This work was funded by SEP-PROMEP and by CONACYT (J51582-R).

References

[1] Kimel, J., et al.: Exploring the Nuances of Murphy's Law Long-term Deployments of Pervasive Technology into the Homes of Older Adults. ACM Interactions, 38–41 (2007)
[2] Rodriguez, M.D., Favela, J., Preciado, A., Vizcaino, A.: Agent-based ambient intelligence for healthcare. AI Communications 18(3), 201–216 (2005)
[3] Chen, H., Finin, T., Joshi, A.: An ontology for context-aware pervasive computing environments. The Knowledge Eng. Review 3(18), 197–207 (2003)
[4] Ranganathan, A., Campbell, R.H.: An infrastructure for context-awareness based on first order logic. Personal and Ubiquitous Computing 7, 353–364 (2003)
[5] Haigh, K.Z., Kiff, M.: The independence LifeStyle Assistant (I.L.S.A): AI Lessons Learned. In: Proceedings of IAAI 2004, San Jose, CA, July 25-29 (2004)

Risk Patient Help and Location System Using Mobile Technologies

Diego Gachet, Manuel de Buenaga, José C. Cortizo, and Víctor Padrón

Grupo de Sistemas Inteligentes, Universidad Europea de Madrid
C/ Tajo s/n, 28670 Villaviciosa de Odón, Madrid, Spain
gachet@uem.es, buenaga@uem.es,
josecarlos.cortizo@uem.es, victor.padron@uem.es

Abstract. This paper explores the feasibility of the inclusion of information and communications technologies for helping and localizing risk and early discharge patients, and suggests innovative actions in the area of E- Health services. The system will be applied to patients with cardiovascular or Chronic Obstructive Pulmonarry Diseases (COPD) as well as to ambulatory surgery patients. The proposed system will allow to transmit the patient's location and some information about their illness to the Hospital or care centre.

Keywords: Ubiquitous computing, ambient intelligence, telecare, mobile location.

1 Introduction

The model for delivery of health services in our country is organized into a hierarchical structure of health facilities centres, where professionals provide services to their patients. The health professional contact with the patient begins in primary care centres, while the specialized care and management of the disease are performed in hospitals of first, second and third level, according to the level of expertise in the treatment and patient care.

In the hospital environment, given the high volume of activity and mobility, both medical and nurses need technological tools that facilitate access to patient medical information and aid them to get clinical decision, with a new form of interaction necessary among professionals, most natural and close to the user, in other words embedded in the working environment.

Moreover treating chronic illnesses, or special needs such as cardiovascular disseases are conducted in long-stay hospitals and in some cases in homes of patients with a follow-up from primary care centre. The evolution of this model is following a clear trend: trying to reduce the time and the number of visits by patients to health centres and derive tasks, so far as possible, toward outpatient care. Also the number of Early Discharge Patients (EDP) is growing permiting a saving in the resources of the care center [1].

This is largely due to certain structural aspects that encourage this trend [2] on one hand, the demographics of the population have changed, not only because of its growing aging but also by the increase in chronic diseases. Moreover, these developments

J.M. Corchado, D.I. Tapia, and J. Bravo (Eds.): UCAMI 2008, ASC 51, pp. 173–179, 2009.
springerlink.com

have an impact on the geographical characteristics of our environment, to do that there is a need to bring health services to rural populations and disconnected from the cities on the one hand and the need to decongest and decentralize health services.

2 Early Discharge Criterion for Risk Patients

The early discharge permits to the patients return to their home for a better recovery, telecare and telemedicine aims to accelerate discharge home from hospital and provide rehabilitation / support in the home setting and provide services to the patients as shown in figure 1:

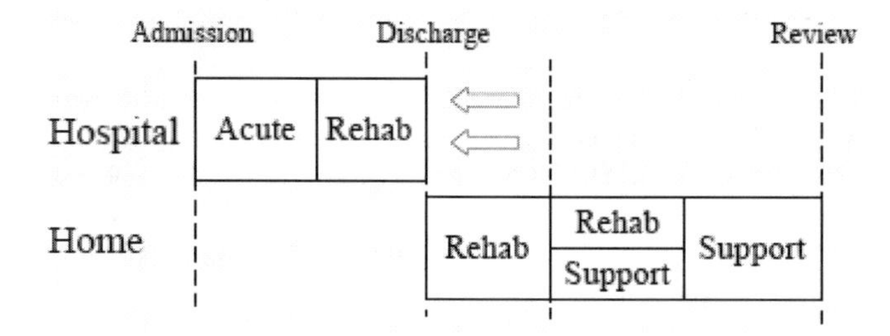

Fig. 1. Services at patient's home for a early discharge

The early discharge service (EDS) enables health professionals to deliver more proactive care and support for patients, who can remain living independently with the reassurance that help is at hand should they need it. Telemedicine units enable they to make early intervention, helping to adopt a more preventtive approach to care, which has helped to reduced hospital readmissions.[3]

The EDS is applicable to a large list of diseases as long as cardiovascular, chronic obstructive pulmonary (COPD) or stroke disseases. The criteria used for patient inclusion in an Early Discharge Program is highly dependent of medical team but in general we can take in account the following information:

- Age no more than 70 years old deppending on the grade of dissability
- Living in a distance no more than 100 km from hospital
- Stable clinical condition
- Sociofamiliar suitable environment
- Informed consent
- Capability to understand the program

The advantages for an early discharge program include home better rehabilitation for the patient, free hospital beds and reducing costs in 85% for care centre. [4]

3 System Architecture

Recent developments in ICT have, however, opened up the prospect of creating a far more customised and integrated approach to care provision, where telecare is used to monitor people at risk within the wider, non-institutional built environment. Two main types of telecare can be distinguished: systems designed for information provision and those designed for risk management [5]. The first approach aims to provide information about health and social care issues more effectively to individuals who need it. Our focus is on telecare for risk management, i.e. supporting activities for daily living.

In our proposal we take measurements from surrounding environment (mainly patient's location) and then transmit it along with other information to a health centre which will then act appropriately, permiting a care professional provide support remotely, notify the patient about the medication, ask patient about some special condition, trigger an emergency response or alert a relative or neighbour.

In effect, telecare can form part of the risk management that every clinician undertakes, by transforming a previously unsuitable environment into one that is sufficiently safe for a patient to be discharged to. A high level vision of the proposed system is depicted in figure 2.

As we can see from figure 2, from the hardware point of view the systems has a mobile device with process and location capability, we think that this device may be a PDA with mobile phone characteristics with an integrated GPS receptor and Wifi capability, also, it will be interesting if the device has an RFID (Radio Frequency Identification) reader for indoor location. From the communications networks point of view we are asuuming at least GSM coberture in the pàtient's area of movement.

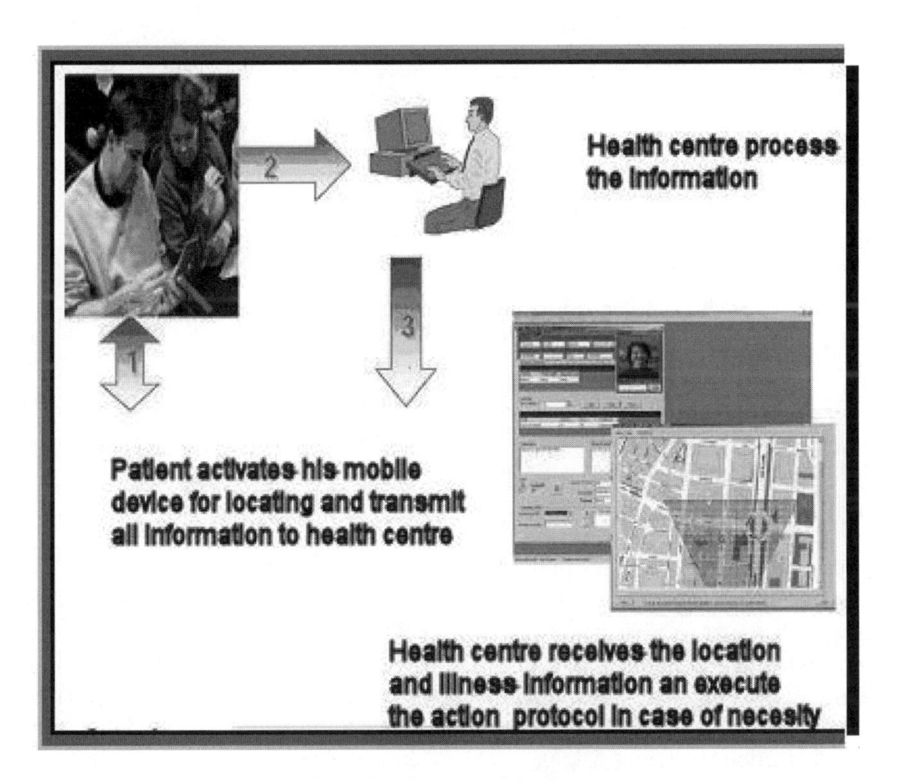

Fig. 2. Global Concept of the System

The illness information to be transmited to the health centre would include a questionnaire with information about the global condition of the patient, taking as example patients with stomach ambulatory surgery, the questionnaire may contains among other the information shown in Table 1. This medical information will change deppendng on the type of ambulatory surgery and must be dertermined by the medical personnel.

As communication system to ensure a correct transmission of information between patient and health centre we will use the GSM/GPRS public mobile networks that also may provide the possibility for patient's location trough the mobile device in case of necesity, i.e. when GPS service is not available.

The servers at the health centre must include the capability for visualizing map and location information and also the possibility for sending and receiving information to and from the patient, also this infrastructure has the possibility to contact inmediatly and in an automatic form with emergency and ambulancy teams in case of necesity.

Table 1. Questionarie about illnes condition to be transmitted to the care center

Surgery Problems	Answer
Pain in the afected body area	Nothing/ Tolerable / Much
Needed to take painkillers	Yes / No
Has bleeding	Yes/ No
Fever	Yes/ No
Deposition	Yes/ No
Vomited	Yes/ No
Headache	Yes/ No
Feeling dizzy	Yes / No
Tingling	Yes/ No
Blurred vision	Yes/ No
Has urine	Yes/ No
Pruritus or rash skin	Yes/ No
Dyspnea	Yes/ No
Difficulty swallowing	Yes/ No

3.1 Location Subsystem

In our case, the patient's location is very important due to medical personnel necesity to know all time where is the patient, this is specially necessary for patients with an early discharge as a consecuence of ambulatory surgery.

For obtaining location information we might mention the existence of mature technology such as the outdoor location through mobile phone networks that are based on triangulation The possibility of locating a mobile terminal has already led to numerous information services, tracking, route selection and resource management.

Because of the nature of the signal and network used, there are basically four types of mobile location systems: location through mobile pone networks, satellite networks, wi-fi networks and RFID devices.

There are several location techniques using the cells of mobile telephone systems, but they are based mainly in two principles: the measurement of the transmission time of the signal or the measure of the reception angle of the signal [6]. On the other hand the most used method for outdoor location is GPS (Global Positioning System). It is a fact that GPS is embedded in more and more devices: GPS receivers, PDAs and last generation mobile phones.

Nevertheless, the application for of GPS for indoor location is not so efficient. It is due to the lost and attenuation of the GPS signals because the lack of direct vision with the satellites and the attenuation that these signals suffer through building walls and roofs. The location using mobile phone terminals presents similar drawbacks. For these reasons in this proposal other variants of location have been taken into account: location using Wi-Fi networks and RFID-based location.

Location using Wi-Fi networks is obtained using several techniques. The simpler of which is determining the network access point close to the device to be located. Other very used technique consists in storing the measurement of the power of the network signal in different points of the indoor environment. The technique known as Wi-Fi mapping is more precise than the cell triangulation, reaching a precision in the range from 1 to 20 meters.

RFID-based location can be used in a big number of applications, because the data transferred from the label can provide information not only about location, but also identification or any other information that had been stored in the label.

4 Interface Elements and Methods

The system we are proposing consists in two main parts, a patient part and a health centre part, each with its own interface: the mobile device interface used by the patient and the hospital information system interface used by the medical personnel to access the information about the patients and take decisions about them.

Figure 3 shows a simplification on the interactions between users and subsystems. Figure 3 also shows the interface needs for each subsystem, where we can see that the mobile device's interface is pretty much simpler that the Hospital information system interface, that should provide the medical personnel to access and visualize patient's medical records and other related information about them.

The Hospital information system interface will be implemented as standard web interface to access the data, but the interface in the mobile device used by the patient needs more attention, it must be simple and usable by any possible risk patient as for example old people with little knowledge about technology. The patient's location tracking subsystem at the Hospital Information Servers could be based on Google Maps API, as it can be seen in the right part of Figure 4.

This interface would allows an easy location of patients showing the most important data for managing a possible emergency: the name of the patient, the state of the emergency and a resume of the clinical history, which can give some ideas about the problem. This interface allows the medical personnel to access more data about

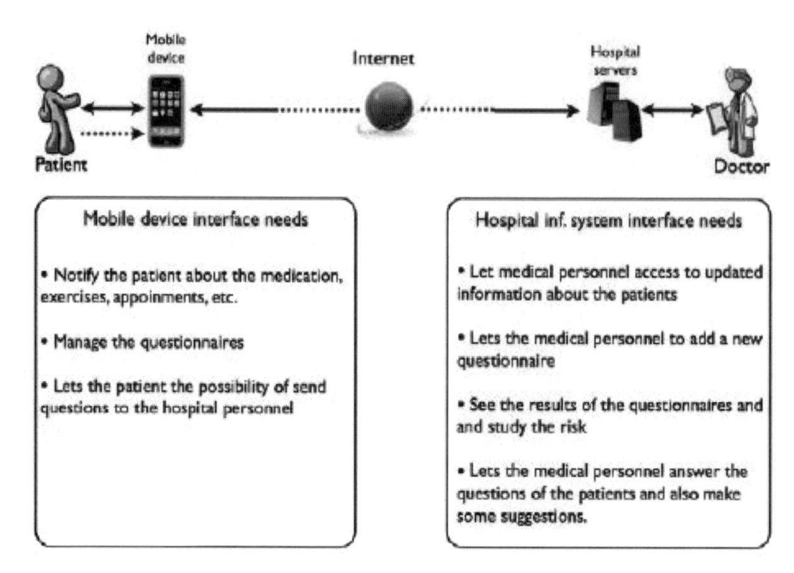

Fig. 3. Interaction between patient's mobile device the hospital infrastructure in order to let the medical personnel access the data

the patient in a very simple way, by two links to the last recorded patient's questionnaire and to all the data about the patient recorded in the system.

Figure 4 also shows (left image) a screenshot of the possible interface, it shows all events and reminders in an ordered list, which is very simple to manage by any possible patient. To fill out the questionnaires or get more information about an event or a reminder the user only needs to select the desired option.

Fig. 4. Screenshots of the interfaces of the system. The left image shows the patient's interface on a mobile device showing all the possible options. The right image shows a detailed view of the interface of the location subsystem in the Hospital Information Server.

5 Future Work

The ideas presented in this paper are parts of a feaseability study and not a final system, but shows interesting questions about the problem that we are trying to solve. The final implementation of the system must be designed to work on a real medical environment, which presents some differences and challenges:

- It must be able to manage different security profiles as the patient's information is confidential and only authorized personnel should access to it.
- It must be able to manage a huge volume of information, allowing a lot of connections between the HIS and the patient's mobile devices.
- The mobile device's interface should be standard or implemented for a large variety of systems.

The final system should also be evaluated in a real environment in a qualitative approach. We plan to evaluate the system with the final users: real patients and medical personnel. In this evaluation, we will study the usability and the satisfaction of the final users with the final system.

References

[1] Hailey, D., Roine, R., Ohinmaa, A.: Systematic review of evidence for the benefits of telemedicine. J. Telemed Telecare 8, 1–30 (2002)

[2] Mechanic, D.: Improving the quality of health care in the United States of America: the need for a multi-level approach. Journal of Health Services & Research Policy 7 (suppl.1), 35–39 (2002)

[3] Torp, C.R., Vinkler, S., Pedersen, K.D., Hansen, F.R., Jorgensen, T., Olsen, J.: Model of Hospital-Supported Discharge After Stroke. Stroke, 1514–1520 (June 2006)

[4] Iakovidis, I., Bergström, R.: European Union activities in e-Health (2006), `http://europa.eu.int/information_society/ehealth`

[5] Barlow, J., Bayer, S., Curry, R.: Integrating telecare into mainstream care delivery. IPTS Report (2003), http://www.jrc.es

[6] Aranda, E., De la Paz, A., Berberana, I., González, H.: Sistemas de localización en redes móviles: el servicio de emergencias 112. Comunicaciones de Telefónica I+D, número 21, 117–131 (2001)

[7] Barlow, J., Venables, T.: Smart home, dumb suppliers? The future of smart homes markets. In: Harper, R. (ed.) Home Design. Social perspectives on domestic life and the design of interactive technology. Springer, Heidelberg (2003)

[8] Brownsell, S., Bradley, D., Bragg, R., Catling, P., Carlier, J.: An attributable cost model for a telecare system using advanced community alarms. Journal of Telemedicine and Telecare 7, 63–72 (2001)

[9] Bull, R. (ed.): Housing Options for Disabled People. Jessica Kingsley, London (1998)

[10] Cabinet Office E-government. A strategic framework for public services in the information age, London, Cabinet Office (2000)

ATLINTIDA: A Robust Indoor Ultrasound Location System: Design and Evaluation

Enrique González[1], Laura Prados[1], Antonio J. Rubio[1], José C. Segura[1], Ángel de la Torre[1], José M. Moya[2], Pablo Rodríguez[2], and José L. Martín[2]

[1] Dpto. Teoría de la Señal, Telemática y Comunicaciones, Universidad de Granada. ETSIIT, C/ Periodista Daniel Saucedo Aranda s/n 18071 Granada, España
{lapm,emg1985}@correo.ugr.es, {rubio,segura,atv}@ugr.es

[2] Telefónica Investigación y Desarrollo. Centro de Granada. C/ Recogidas 24 - Portal B, Escalera A, Planta 1, Puerta A, 18002 Granada, España
{jmml,pra,joselu}@tid.es

Summary. We describe the ATLINTIDA indoor ultrasound location system, which provides positioning services in the context of Ambient Intelligence. ATLINTIDA makes use of spread spectrum techniques in order to improve robustness. In this paper we describe the design of ATLANTIDA, similarities/differences with respect to a commercial ultrasound location system (the CRICKET) and a comparison of the performance provided by both systems. According to experimental results, ATLINTIDA significantly improves the performance of CRICKET in positioning accuracy (mean error in location reduced from 37.74cm to 7.37cm), and rate of invalid measurements (reduced from 39.18% to 4.27%), for a similar distribution and density of nodes.

Keywords: ATLINTIDA, CRICKET, spread spectrum, ultrasound location.

1 Introduction

In the context of Ambient Intelligence, a number of systems oriented to location of the actors or objects have recently been developed. Some of them are based on ultrasound or acoustic techniques, while others make use of radio, optical or artificial vision methods [1, 2]. Among the different proposed technologies, ultrasound-based (and acoustic-based) systems are shown to provide the best results regarding accuracy in the location results, due to the extremely low velocity of propagation of the acoustic wave (compared to the propagation of radio signals).

Most ultrasound location systems are based on the transmission of a short signal from a transmitter to a receiver. The time of arrival is obtained by comparing the delay of the ultrasound acoustic signal with respect to a synchronization signal (which can be transmitted by radio or cable). From the time of arrival and taking into account the propagation velocity of sound, the distance between the transmitter and the receiver can be computed, and using the distances from several nodes to the object to be located, its position can be estimated.

J.M. Corchado, D.I. Tapia, and J. Bravo (Eds.): UCAMI 2008, ASC 51, pp. 180–190, 2009.
springerlink.com

Ultrasound location systems like the CRICKET [3, 4] use ultrasound ceramic transducers for the transmitter and the receiver, and send a short pulse of signal modulating a carrier at the transmission frequency (40 kHz in the case of CRICKET). This provides a narrow-band signal that can be appropriately transmitted using ceramic transducers (which present a bandwidth of 2 kHz in the case of CRICKET). Even though these systems provide an accurate location, due to the relatively short duration of the pulses and the characteristics of the transmitted signal, the transmission is not robust enough in the presence of noise in the transmission band, or for a low density of nodes in the transmission system (i.e., for low signal-to-noise ratio in the transmission). In order to minimize some inconveniences of ultrasound location systems based on ceramic transducers, locations systems based on piezofilm transducers have been proposed [5, 6]. These transducers provide wider bandwidth, which allow the application of spread spectrum techniques [7], in order to increase robustness of the transmission. Additionally, piezofilm transducers present wider radiation diagrams than ceramic transducers, which in principle would provide better coverage of the room to be located. However, the maximum acoustic power that can be radiated by a piezofilm transducer is significantly lower than that radiated by a ceramic transducer, and similarly, the sensitivity of piezofilm transducers is significantly lower. This reduces the maximum distance between transmitter and receiver when using piezofilm transducers.

In order to benefit from advantages of spread spectrum techniques (robustness in transmission) and those of ceramic transducers (high radiation power and sensitivity) we have adapted the spread spectrum method to the bandwidth available in the ceramic transducers, by transmitting a pseudo-random sequence of N bits, each bit with the required duration to accommodate into the bandwidth of the transducer. This leads to an expansion in time duration (by a factor N) of the transmitted signal with respect to the transmission of a single pulse. The transmission of a pseudo-random sequence provides robustness against noise and multi-path effects. The use of ceramic transducers allows transmission with high power and reception with high sensitivity. Finally, the expansion in the total duration of the transmitted signal increases the total energy involved in each transmission, which also increases the robustness of the positioning system. In this work we describe the ATLINTIDA indoor ultrasound positioning system, which has been designed according to de above described criterion. This system uses Kobitone 40 kHz ceramic transducers (like the CRICKET system) but instead of transmitting a single pulse, our system transmits a pseudorandom sequence of bits for the estimation of the distance between the nodes and the object to be located.

In the next sections we describe the design of the ATLINTIDA system (including principles of design, hardware and software), and the results of some experiments carried out to compare the ATLINTIDA and the CRICKET ultrasound location systems. Finally, we present some conclusions and outline future improvements of the ATLINTIDA system.

2 Description of ATLINTIDA Ultrasound Location System

The ATLINTIDA robust ultrasound location system has been developed in collaboration between the University of Granada and Telefónica Investigación y Desarrollo. ATLINTIDA project has included the design, implementation and evaluation of the specific hardware and software necessary to perform accurate and robust location in the context of Ambient Intelligence and Teleassistance Services.

2.1 Description of Hardware and Software

ATLINTIDA specific hardware consists of 7 ultrasound transmitters, an interface and one ultrasound receiver. The interface connects the transmitters to an AD/DA acquisition board. The interface also provides DC supply for the transmitters. The AD/DA board is connected to a laptop running the control and positioning software. The software sequentially activates the different transmitters and prepares the pseudorandom sequence to be transmitted. The sequence generated by the AD/DA board is sent to the interface, where a demultiplexer sends it to the appropriate transmitter. The receiver is also connected to the AD/DA board, and the software acquires and processes the received signal in order to compute the time of arrival of the ultrasound signal. Using the times of arrival from several transmitters to the receiver, the position of the receiver is computed by a trilateration algorithm. Therefore, the ATLINTIDA is a centralized system where transmission and reception is managed by the same software. However, the architecture of the system could easily be modified in order to allow the receiver to independently compute the location.

The architecture of each transmitter is described in the block diagram in figure 1. Each transmitter includes a 40 kHz square wave oscillator. A TTL signal coming from the AD/DA board is used to modulate the 40 kHz carrier provided by the oscillator. Modulation is performed with an AND gate, providing ASK modulation. Finally a power amplifier is included in order to provide enough amplitude to the transducer.

On the other hand, the receiver obtains the transmitted sequence by demodulating the received signal. Demodulation is based on envelope detection, using a half-wave precision rectifier. The basic architecture is presented on figure 2.

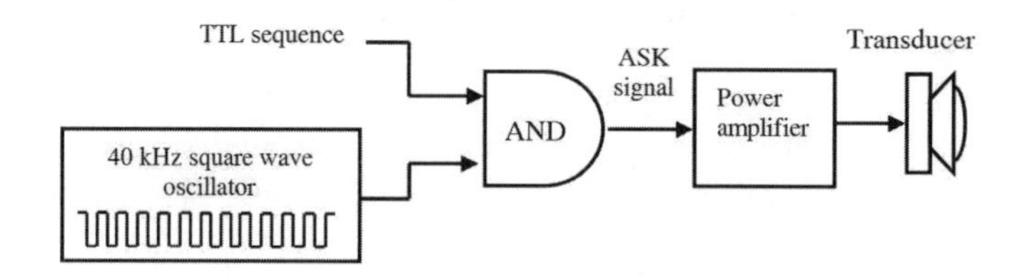

Fig. 1. Block diagram of the ultrasound transmitter

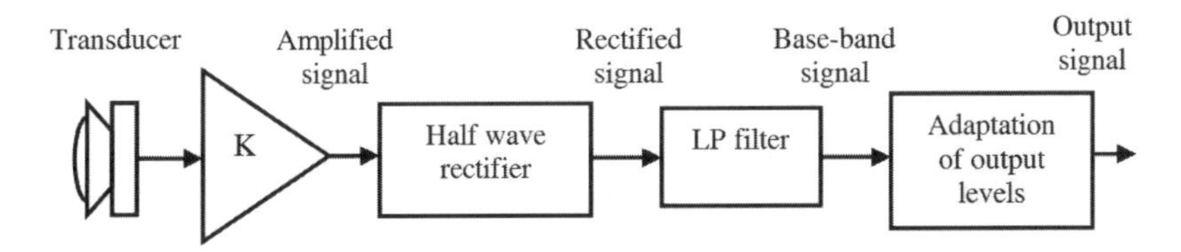

Fig. 2. Block diagram of the ultrasound receiver

Fig. 3. Pictures of the ATLINTIDA ultrasound transmitter (left side) and receiver (right side)

The receiver includes a receiving ultrasound transducer, an amplifier, a rectifier, a low-pass filter and an adaptation of output levels.

In order to complete the description of the hardware designed for the ATLINTIDA system, figure 3 shows pictures of the implemented transmitter (left side) and receiver (right side), respectively.

2.2 Location Procedure

The ATLINTIDA system obtains a 3D location of the receiver by spherical trilateration, using the estimated distances between the receiver and several transmitters. Each distance is estimated from the time of arrival of the ultrasound signal, like in the CRICKET system. The main difference between both systems is that CRICKET sends a single pulse when transmitting, while ATLINTIDA transmits a pseudorandom sequence of pulses, in order to benefit from advantages of spread-spectrum technique. In order to accommodate the bandwidth of the transducers, instead of inserting the sequence into the pulse duration (which would increase the bandwidth), the sequence contains N bits, each one with the duration of the pulse. This way, the bandwidth is kept constant, but duration of the sequence increases by a factor N. So this method provides a time expansion rather than a spread spectrum.

Like in the case of spread spectrum, the proposed time expansion approach provides immunity against noise in the transmission, thanks to the redundancies inherent to the use of a pseudorandom sequence of bits (that is applied for

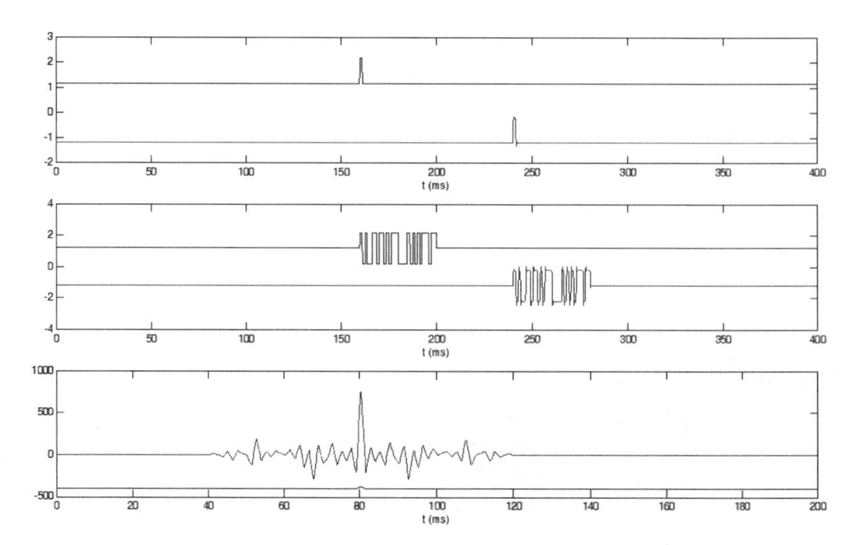

Fig. 4. Simulation comparing transmission and reception of a single pulse (top) and a pseudorandom sequence (center). The cross-correlation functions between transmitted and received signals (bottom) have been computed for single pulse (blue line) and pseudorandom noise (green line).

both, transmission and reception) [7]. Compared with spread spectrum, the time expansion technique increases the duration of the transmission (because of the increment in the duration of the transmitted sequence). Similarly, if transmission power is kept constant, the total energy transmitted increases in the case of time expansion, due to the increment in the duration of the transmitted signal. This also increases the robustness of the transmission.

Figure 4 shows a comparison of transmission of a single pulse (like in the CRICKET system) and a pseudorandom sequence (like in ATLINTIDA). The plot in the top shows, respectively, the transmitted and received signals for a single pulse. The plot in the center shows the transmitted and received signals for a pseudorandom sequence. The plot in the bottom represents the cross-correlation between transmitted and received signals in the case of single pulse (in blue) and in the case of pseudorandom sequence (in green). In both cases, the peak in the cross-correlation function provides the delay between the transmitted and received signals (80 ms in the simulation), and can be used to estimate the time of arrival. However, the peak of the cross-correlation function is significantly higher in the case of the pseudorandom sequence, because the total transmitted energy is higher in this case.

In absence of noise, both single pulse and pseudorandom sequence provide similar accuracy in the estimation of the delay. However, in presence of noise, pseudorandom sequences perform better due to immunity associated to spread spectrum method. Figure 5 shows a simulation similar to that in figure 4, but for a signal to noise ratio of 0dB in the received signals. As can be observed, in the

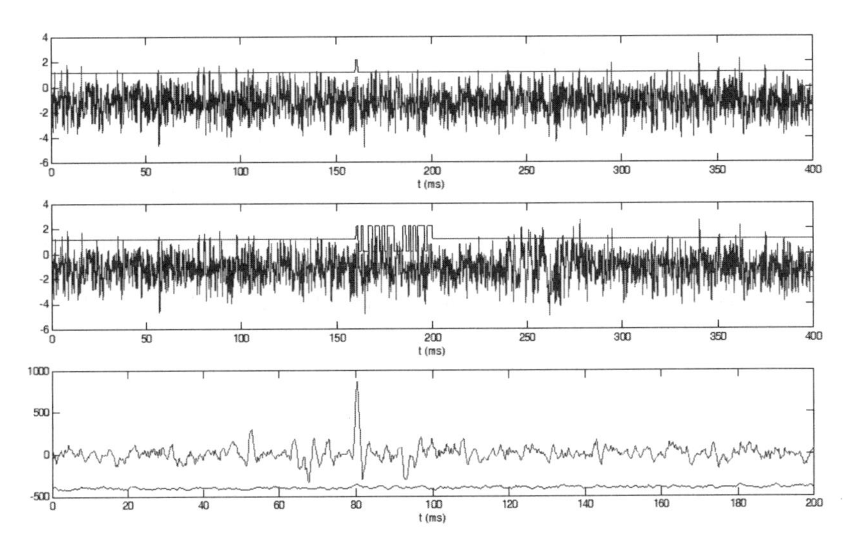

Fig. 5. Simulation similar to that in figure 4, but contaminating the received signal with a white noise at a signal to noise ratio of 0 dB. The cross-correlation function between transmitted and received signals (bottom) shows better immunity in the case of pseudorandom sequences.

case of pseudorandom sequence, the cross-correlation function provides a clear peak, corresponding to the correct delay, while in the case of single pulse, peaks associated to noise are higher than that corresponding to the correct delay.

The time of arrival of the ultrasound signal corresponds to the position of the peak of the cross-correlation function. The distance between the transmitter and the receiver can be estimated taking into account the propagation velocity of sound (340 m/s).

Compared with CRICKET, ATLINTIDA presents the advantage of robustness, but the disadvantage of a longer duration in the transmission, due to the use of a longer signal to be transmitted. This could affect the updating period of the location system, which would be critical for the location of moving objects. On the other hand, since management of the transmission is centralized in ATLINTIDA system, the interval among transmissions can be minimized without risk of collisions. This is an important difference with respect to CRICKET. In the CRICKET, each node autonomously transmits, and therefore transmission is configured with a random interval that must be large enough to minimize the risk of collisions, which significantly increases the updating period.

In the case of distributed systems like CRICKET, synchronization between transmitter and receiver is also necessary. A radio link between transmitter and receiver provides synchronization for the ultrasound signal. In the case of ATLINTIDA, since the management of transmitters and processing of the received signal are performed in the computer, synchronization is also centralized and no radio link between the transmitters and the receiver is necessary.

3 Experimental Results

In order to evaluate the developed location system and also to compare the performance of ATLINTIDA and CRICKET, location experiments were carried out in similar conditions for both systems. The same number of transmitters were used (7 in both cases) and they were placed at the same positions at the ceil of the testing room. Location was measured for a set of 19 positions and 3 different altitudes in the room (i.e., a total of 57 positions). For each position, 300 location measurements were recorded with both, ATLINTIDA and CRICKET location systems. Figure 6 shows a diagram of the testing room, including dimensions, the position of the 7 transmitters and the 57 different positions where the receiver was situated.

Three aspects have been considered when comparing the developed location system (ATLINTIDA) and the reference one (CRICKET). Taking into account that the coverage provided by the location systems strongly depends on the dimensions of the room and the distribution and density of nodes, for some positions of the receiver the received signal could be insufficient to provide a valid location. Therefore, the first aspect to be compared is the rate of invalid location measurements. The second aspect to be compared is the accuracy in the location results for the valid location measurements. Finally, the updating period in the location results is an aspect to be compared that becomes important in the case of location of moving objects.

3.1 Rate of Invalid Measurements

The number of invalid measurements for each receiver position is shown in figures 7 (CRICKET system) and 8 (ATLINTIDA system). In these figures, each bar corresponds to one of the 57 receiver positions. The results are grouped for the different altitudes of the receiver, and each group of results contains measurements for 19 positions at a given altitude. The total number of measurements at each receiver position is 300.

In the case of CRICKET system, a high rate of invalid results was observed, particularly in the case of receiving positions in the limit of the room (where the signal is not appropriately received for the minimum number of transmissions

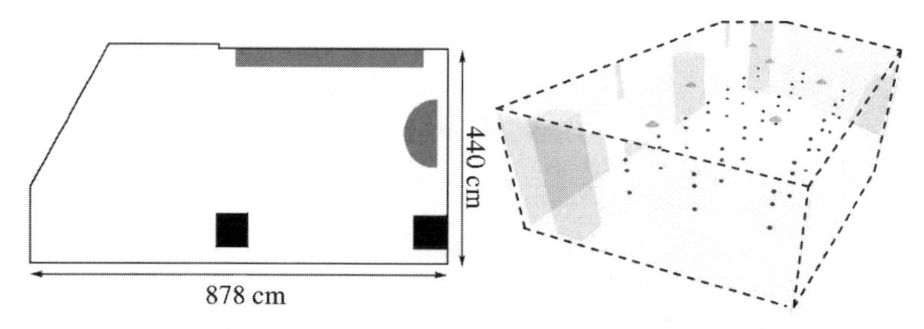

Fig. 6. Description of the test configuration for comparing the ATLINTIDA and the CRICKET ultrasound location systems

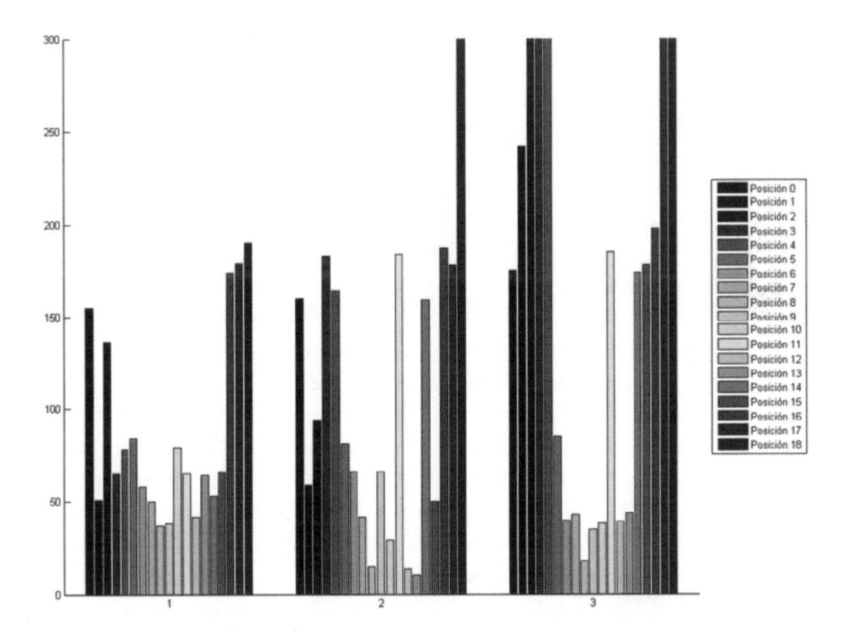

Fig. 7. Distribution of invalid measurements for the CRICKET location system, considering different positions and altitudes of the receiver, for a total of 300 measurements at each position

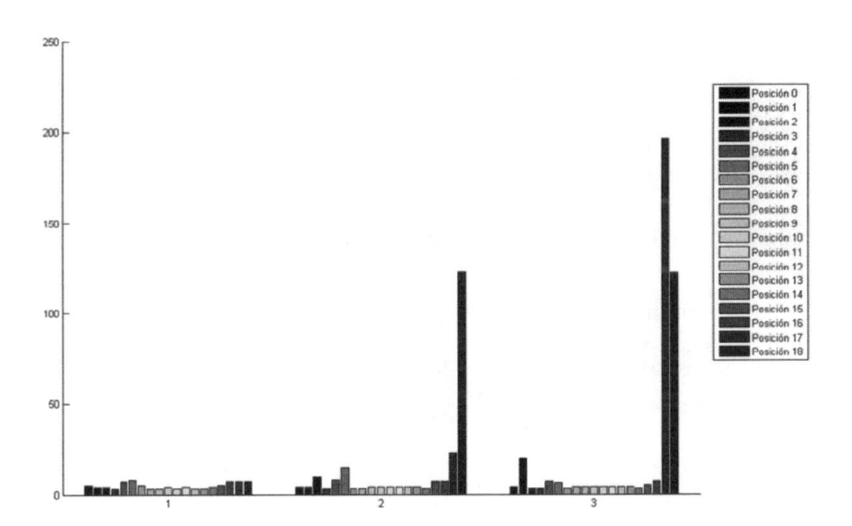

Fig. 8. Distribution of invalid measurements for the ATLINTIDA location system, considering different positions and altitudes of the receiver, for a total of 300 measurements at each position

necessary for trilateration), and for positions close to a column (due to multipath interference). The rate of invalid measurements increases for altitude 3 because in this case the angles of arrival in the transmission direction with respect to the axis of the transducers increases, and this causes attenuation due to the relatively narrow radiation diagram of the transducers. Note that for 23 (of the 57 receiving positions) the CRICKET presents a rate of invalid measurements higher than 50%, and for 6 positions, CRICKET is unable to provide any location result.

Figure 8 presents the number of invalid measurements for the ATLINTIDA location system. As in the case of CRICKET, the positions providing the highest rate of invalid measurements are those in the limit of the room, those closer to the columns and the ones for the highest altitude. However, in this case the rate of invalid measurements is significantly lower than in the case of CRICKET. The rate of invalid measurements is 29.19% for altitude 1, 35.81% for altitude 2, 52.53% for altitude 3, and the global rate is 39.18% for the CRICKET system, while in the case of ATLINTIDA it is reduced to 1.56% for altitude 1, 4.14% for altitude 2, 7.12% for altitude 3, and 4.27% the global rate. The rate of invalid measurements is higher than 50% only for position 17 - altitude 3, and it remains below 15% for 54 of the 57 receiving positions. This reveals a better coverage of the ATLINTIDA system for the same distribution of transmitters. This advantage of the designed system with respect to the reference one is obtained thanks to the improvement in the robustness of the transmission.

3.2 Accuracy in Location

The accuracy in location has been evaluated by comparing the distribution of the location error for CRICKET and ATLINTIDA location systems. The location error is defined as the Euclidean distance between the position of the receiver and the position estimated by the location system. The distribution of location error has been analyzed only for valid measurements. Table 1 presents a comparison of the location error observed for both location systems. The mean and standard deviation of the location error is shown, as well as 50%, 80% and 95%

Table 1. Comparison of the location error provided by CRICKET and ATLINTIDA location systems. Mean, standard deviation and 50% 80% and 95% percentiles are shown for different altitudes and total valid measurements.

System	altitude	mean error	std. dev.	50% percent.	80% percent.	95% percent.
CRICKET	1	39.6 cm	29.3 cm	31.8 cm	35.7 cm	94.7 cm
	2	35.7 cm	44.9 cm	23.4 cm	26.5 cm	151.9 cm
	3	37.7 cm	50.1 cm	20.4 cm	27.7 cm	131.5 cm
	TOTAL	**37.7 cm**	**41.2 cm**	**26.3 cm**	**33.9 cm**	**123.7 cm**
ATLINTIDA	1	5.73 cm	9.75 cm	4.56 cm	8.05 cm	11.9 cm
	2	7.52 cm	6.68 cm	5.85 cm	9.68 cm	19.4 cm
	3	8.94 cm	10.6 cm	5.60 cm	10.6 cm	37.7 cm
	TOTAL	**7.37 cm**	**9.27 cm**	**5.47 cm**	**9.05 cm**	**16.6 cm**

Table 2. Comparison of the updating period provided by CRICKET and ATLINTIDA location systems. Mean, standard deviation and 50% 80% and 95% percentiles are shown for valid measurements.

System	mean time	std. dev.	50% percent.	80% percent.	95% percent.
CRICKET	187.3 ms	175.1 ms	169.9 ms	305.9 ms	633.8 ms
ATLINTIDA	188.8 ms	176.8 ms	154.5 ms	155.5 ms	203.1 ms

percentiles. The comparison shows a clear improvement of the accuracy provided by the designed system with respect to CRICKET. This improvement is again associated to the increased robustness in the ultrasound transmission due to the use of a pseudorandom sequence.

3.3 Updating Period

Table 2 shows a comparison of the updating period for CRICKET and ATLINTIDA systems. Mean time of updating, standard deviation, and percentiles 50%, 80% and 95% are shown in the table. It can be observed that both systems provide similar average updating period, because the increment of the updating period due to the use of a pseudorandom sequence (in ATLINTIDA) is partly compensated by the increment of updating period associated to the procedure of the CRICKET for reducing the probability of collisions in a not centralized management of transmissions.

4 Conclusions and Future Work

This work describes the ATLINTIDA ultrasound location system. This system has been designed in order to improve robustness in location. Robustness is achieved by applying spread spectrum technique, but due to restrictions of the transducers bandwidth, the implementation of the technique leads to a time expansion rather than a spread spectrum. The ATLINTIDA location system has been evaluated and compared to the CRICKET system. The transmission of a pseudorandom sequence of bits increases the immunity of the transmission and improves performance of ATLINTIDA with respect to CRICKET by reducing the rate of invalid measurements (by a factor 8) and reducing the average error in location (by a factor 5) without an increment in the updating period. Some improvements are considered for future versions of the ATLINTIDA prototype. Including a radio link between the control computer and the receiver would allow local and autonomous estimation of location results. Changing ASK modulation by PSK modulation would increase immunity in transmission. The effect of some parameters (like properties of the pseudorandom sequence and its length, sampling frequency, etc.) over the accuracy should also be studied in order to optimize the performance of the location system.

References

1. San, S.: Location Authorities for Ubiquitous Computing. In: Proceedings of the Workshop on Location-Aware Computing (2003)
2. Jiménez, A.R., Seco, F., Prieto, C., Roa, J.: Tecnologías sensoriales de localización para entornos inteligentes. In: Congreso Español de Informática (2005)
3. CSAIL, Massachusetts Institute of Technology (MIT), EE.UU. The Cricket Indoor Location System, `http://cricket.csail.mit.edu/`
4. Balakrishnan, H., Baliga, R., Curtis, D., Goraczko, M., Miu, A., Priyantha, B., Smith, A., Steele, K., Teller, S., Wang, K.: Lessons from Developing and Deploying the Cricket Indoor Location System. MIT Computer Science and Artificial Intelligence Laboratory (CSAIL) (2003)
5. Hazas, M., Hopper, A.: Broadband Ultrasonic Location Systems for Improved Indoor Positioning. IEEE Transactions on Mobile Computing 5 (2006)
6. Hazas, M., Ward, A.: A Novel Broadband Ultrasonic Location System. In: Proceedings of the Fourth International Conference on Ubiquitous Computing (2002)
7. Torrieri, D.: Principles of Spread Spectrum Communication Systems. Springer, Heidelberg (2005)

Standard Multimedia Protocols for Localization in "Seamless Handover" Applications

Jorge Parra, Josu Bilbao, Aitor Urbieta, and Ekain Azketa

Software Technologies & Communications Area, Ikerlan Technological Research Centre, 20500 Mondragon, Spain
{jparra,jbilbao,aurbieta,eazketa}@ikerlan.es

Summary. The increasing number of networked devices with multimedia rendering capabilities comes down to the deployment of user-centered applications like "Follow Me" multimedia services. In these scenarios, the multimedia content comes along with the user because it is conveniently distributed over the most suitable device as he/she moves. In order to support user mobility, a key issue is to know the localization of users and devices. Nowadays, a great number of technologies are being used for outdoor and indoor localization, such as GPS, RFID, WiFi and UWB-based systems. However, all of them require the utilization of their own API with their own protocols (not compliant with well-known or standard multimedia protocols). This can seriously hinder efforts to develop heterogeneous scenarios where different localization systems have to be used for multimedia seamless handoff. This paper presents and analyzes an approach that makes use of multimedia standard protocols for heterogeneous localization in seamless handoff scenarios. This is achieved by using the RTP/RTCP protocol that hides underlying technologies for localization.

Keywords: RTP/RTCP, localization, multimedia seamless handoff or handover, ubiquitous computing, mobility awareness.

1 Introduction

Multimedia services are no longer the same since the digital era began. The digitalization of broadcasting audiovisual contents and the advent of IP services have supported the generation of a wide range of multimedia services that are offered to users in indoor and outdoor situations [1]. This has been a reality thanks to the growing advances in wired and wireless technologies and in portable and more capable devices.

Nowadays, the tendency is to redistribute these multimedia services for mobile users [2]. For example, the concept of *extended home* has been coined to indicate an environment where mobility aspects are considered. In an extended home environment many devices are connected together, both wired and wireless, to support user and service mobility while moving (indoor and outdoor) [3]. Therefore, currently there is increasing interest in deploying user-centered services and contents [4]. The user no longer follows or searches for services, but

J.M. Corchado, D.I. Tapia, and J. Bravo (Eds.): UCAMI 2008, ASC 51, pp. 191–200, 2009.
springerlink.com

they go with the user whilst on move in the environment limited by the extended home (i.e. in a car, in a city).

Advances in knowing position, location and capacities of the different agents that participate in a streaming session unfold new avenues of applications/services that can follow users [5]. In this vast field of applications, this research aims at multimedia streaming systems where the multimedia content goes with the user and, if required, the user chooses where to display it.

In this kind of scenarios, there must exist at least: a mobile service that should go with a user, one or more localization mechanisms for user or system location, a unified communication networking (i.e. IP network), and a system to indicate servers where to send the content. The novelty of this research work is that the localization information is given by means of multimedia standard protocols. According to the authors, the most optimal conditions to be successful in this kind of applications are: server availability and the number of clients requesting a particular multimedia content should not maximize the number of clients that can be serviced simultaneously by the server.

The rest of the paper is organized as follows. In Section 2, a reference scenario is described. This is used to show challenges when using different localization systems and the main research goal. In Section 3, the state-of-the-art in seamless handoff is shown, followed by a description of the main localization systems available (Section 4). Section 5 and 6 describe, respectively, the novelty and its advantages and drawbacks. Finally, a proof-of- concept study is outlined in Section 7 and Section 8 concludes.

2 Reference Scenario

A reference scenario is depicted in Figure 1. It shows all scenario types that can be possible in outdoor and indoor situations for multimedia seamless handoff. Throughout this paper, this layout is used to describe challenges for each particular situation, as well as to explain the solution addressed by the authors.

The scenario consists of a user provided with a PDA. He moves both indoor (i.e. at home, at his/her office) and outdoor. In general, the most typical situation, which is studied in this research, is when a user moves at home and multimedia content goes with the user.

Basically, four elements are shown in this scenario: a user, user location (i.e. home or outdoor places), display devices and lastly, multimedia content servers, In this scenario the user always carries a personal display device with him/her, a PDA. The remaining display devices are shown like TVn (n varies from 0 to 6). The devices used for user localization are denoted by Ln (n varies from 0 to 8). L0 denotes the user location provided by the PDA. Sn (n from 0 to 3) denotes content servers that could be available or not. For example, S1 could be a Satellite TV tuner. Generally and without loss of generality, the key elements to be considered are the user, the multimedia display devices and the localization system used depending on user location.

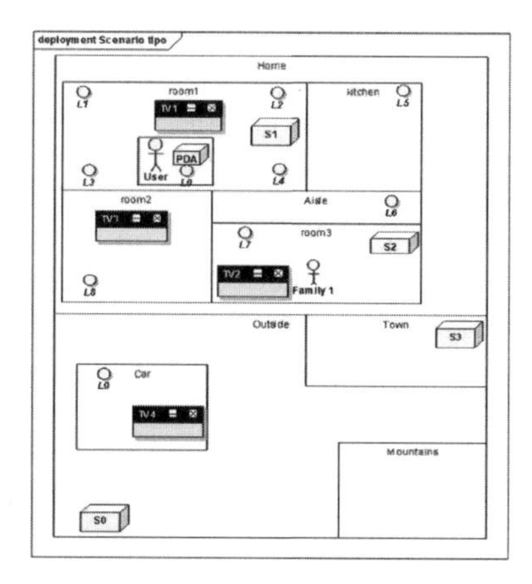

Fig. 1. Reference Scenario

This reference scenario allows easily describing many cases where multimedia content follows a user as he/she moves. In today's real world environments, most of these cases imply situations wherein the person is either in a building or outside, although it can be possible a mixture of indoor and outdoor situations, especially in the years to come.

3 State of the Art and Challenges

One of the key points in providing user-centered services is the fact that such services are given through a wide range of heterogeneous networks based on different physical means and protocols. Therefore, user mobility will generally imply the transition or migration across different networks in the extended home. When it comes down to describing this migration based on the same or different network technologies, handoff is an important issue. Handoff (also called Handover) is a fundamental concept in the provision of the required QoS in multimedia services while a user moves [6].

A striking research work is being developed by MIH (Media Independent Handover) or the 802.21 Working Group [7]. They classified handoff into two types: homogeneous (or horizontal) and heterogeneous (or vertical). The first term refers to interoperability within a single network based on the same technology (localized mobility), and the latter one refers to global mobility, that is, across different networks. There is another fundamental classification for handoff processes: hard handoff and soft handoff [6]. A convergence of these handoff mechanisms is of paramount interest in order to enable media streaming moved between access networks within the same device as a user moves or to adjust seamlessly the service to a new device available depending on the user's position. Some related research has been reported using the Session Initiation Protocol (SIP) [8], which provides mobility support [9, 4]. Although SIP-based handoff mechanisms solves problems related to mobility, they suffer from undesirable delays causing packet loss in detrimental to reducing the QoS of the multimedia service [9]. Therefore, some SIP-based mechanisms have been proposed to improve the QoS of the streaming service. In [9] a SIP-based technique is studied to support soft handoff in heterogeneous wireless networks whereas in [4] SIP-based vertical handoff is addressed.

In general, the use of handoff has been based on the detection of SNR (signal to noise ratio) to select which access network should be used. There are many situations where handoff is an important issue. For instance, handoff has been used in the case of network congestion [6]. Another research direction focuses on the usage of SNR and user localization for seamless handoff [10, 11]. In [12] soft handoff adaptive methods based on signal strength and RF propagation statistics are described. An interesting work is also described by Sulander and his colleagues to provide a new method for faster handoff for mobile IPv6 networks [13].

Our main goal is to provide a mechanism based on standard and widely-used multimedia protocols that support continuity on multimedia services over heterogeneous networks and devices as users move. In "Follow Me" scenarios where multimedia content has to be adapted to user location, different localization technologies are sometimes required [5]. That means that the localization system used may vary or be comprised of heterogeneous devices, which negatively contributes to the application complexity. The existence of different localization systems requires the API device integration to each application.

4 User Location Methods

In almost any multimedia distributed scenarios, both wired and wireless, protocols, such as RTP/RTCP or RTSP are mainly employed [2, 10].

In ubiquitous computing environments, it is needed to know a user's and objects' location with certain degree of accuracy in order to adapt services to user requirements or desires or provide location-awareness information. Currently, many technologies are being used for location-aware systems.

- GPS - Global Positioning System [14]: a satellite navigation system traditionally used for vehicle navigation. It is most widely used for outdoor positioning information [5].
- RFID - Radio-Frequency Identification: Due mainly to its simplicity and low cost, RFID technology has recently gained popularity for use in real-time user localization [15]. RFID-based wearable or handheld devices allow determining where a user is located. One of the most interesting features is that RFID tags for positioning can be incorporated into everyday objects, such as, pieces of clothing.
- Ultra-Wideband (UWB) Radio Technology [16, 17]: provides a useful means for wireless positioning due to its high resolution. Some companies have made use of UWB to provide real-time location systems with high precision, responsiveness, reliability and scalability [18].
- Wi-Fi technology [19]: is one of the most popular wireless technologies at home and in mobile devices. The idea is to determine the precise location of

any Wi-Fi enabled device based on the Wi-Fi networks. In [20] there is an example of using the WiFi technology for location-aware environments.
- Zigbee [21, 22]: is a wireless technology to address the needs of low cost and low power consumption sensor networks. The Zigbee technology can be used for determining user location; it can be estimated according to their received signal strength samples from Zigbee sensors [23]. Presently, it is increasingly being used in indoor environments.

Many systems have utilized a hybrid combination of these technologies. In [20], a location-aware computing system is studied where GPS is used for outdoor cases and Wi-Fi for indoor situations, such as a laboratory. In [24], RFID technology is incorporated into a navigation system to improve the accuracy. A survey of commercial products and research in the area of localization is reported in [5]. Therefore, the varying number of localization methods provides evidence for the need of vertical and soft-handoff over heterogeneous networks for locating users. However, in related research works different localization systems are used for providing user location coordinates to content providers and services. This paper addresses this issue by describing a homogenous mechanism based on multimedia standard protocols to provide user location. In that way, any of the handoff processes could be carried out.

5 Novelty Description

Real-time Transport Protocol (RTP) provides end-to-end network transport functions for applications streaming or transmitting real-time data, such as, mainly audio and video data, over unicast or multicast network services [25]. The Real-Time Transport Control Protocol (RTCP) is a control protocol that runs beside RTP for monitoring data delivery and for providing control and identification functionality. It is based on a periodic transmission of control packets to all participants in an on-going session. The control packets use the same distribution mechanism as the data packets. Therefore, RTP is designed to send real-time data; whereas, RTCP monitors the quality of service and conveys information about the participants in the same session.

RTCP performs four functions:

- (Required) Provide feedback on the quality of the data distribution.
- (Required) Carry a persistent transport-level identifier for an RTP source.
- (Required) Control the RTCP rate in order to allow scaling up to larger number of participants.
- (Optional) Convey minimal session control information, such as participant identification, or localization.

This research direction is to exploit the last function of RTCP to simplify the development of seamless handoff applications and decouple such applications from any localization technology. The Source Description RTCP Packet (SDES) allows the participants to provide their own description, by means of some optional well-known items:

- CNAME: user and domain name
- NAME: common name of source
- EMAIL: email address of source
- PHONE: phone number of source
- LOC: geographic user location
- TOOL: name/version of source application
- NOTE: notes about the source
- PRIV: private extensions

The main contribution of this study is that sensors, technologies or any other devices for user localization become participants in a RTP session. These localization devices convey user location to the rest of participants (content servers, receivers and other locators) by means of RTCP packets. User location is conveyed by using the LOC item in a SDES packet that describes a user.

User location or even mobile receivers are published in RTCP sessions, without prior knowledge of the localization technology that is being used. Hence, seamless handoff applications are decoupled from underlying localization technologies. The application only needs to listen to SDES packets from the participants (senders, receivers and users) to know the presence, identification and location of each participant and acting in consequence. The RTP/RTCP protocol is currently being used in many applications that involve the transmission of real-time data (mainly audio and video), such as applications for IP phones, IP TV, IP TV Set Top Boxes, videoconferencing systems, mobile TV, mobile Handsets and so on.

6 Conclusions and Future Work

This section provides a better understanding of the novelty by outlining the advantages and drawbacks of our research.

Advantages:

1. By abstracting the localization technology, seamless handoff processes become simpler. In this case, different technologies and protocols do not need to be integrated. (i.e. NMEA, UWB, RFID).
2. The use of different localization technologies makes it easier seamless handoff. It makes feasible continuous user positioning for indoor and outdoor movements (Figure 1), for example, the use of GPS for outdoor situations and when users roam through indoor places, the employment of RFID, UWB and so on.
3. The handoff is not centered at any agent because any of the participants (or an external monitor) can implement the handoff mechanism.
4. It is possible to employ the same localization devices in different applications that can be simultaneously executed.
5. Location awareness is transparent for the handoff system, only the localization methods are involved.

Possible drawbacks:

1. The transmission conditions to maintain scalability defined in RFC 3550 - RTP [25] can be an inconvenient when the number of participants in RTP sessions is high.
2. The degree of detail and format of positioning data in SDES packets is not specified in RFC 3550 - RTP [25]. Therefore, it is left to the implementation and/or developer, although format and content is recommended to be prescribed by a profile.
3. It is required an intermediate software layer, which may reside between the application and the localization system, being responsible for: RTP management establishing RTP sessions and publishing RTCP packets.
4. Furthermore, as an extension of 3, the software layer should be able to simultaneously establish more than a RTP session to support several applications. This complicated the implementation of this software layer.

7 Proof of Concept

To evaluate this approach, a proof-of-concept implementation of RTCP supporting various localization systems has been built. The main objective is to show the feasibility of the proposed approach and to provide an adequate testbed for analysis and further improvement.

7.1 The Prototype Scenario

As outlined in Section 2 and illustrated in Figure 1, a typical scenario consists of users that move at home/office, although they can go out at any time. For example, the home and the user can be equipped with the following devices:

- Display appliances:
  - PDA: a user's PDA
  - TV1: IP TV set
  - TV2, TV3: legacy TVs connected to an IP Set Top Box
- Content Servers:
  - S1: DVB-S tuner
  - S2: DVB-T tuner
- Localization systems:
  - L0: User's PDA equipped with GPS
  - L1, L2, L3, L4: UWB-based indoor locators (Ubisense)
  - L5, L6, L7, L8: RFID readers
  - L9: car with a GPS navigator

According to a user's indoor location provided by either the Ubisense system (L1-L4) or RFID readers (L5-L8) the multimedia content selected by the user from any of the servers (S1, S2) will "follow" the user, being rendered in the most appropriate displayer: PDA, TV1, TV2 or TV3. The system might trigger the nearest one or not depending whether they are being used by other participants.

In case of outdoor movement, a user's location could be provided by the PDA equipped with GPS; the PDA itself could be the rendering device; and some external multimedia servers (S0, S3) could also participate in the scenario. In this case, we will focus on an indoor scenario, although the proposed approach can be easily extended to outdoor movement.

7.2 The Prototype Components

The following software components have been designed and implemented for the prototype scenario: three RTCP wrappers for the Ubisense, GPS and RFID localization APIs and an RTCP-based multimedia handoff application. In the case of using GPS as a localization system, a RTCP wrapper for NMEA 0183 based on GPS receivers was performed [26]. These wrappers deal with communications to each localization technology API. They send RTCP SDES packets with the identification and location information of all users at home, no matter which localization technology is being employed.

The RTCP-based multimedia handoff application monitors the RTCP traffic in sessions. It collects user location information and redirect multimedia streaming over the network from servers to the most convenient displayers.

8 Conclusions and Future Work

Since multiple wired and wireless technologies and more capable multimedia appliances are increasingly present in our lives, a wide range of mobile multimedia services for users are gradually emerging. The seamless handoff of advanced multimedia services decoupled from localization technologies and over indoor and outdoor situations is an important challenge to move beyond today's state of the art. In this paper, an approach based on multimedia standard protocols is described. This approach allows adopting different techniques for seamless handoff and is independent from the localization system in use.

With a user case example, it is shown the potential of using standard multimedia protocols to standardize 'user location' information and thus decouple these data from the application or system. Therefore, the system becomes highly independent from the localization system; meaning that any localization system compatible with RTP/RTCP standard will be able to be used in any system.

As per future work, we plan to create a standard location information format (compliant with RTCP) to be used by any participant in a RTP session. Moreover, extending the use of RTCP to other localization/identification technologies is one of our immediate future works. Finally, we plan to explore the potential of this approach in highly dynamic and populated environments (users and devices).

References

1. Bilbao, J., Armendariz, I.: Convergence in Digital Home Communications to redistribute IPTV and High Definition Contents. In: Proc of IEEE CCNC, pp. 885–889 (2007)

2. Kashibuchi, K., Taleb, T., Jamalipur, A., Remoto, Y., Kato, N.: A new Smooth Handoff scheme for mobile multimedia streaming using RTP dummy packets and RTCP explicit handoff notification. In: IEEE Wireless Communications and Networking Conference - WCNC, pp. 2162–2167 (2006)
3. The extended home environment concept, `http://cordis.europa.eu/ist/audiovisual/neweve/e/ws280606/ws280606.htm`
4. Izumikawa, H., Fukuhara, T., Matsunaka, T., Sugiyama, K.: User-centric seamless handover écheme for realtime applicacions. In: IEEE 18th International Symposium on Personal, Indoor and Mobile Radio Communications (PIMRC), September 2007, pp. 1–5 (2007)
5. Muthukrishnan, K., Lijding, M., Havinga, P.: Towards Smart Surroundings: Enabling Techniques and Technologies for Localization. In: Strang, T., Linnhoff-Popien, C. (eds.) LoCA 2005. LNCS, vol. 3479, pp. 350–362. Springer, Heidelberg (2005)
6. Cunningham, G., Murphy, S., Murphy, L.: Seamless Handover of Streamed Video over UDP between wireless LANs. In: Second IEEE Conference on Consumer Communications and Networking, January 2005, pp. 284–289 (2005)
7. IEEE 802.21, `http://ieee802.org/21/`
8. Rosenberg, J., Schulzrinne, H., Camarillo, G., Johnston, A.R., Peterson, J., Sparks, R., Handley, M., Schooler, E.: 2002. SIP: Session initiation protocol, RFC 3261 (June 2002)
9. Banerjee, N., Das, S., Acharya, A.: SIP-based Mobility Architecture for Next Generation Wireless Networks. In: IEEE International conference on Pervasive Computing and Communications (PerCom), March 2005, pp. 181–190 (2005)
10. Dutta, A., Chakravarty, K., Taniuchi, V., Fajardo, V., Ohba, Y., Famolari, D., Shculzrinne, H.: An experimental study of location assisted proactive handover. In: IEEE GLOBECOM 2007, November 2007, pp. 2037–2042 (2007)
11. Ohta, K., Yoshikawa, T., Nakagawa, T., Isoda, Y., Shoji Kurakake, S., Sugimura, T.: Seamless service Handoff for ubiquitous mobile multimedia. In: Chen, Y.-C., Chang, L.-W., Hsu, C.-T. (eds.) PCM 2002. LNCS, vol. 2532, pp. 9–16. Springer, Heidelberg (2002)
12. Wang, S.S., Sridhar, S., Green, M.: Adaptive soft handoff method using mobile location information. In: IEEE 55th Vehicular Technology Conference, May 2002, vol. 4, pp. 1936–1940 (2002)
13. Sulander, M., Hamalainen, T., Viinikainen, A., Puttonen, J.: Flow-based Fast Handover method for Mobile IPv6 network. In: IEEE Semiannual Vehicular Technology Conference (VTC'S 2004), May 2004, vol. 5, pp. 2447–2451 (2004)
14. Standard Positioning Service specification, `http://gps.afspc.af.mil/gpsoc`
15. Manapure, S., Darabi, H., Patel, V., Banerjee, P.: A comparative study of Radio Frequency-based Indoor Location Sensing Systems. In: Proceedings of IEEE ICNSC, March 2004, vol. 2, pp. 1265–1270 (2004)
16. Porcino, D., Hirt, W.: Ultra-wideband radio technology: potential and challenges ahead. IEEE Communications Mag. 41(7) (July 2003)
17. Roy, S., Foerster, J.R., Somayazulu, V.S., Leeper, D.G.: Ultrawideband radio design: the promise of high-speed, short range wireless connectivity. Proc. IEEE 92(2), 295–311 (2004)
18. Ubisense, `http://www.ubisense.net/`
19. Wi-Fi Technology, `http://www.wi-fi.org/`

20. Schilit, B., Lamarca, A., Borriello, G., Griswold, W., McDonald, D., Lazowska, E., Balachandran, A., Hong, J., Iverson, V.: Challenge: Ubiquitous Location-Aware Computing and the Place Lab Initiative. In: ACM International Workshop on Wireless Mobile Applications and Services on WLAN, pp. 29–35 (2003)
21. Zigbee Alliance, http://www.zigbee.org/
22. Carcelle, X., Heile, B., Chatellier, C., Pailler, P.: Next WSN applications using ZigBee. In: IFIP International Federation for Information Processing, vol. 256, pp. 239–254. Springer, Boston (2007)
23. Lin, C., Song, K., Kuo, S., Tseng, Y., Kuo, K.: Visualization Design for Location-Aware Services. In: IEEE International Conference on Systems, Man and Cybernetics- SMC 2006, October 2006, vol. 5, pp. 4380–4385 (2006)
24. Chon, H.D., Jun, S., Jung, H., An, S.W.: Using RFID for Accurate Positioning. Journal of Global Positioning Systems 3(1-2), 32–39 (2004)
25. RTP: A transport Protocol for Real-Time Applications, http://www.ietf.org/rfc/rfc3550.txt
26. NMEA 0183 Standard, http://www.nmea.org/pub/0183/index.html

Location Based Services: A New Area for the Use of Super-Resolution Algorithms

Raúl O. González-Pacheco García[1] and Manuel Felipe Cátedra Pérez[2]

[1] Dpto. Ciencias de la Computación, Universidad de Alcalá 28806
Alcalá de Henares, Madrid, Spain
Fax: +34 91 7461607
rgonzalez@iberia.es
[2] Dpto. Ciencias de la Computación, Universidad de Alcalá 28806
Alcalá de Henares, Madrid, Spain
Fax: +34 91 8856646
felipe.catedra@uah.es

Abstract. This article focuses on the location of radiating sources that are in line of sight, using calculations encompassing wide ranges of an electromagnetic signal. The purpose of using the MUSIC algorithm is to consider the distance at which the transmitting antennas are, implementing a measurement procedure throughout simulations. In order to do these simulations we use a tool that determines the value of the real surroundings electromagnetic field under study and that is validated to provide trustworthy results in a reduced runtime. Thus the signal spatial signature is observed.

The algorithm is based on the UTD (Uniform Theory of Diffraction) and/or ray tracing.

1 Introduction

A review of the works published over the last years shows an increasing interest in the location of electromagnetic emission sources. Talking about radar, sonar, radio astronomy and seismology, an important problem to be solved is the spatial source location through passive sensors.

In essence, a radio location system can operate measuring and processing physical amounts related to the radio signals travelling between a mobile terminal (MT) and a set of base stations (BSs), like the time of arrival (ToA), the angle of arrival (AoA) or the signal strength.

The use of the signal strength magnitude is based on the fact that the average power of a radio signal decays over distance, following a well known law.

The disadvantage of this method is a significant, random, deviation of the average received signal strength, caused by the small scale effects of the channel and the shade, which can give an average variation of 10dB or even greater in the results, which translates into an important inaccuracy of the estimated position.

As it is clear enough reviewing the state-of-the-art on technology, traditionally the search algorithms have been based on estimation of the direction of arrival (DoA), by means of digital signal processing techniques [1].

An area of important investigation here is the security improvement through the use of wireless communications. A good example is the regulation created by the

J.M. Corchado, D.I. Tapia, and J. Bravo (Eds.): UCAMI 2008, ASC 51, pp. 201–209, 2009.
springerlink.com

FCC (USA) in June of 1996, indicating that the suppliers of wireless services must offer information about the location of the mobile users who make use of the emergency system 911 (E-911) [2], which represents a tremendous technical challenge for the service suppliers, due to the channel conditions and to the interference due to multiple user access associated to wireless systems. Other potential necessities have to do with applications related to the position based pricing services, resource management, development of advanced management strategies (such as the control handover based on position), intelligent transport systems (ITS), fleet management and others like, for example: electronic yellow pages, applications such as "where am I?", navigation & customized traffic systems, algorithms of Radio Resource Management (RRM) based on location and acoustic arrays [3], fighting against the fraudulent use of wireless emissions.

This paper presents an approach for locating radiating sources distributed in urban as well as indoor scenarios through the application of super-resolution algorithms. An important research effort in the development of a solution, that uses only one base station and various mobile terminals that do not require modification, has been undertaken. This means that market penetration is 100% from the day the system is launched, reducing the complexity of implementation.

To estimate the ToA/DoA in urban propagation scenarios (including the shaded ones) in the past were used methods like the one of time-frequency Maximum Likelihood (t-f ML) [4] or the space-time of hyperbolic channel model [5], which were better adapted for the case of no stationary scenarios, where the number and delay of the multiple ways changes with time. The classic method for the search of spatial signatures is based on the direct application of the Fourier transform. Its limitation is that it is not possible to distinguish between sufficiently close signals. This gave way to the great resolution methods, whose philosophy is based on the underlying signal model, that is, they assume that the data is adjusted to a model which parameters contain the information to be considered. The models based on analysis methods increase the spectrum resolution. They have nothing to do with power density spectrum (PDS) estimators, since they do not keep the process power being measured, but produces a kind of pseudo-spectrum.

From all the super-resolution algorithms (so called because the estimation of frequencies or the angle of arrival has, under controlled conditions, the capacity to exceed the limitations of the classic methods based on Fourier), the most important is MUSIC (MUltiple SIgnal Classification), published by R. O. Schmidt in 1979 [6], partly due to its adaptation to arbitrary orientation and response arrays.

In the scenario presented in this paper, the energy contained in the field of waves, created by the emitter, is gathered by a sensor. A receiver measures the response of this sensor, extracting the emitter location through the measurement of the phase difference of the waves that affect the element. The approach formulates signal energy estimation as a function of the angle, presenting dominant peaks in the sources direction. Propagation scenario is represented by a flat faceted model.

The paper is organized as follows: point 2 describes the signal model; section 3 presents the algorithm. Point 4 talks about the simulation that allows obtaining the values of magnitude and phase of the electric fields from the models; point 5 explains the presentation of the results and, finally, point 6 details the conclusions.

2 Signal Model

A sensor array is generally used in signal process, where there are a very small number of sensors. In our case the intention is not to modify the elements, using a sweeping of frequencies in the emitter and only one element in the receiver.

Beginning with the coverage analysis, in order to lead the study of answer to the vectorial impulse / spatial signature, a coverage calculation program is used. The selected tool for outdoors is FASPRO (Fast Computer Tool for the Analysis of Propagation in Personal Communication Network) [7], whereas FASPRI is being used for indoor environments; both of them being developed by the Electromagnetic Computational Group of the University of Alcalá. The codes are based on the uniform version of the Geometric Theory of Diffraction, and take into account the effects from direct, reflected, diffracted in the edges, double reflected, reflect-refracted, diffracted-reflected fields as well as the influence of the ground.

3 Algorithm

In order to present the results of this work we made use of MUSIC, enabling us to verify the benefits of the formulation with direct ray using a robust method.

The use of other equally valid super-resolution algorithms like Root-MUSIC, ESPRIT (Estimation of Signal Parameters via Rotational Invariant Techniques) [8], or Matrix-Pencil, was considered, however MUSIC was selected due to its good behaviour in noisy surroundings and direct rays and its ease of deployment. Although ESPRIT, in particular, was developed to bear some of the requirements of necessary calculation and previous information in MUSIC [9], the results that are exposed in point 5 demonstrate that, in this case, none of those determinant factors were important. The most crucial being the simplicity of the method.

The MUSIC algorithm was originally proposed for the estimation of corrupted sine frequencies; these frequencies were corrupted by additive white Gaussian noise. Later it showed its effectiveness in the detection of the signal´s direction of arrival, which was mainly used for the formation of adaptive beams. Here we will use MUSIC for a new purpose: the calculation of distances to radiating sources.

In the previous use of MUSIC and ESPRIT at least three or more sensors were needed to produce suitable results; to increase its resolution the number of sensors had to be raised. These methods determine the range exploding statistical of second order (the covariance matrix and its decomposition in eigenvalues and eigenvectors in particular) In the proposed algorithm it is only necessary a sensor in the receiver, leaving the transmitter to adjust the range (number of frequencies swept and total bandwidth).

The data gathered executing the FASPRO / FASPRI simulation tools, and after having measured the signal power, are contaminated with complex white Gaussian noise, establishing the SNR wished for the test.

Let us consider M uncorrelated and contaminated with Gaussian noise signals. Assuming that M < N:

$$x[n] = \sum_{i=1}^{M} A_i s_i[n] + \eta[n] \tag{1}$$

where: $s_i = e^{j\omega_i n}$; N is the number of discreet samples and *Ai* is the complex amplitude of the i-th signal: $A_i = |A_i| e^{j\phi_i}$

The correlation matrix is expressed like:

$$\mathbf{Rx} = \sum_{i=1}^{M} P_i \mathbf{s}_i \mathbf{s}_i^{\mathrm{H}} + \sigma_o \mathbf{I} \tag{2}$$

where: P_i is the power of the $s_i[n]$ sample, I is the identity matrix of N x N size and σ_o is the variance of the zero average white Gaussian noise. Then the covariance matrix of the spatial signature:

$$R_x = E \wedge E^H = E_{sig} \wedge_{sig} E_{sig}^{\ H} + E_{noi} \wedge_{noi} E_{noi}^{\ H} \tag{3}$$

where E_{sig} and E_{noi} are defined from the eigenvectors of the correlation matrix, and whose columns are the R_x eigenvectors. In the same manner we obtain Λ_{sig} y Λ_{noi} from its eigenvalues. The result is that the pseudo spectrum function adopts the following form:

$$\hat{P}(e^{jw}) = \frac{1}{w^H P_{noi} w} = \frac{1}{w^H E_{noi} E_{noi}^{\ H} w} \tag{4}$$

and it shows peaks for the frequencies in which $w = s_i$.

P_{noi} is the projection operator on the subspace of noise.

4 Simulation

The propagation model used to prove and to validate the location algorithm is full 3D.

The geographic area that was used in outdoor simulations corresponds to downtown Madrid. Its model, created with AutoCAD following the characteristics of the real buildings located in that area, once processed and inserted into the simulation program, form the representation shown below:

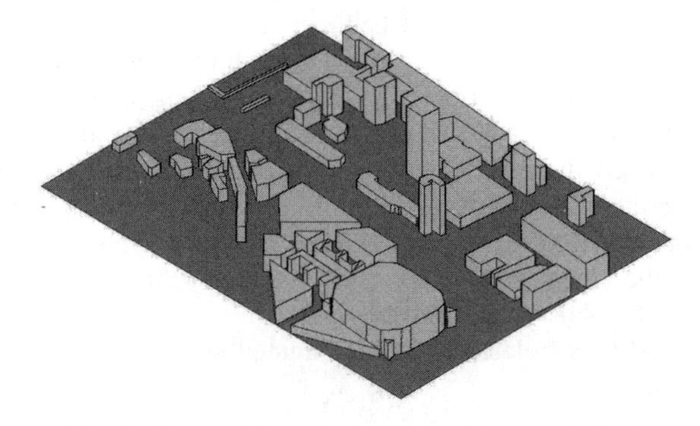

Fig. 1. Geographic area used in outdoor simulations

In the case of interiors the simulation used was the office plan that follows:

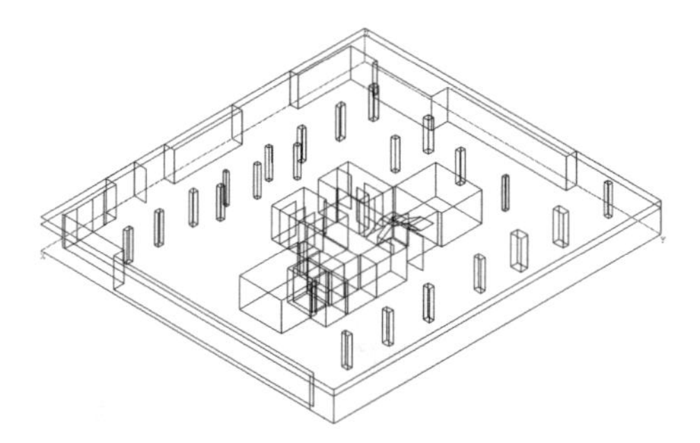

Fig. 2. Office plan used in indoor simulations

The radiating element and the direction of its pattern are located within the scenario. After that, the point where the movable observer positions itself is established, in order to be able, immediately afterwards, to the execution of the emulation itself. Along the emulation the emitter sweeps along a range of frequencies with Δf settle according to the desired distance resolution.

Data obtained from the simulation program are contaminated with white Gaussian noise before proceeding to the autocorrelation of the matrix, which will become the input for the distance calculation program based on super-resolution algorithms and developed in MATLAB:

$$sigPower = \sum \frac{|sig^2|}{length(sig)} \tag{5}$$

$$sigPower = 10 * \log_{10}(sigPower) \tag{6}$$

$$noisePower = sigPower - reqSNR \tag{7}$$

$$x_{noi} = sig + wgn(size(sig,1), size(sig,2), noisePower, 1, 'dB') \tag{8}$$

Having chosen MUSIC, the number of points considered necessary to evaluate the pseudo spectrum function is introduced. Generally the higher the number, the greater the precision, but as it is demonstrated next, this method is robust enough so that the exactitude of the results does not vary significantly with the number of evaluation points.

The application output is the pseudo spectrum function, whose peaks, properly adapted, offer the distances to the radiating elements.

5 Results

Diverse simulations were performed considering different types of incident waves and frequency hopping (always within the margin that allows us to solve the distance to which we work).

The following cases were considered:

1. Noise power variation
2. Variation of the number of samples used to extract the received signal correlation matrix

3. Variation of the number of points used for the evaluation of the pseudo spectrum function

In all the cases we tried to determine the strength of the algorithm with relation to these variables.

The picture represents an upper view of the urban centre for the outdoor simulation, indicating the antenna position and its main lobe direction as well as the location of the mobile unit:

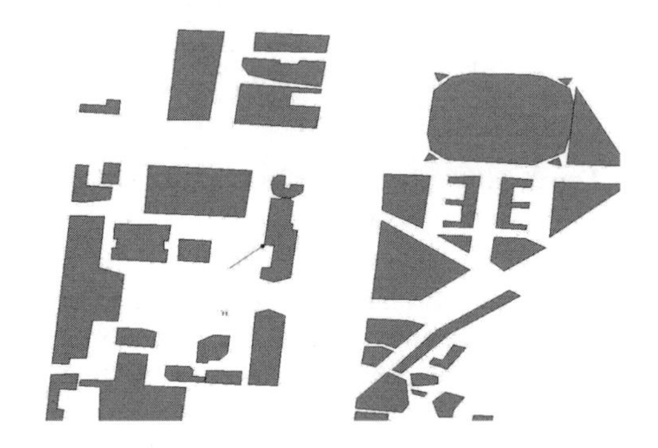

Fig. 3. Upper view of the outdoor simulation scenario

The figures below show the results of these simulations:

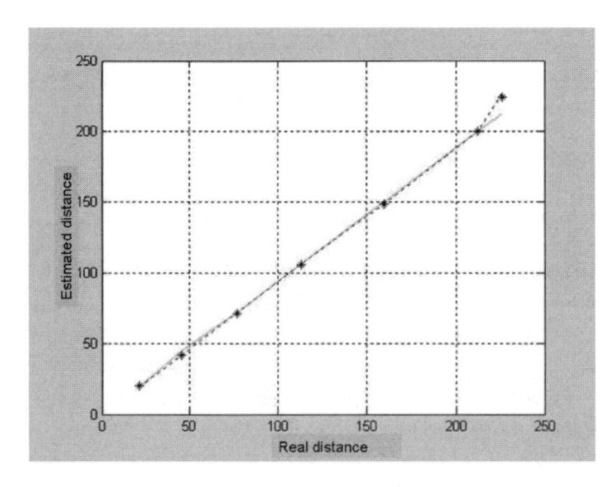

Fig. 4. Results obtained for different noise levels: with - SNR=30dB; with * SNR=3dB; with: SNR=60dB

For the first scenario, the error is negligible in the first third of the resolution window for all the conditions. The same happened when approaching to the window´s limit for small S/N. In other distances the maximum error was 9% in defect, regardless of the S/N.

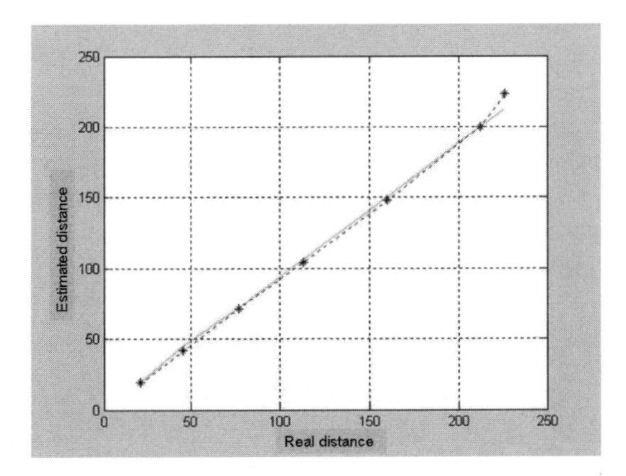

Fig. 5. Results obtained for different data volumes: with - 100 samples; with * 10 samples; with: 3 samples

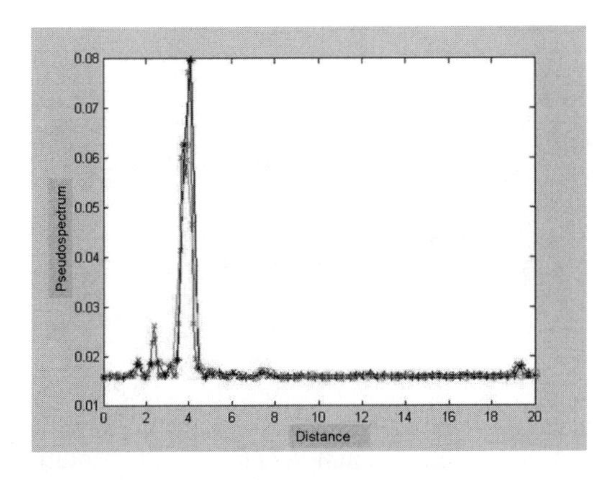

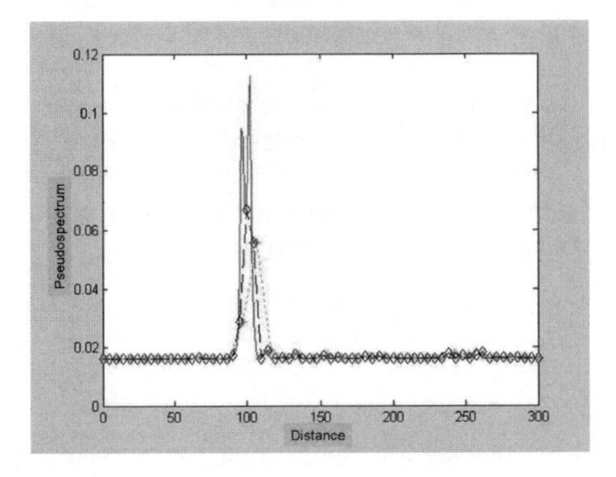

Fig. 6. Effect produced varying the number of points used to calculate the pseudo spectrum for different distances from the mobile unit to the antenna: with x -> 256, with * -> 64, with o -> 32

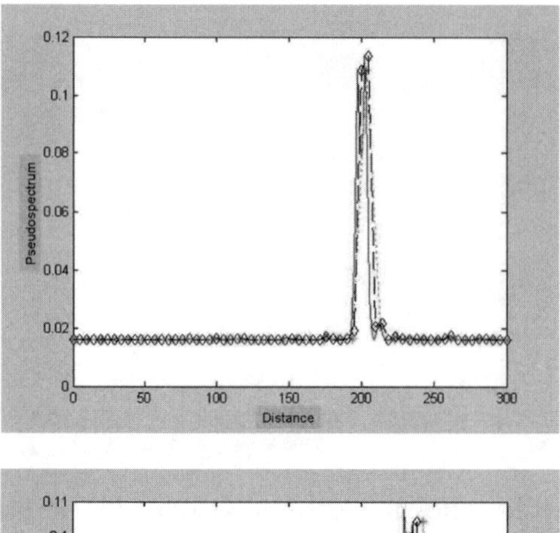

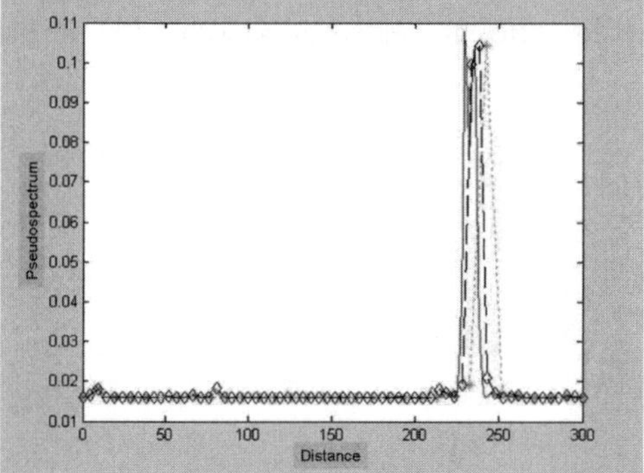

Fig. 6. (*continued*)

Varying the number of samples of the electromagnetic field, the answer is similar to that of the S/N. Not taking into account the ends of the resolution window, the majority of the samples offer accuracy in the first third. Then the results are degraded until 6% in defect, but they remain slightly more accurate than for the lesser number of sampling.

For the last assumption, the variation of the number of points used to evaluate the pseudo-spectral function acts as follows: the lesser the number of samples, the greater the protection against false positives, however the imprecision of the radiating point location increases. On the contrary, the greater the number of samples, the greater is the precision. In all the cases the precision is over 10%.

6 Conclusions

The super-resolution techniques have allowed improving the ToA estimations, transforming the data of the frequency corresponding to the correlation peak, to the temporary domain, offering an image of the times of arrival of the multi-path components of the signal [10].

From the experimental results we can extract that the analytical capacity of the algorithm does not diminish significantly when the SNR does.

In the same way it is demonstrated that the amount of data or number of samples in time which are used to consider the correlation matrix does not improve considerably the spatial resolution of the algorithm, demonstrating the robustness of the method with small amounts of data.

The run time is short enough to be used in real time scenarios.

The experimental analysis on simulated data demonstrates, therefore, that the technique proposed offers a great resolution, with minimum levels of lateral lobes and robustness to the noise and to the limitation in the data volume.

Simulation results exceed the FCC requirements that call for a service supplier to be able to locate sources within a precision of 125m in 67% of the cases.

A continuation of this work sets out the use of Matrix Pencil for scenarios of multiple reflections.

Acknowledgements. This work has been supported in part by the Madrid Community Project S-0505/TIC/0255.

References

1. Godara, L.: Applications of Antenna Arrays to Mobile Communications, Part II: Beamforming and Direction-of-Arrival Considerations. Proceedings of the IEEE 85(8), 1195–1245 (1997)
2. Federal Communications Commission, Revision to the commission's rules to ensure compatibility with enhanced 911 emergency calling systems, CC Docket Nº, pp. 94–102 (July 1996)
3. Cheng, J.C., Kung, Y., Hudson, R.E.: Source localization and beamforming. IEEE Signal Processing Magazine 19(2), 30–38 (2002)
4. Zhang, Y., Mu, W., Amin, G.: Time-frequency maximum likelihood methods for direction finding. Journal of the Franklin Institute 337, 483–497 (2000)
5. Mahmoud, S., Hussain, Z., O´Shea, P.: Space-time model for mobile radio channel with hyperbolically distributed scatterers. IEEE Antennas and Wireless Propagation Letters 1(12), 211–214 (2002)
6. Schmidt, R.O.: A signal subspace approach to emitter location and spectral estimation., Ph. D thesis, Stanford University (November 1981)
7. Cátedra, M.F., Pérez, J.: Cell Planning for Wireless Communications. Artech House (1999)
8. Roy, R., Kailath, T.: ESPRIT – estimation of signal parameters via rotational invariance techniques. IEEE Trans. on Acoustics, Speech and Signal Proc. 37, 984–995 (1989)
9. Roy, R., Paulraj, A., Kailath, T.: Comparative performance of ESPRIT and MUSIC for direction-of-arrival estimation. In: Proc. 20th. Asilomar Conf. Circuits (November 1986)
10. Dumont, L.R.: Super-resolution of discrete arrivals in a spread spectrum system. M.S. thesis, University of Calgary (1994)

Developing Ubiquitous Applications through Service-Oriented Abstractions

J. Antonio García-Macías and Edgardo Avilés-López

Computer Science Department, CICESE Research Center
Km. 107 Carretera Tijuana-Ensenada, Ensenada, Baja California, Mexico
jagm@cicese.mx, aviles1@cicese.mx

Abstract. Many infrastructural elements required to make pervasive computing a reality are now in place: some have been tested in the labs, while others are already commercially available. However, the integration of these elements to create ubiquitous applications is still a daunting task, as the developers have to delve into the low-level details of each of them. With this in mind, we present UbiSOA, a service-oriented architecture and middleware that allows developers to integrate RFID tags, wireless sensor networks, and other elements, into ubiquitous applications through the well-known service-oriented paradigm, using the language of their choice.

Keywords: Ubiquitous computing, service-oriented architectures, RFID, wireless sensor networks.

1 Introduction

Everyday, the vision of ubiquitous computing is getting closer to being realized. Remember the marketing mantra of "anytime, anywhere computing" of some ten years ago? It is closer than ever to being realized as wireless networks complement wired infrastructure to extend the reach of communications. Many commercially available products can make intelligent spaces out of our homes and offices, as envisioned in the past by domotics and by several sci-fi films. RFID infrastructures are being used for several applications, from tracking goods along the supply chain, to keeping children secure in playgrounds, to enabling better healthcare provision in medical institutions. Sensor networks have also found a wide variety of applications, from intelligent transportation systems, to precision agriculture, to habitat monitoring, etc. The vision statement of Sun Microsystems in the 1980s was "the network is the computer", and today we can assess that this vision has been realized; the desktop (or laptop, for that matter) is no longer the center of computing and communications, as distributed infrastructures, handheld devices, and embedded networks are omnipresent and keep gaining momentum.

As new infrastructures are being constructed, the vision of having computing services readily available as they "weave themselves into the fabrics of everyday life until they are indistinguishable from it" [1] is becoming more real. However, even when the different infrastructural elements are already available and we can use them to create new applications, integrating all the needed elements is no easy task. Even a

J.M. Corchado, D.I. Tapia, and J. Bravo (Eds.): UCAMI 2008, ASC 51, pp. 210–218, 2009.
springerlink.com

simple task such as reading a set of RFID tags would imply learning the API of the tag reader (if there is any), and probably programming in a specific language to comply with the API or binary library; changing to a different hardware would probably imply porting the application to the new platform. If you want to add sensing capabilities to your application, you will have to learn how to program the sensing nodes, which typically implies: learning the details of the embedded operating system (*e.g.*, TinyOS), learning the programming model (*e.g.*, component-based), and the programming language (*e.g.*, NesC) [2], and then learning how to network these sensor nodes with a multihop routing protocol and interfacing everything to a gateway to communicate all sensor data to an external network such as the Internet. Of course, dealing with some other hardware platform would imply repeating the process of learning low-level details about the hardware and developing specialized software to access it and making its services available for the application.

We believe that the current state of development process for ubiquitous applications is unnecessarily primitive, as there are already many abstractions and facilities that can make the process much easier, allowing application developers to concentrate only on applications, not on low-level tasks. One of such facilities is the service-oriented model, which we believe is very suitable for ubiquitous applications development. In the following section we give a brief discussion on the advantages of service orientation and its applicability to ubiquitous computing applications development, overviewing some related work. Then, on section 3, we present UbiSOA, a service-oriented architecture and middleware for providing better abstractions to developers. Section 4 introduces some applications that have been developed using UbiSOA, showing how easy it is to integrate diverse platforms into a single application. We then conclude in section 5 with some final remarks and outlining future work.

2 Related Work

If the infrastructural elements to make ubiquitous computing a reality are already present, why hasn't it been fully realized? Some authors argue that the main barrier is the integration between all these technologies [3]. Many software infrastructures and frameworks have been proposed to diminish these barriers [4].

GSN [5] is one of the projects that exemplify the integration of different technologies, such as WSN, RFID and wireless cameras. However, it doesn't provide an external API to allow the development of new applications. Another integration effort between WSN and RFID is the EPC sensor network [6] where they propose an integration model based on a standardized framework to share information between the two technologies.

Since it allows high-level abstractions, the service-oriented paradigm has received attention and has also been proposed for these purposes, especially via Web services due to the near-ubiquity of the Internet. Just recently the World Wide Web Consortium created a group dedicated to Ubiquitous Web Applications (UWA) that "focuses on extending the Web to enable distributed applications of many kinds of devices including sensors and effectors. Application areas include home monitoring and control, home entertainment, office equipment, mobile and automotive" [7].

3 The UbiSOA Platform

UbiSOA is a software architecture and middleware [8] based on a service-oriented model to make easier the use of WSN, RFID, embedded systems and other elements typically found in ubiquitous computing applications. Its implementation provides a set of Web services and tools that make possible to send and receive information from several data sources. UbiSOA (formerly known as TinySOA) has been released as an open-source project and its current version, and some documentation, can be found at http://www.tinysoa.net. We will overview its architecture and how developers can use it.

3.1 UbiSOA Architecture

The architecture consists of 5 main components (see Fig. 1):

- Sensor Node. This is the lowest-level component of the architecture. It represents software running on each of the wireless sensor nodes in a network. The functionality relies of the hardware abstraction layer provided by a special embedded operating system (e.g. TinyOS).
- RFID Reader. This component represents the RFID hardware reader. UbiSOA uses the API provided by the manufacturer to interact with it.
- Gateway. It receives and processes all the incoming data from the end devices. It also sends configuration parameters to the devices on request.
- Registry. It stores all the data for the infrastructure. It keeps track of all the services and specific information (readings, capabilities, events, and management parameters) for each of the wireless sensor networks or RFID readers available.
- Server. Its main purpose is to serve as the external interface to developers. It provides SOAP Web services, which can be used to request data or to control the end devices.

A typical working scenario is as follows: for a sensor network everything starts when one of its nodes (Sensor Node component) turns on. After hardware initialization, the Discovery subcomponent identifies the capabilities of the hardware, specifically the services it can provide (sensing of humidity, light, temperature, actuators, etc.), and sends a registry message to the Gateway component containing this information. The Gateway component using these messages knows what services are available through the whole network which is seen as a single provider, and it proceeds to update the information of the corresponding network in the Registry (Providers Registry subcomponent). Once registered, it subscribes itself to the network acknowledged services. When the subscription message arrives to the nodes, they proceed to activate their Reading subcomponent to start periodically communicating collected data to the Gateway. In a similar way, the Gateway looks for any RFID reader and then updates its status in the Registry. There is one instance of Gateway in charge of gathering, processing and broadcasting messages for each type of end device. When the end devices and their corresponding Gateway instances are initialized and ready, the Server component can be executed. At start up, it looks at the Registry (specifically at the

Providers Registry) and sets up a Web service for each end device. There are two kinds of Web services, Network Service and RFID Service, each with its own specialized API. In addition to these services, the Server sets up an Information Service, which can be used to query the Registry (*e.g.*, get a list of the available sensor networks and their sensing capabilities). When messages arrive at the Gateway, it converts them into higher-level representations (e.g. raw temperature readings to Celsius degrees) before storing them into the Registry component (Historical Registry subcomponent). Once the infrastructure is set up, developers can use the Web services provided to query the Historical Registry, register events (*e.g.*, "when the temperature rises above 30º degrees Celsius") to be detected (based on the incoming data), write information in RFID tags, and send management parameters such as changing the data rate, putting some nodes into sleep mode, and so on. The available methods are discussed in the next section and shown in Figure 2.

Summarizing, UbiSOA's infrastructure is conformed by multiple sensor networks with nodes running Sensor Node, multiple RFID readers, multiple instances of Gateway attached to an end device of which it is responsible, one instance of Registry where all data is stored, and one or more instances of Server. It is worth noting that this is a semi-distributed architecture that overcomes the single point of failure and central overload problems of centralized architectures.

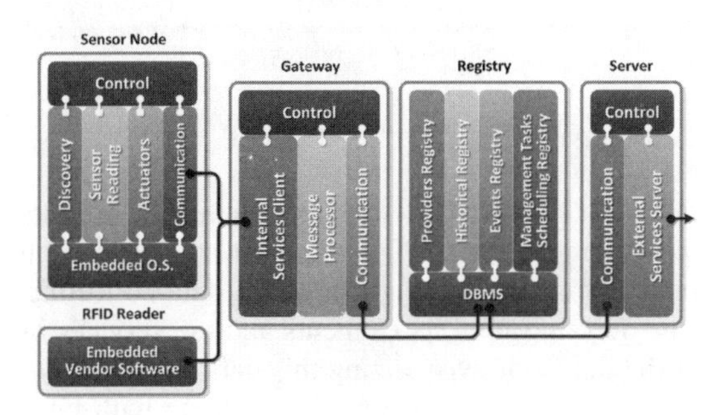

Fig. 1. Architecture components of UbiSOA

3.2 Facilities for Developers

Once installed, the UbiSOA middleware will allow developers to have easy access to the services offered by different hardware elements. Since these elements are abstracted as web services, all the developer has to do is to connect to the Web server that manages these services and use them. Since Web services are used, the developer can use any programming language of its choice, as long as it has libraries or some means to access SOAP Web services. There are 5 sets of methods (shown in Fig 2). Since a Web service is directly associated to a specific type of end-device, all methods end on its corresponding device (*e.g.*, sensor network node or RFID device). The Network Information set features methods to retrieve information concerning the capabilities of a sensor network such as the list of nodes (and their last known state), sensing capabilities of the full network, actuators available through the network, and the time window of the available historical sensing data. RFID Web service methods allow retrieving the reader information, getting tag information and also writing a serialized object into them. The Readings Management set allows querying the historical sensing registry. The Events Management set allows the developer to set criteria to be matched based on the incoming readings (*e.g.*, "when temperature rises above 30° C"). Finally, the Tasks Management set, allows to broadcast a control parameter such as turning on or off a particular actuator, changing sensing rates, entering or leaving sleep mode on particular nodes, and other actions that could be later implemented on the Sensor Node component. Network service features all method groups but the RFID one, and RFID service features the Events, Tasks and the RFID methods.

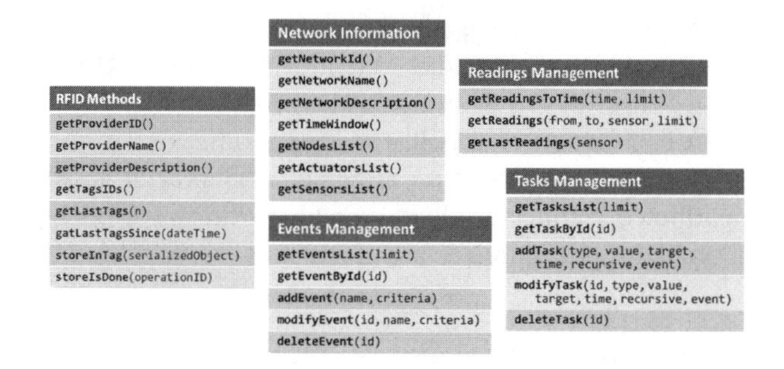

Fig. 2. Methods available through Web Services

In order to provide a set of basic tools that are commonly used in many ubiquitous applications, a data gathering and visualization tool was developed using all the functions mentioned above. This allows connecting to a data-collecting platform (*e.g.*, a WSN) and getting the description of its elements and the services offered, generating historical records of the data read, visualizing this data as charts, and showing the topology of the interconnected nodes in the platform, along with visual representations of data (*e.g.*, a heat-flow map indicating temperature). Samples of screenshots are shown in Figure 3. The idea is then that developers can reuse these commonly needed tools and customize them for their specific applications. All components are implemented as independent pieces and are ready to work with the data format of the Web services outputs.

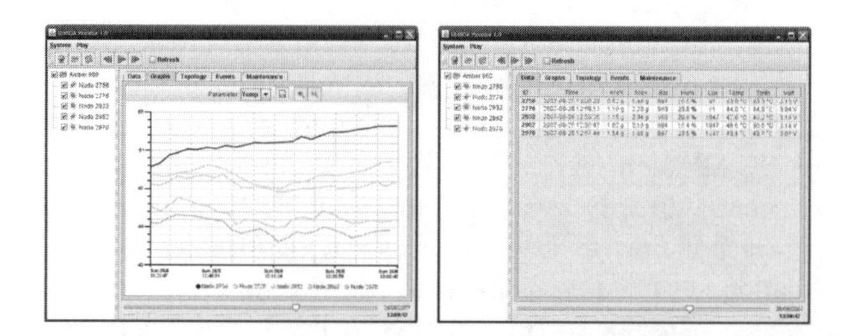

Fig. 3. Visualization tool showing numeric and chart views for data collected with a WSN

4 Developments with UbiSOA

Since its inception, several applications have been developed using UbiSOA. This has allowed us to assess its capabilities and usefulness, and has also shed light on possible enhancements. In order to give a hint of the type of ubicom applications that can be developed with UbiSOA, we will show a couple of them next.

4.1 Environmental Monitoring

Monitoring environmental conditions is essential for many context-driven applications. For instance, intelligent buildings would need to monitor ambient temperature, humidity, CO_2 concentration, along with other parameters, to automatically turn on or off the air conditioning system, open window blinds, play background music, diminish luminosity in certain areas, sound safety alarms, and so on. But these types of applications cannot only be implemented in urban areas, as also rural zones can benefit from them. For instance, for almost a year now, we have been monitoring greenhouses used to grow vegetables in a ranch. Some important variables that affect growth and quality of the plants are solar radiation, ambient humidity, and temperature; they are also important to determine the likelihood of appearance of plant diseases. These variables are being monitored with wireless sensor networks installed in the greenhouses, which connect through a gateway to a monitoring station (a desktop computer) in the farmer's house. An aerial view of the installation (as provided by the Google Earth service) is shown in Figure 4, where we depict the position of the sensing nodes and the monitoring station.

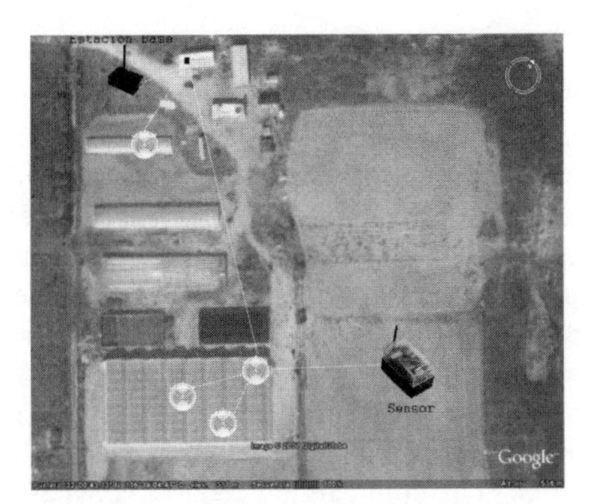

Fig. 4. Aerial view of the agricultural monitoring network

Using the monitoring station, the farm owner can monitor the status of its greenhouses at anytime from the comfort of its own house, through a screen similar to that shown in Figure 3. Moreover, even if the farmer goes to get some groceries to the city, he can still monitor the greenhouses using any computer connected to the Internet (as UbiSOA is web-based) or even through a cellular phone or other mobile device with Internet access. Also, since environmental data in the greenhouses may change frequently, we thought it would be natural to create RSS feeds corresponding to each greenhouse, in a similar fashion as electronic newspapers inform of breaking news. This way, the farmer would just subscribe to a feed and automatically be informed each time new data is available or if any alarm or warning has been issued. Figure 5 shows the RSS client developed for the iPhone, where different feeds can be selected.

Fig. 5. RSS client for the iPhone

The agricultural network was implemented using Crossbow's nodes and gateway; two AA batteries power each one of the nodes, and the gateway uses the power coming from the monitoring station through a USB cable. The monitoring station is an off-the-shelf desktop computer installed at the farmer's house. All the Web services method sets, except the RFID one, were used on the implementation.

4.2 Tracking of Assets

Location-based services, context-aware virtual museum guides, and many other applications have been created using RFID technologies. Food traceability is also an important issue that has received a lot of attention due to E. Coli outbreaks in spinach, mad-cow disease in meats, etc. Thus, food suppliers are increasingly receiving pressure to reliably determine trace the path that their products have followed, from the growing fields to the supermarket. An important technology being used for these purposes is RFID, as electronic tags can store important information along the supply chain. Along this line, we developed a sample application was developed using all the Web services sets to show what can be achieved using WSN and RFID technologies, but most important, how an application developer can easily develop these types of applications using UbiSOA. The application scenario consists of a packing facility in which a conveyor belt carries product boxes that have an RFID tag specifying its content and the optimal temperature range for the product, as well as some other information such as the ID of the farm where the product was collected, etc. Figure 6 shows a typical box with an attached RFID tag and also with a tracking pattern that will be explained later. Wireless sensor networks complement RFID perfectly in this application, as they would provide environmental data to determine if the temperature has gone beyond the upper level or below the lower level of the optimal temperature range for the product. So, ideally there would be a WSN in the growing field, another at the packing facilities, etc. and the local max-min temperature values will be recorded. Along the conveyor belt there will be several RFID readers.

Fig. 6. A typical product box with an RFID tag and a tracking pattern

While the product boxes are passing through the conveyor belt, an operator using video lenses will look at the boxes and trough augmented reality will see the content of the box, which will be stored in the RFID tags. On top of each box a bar will display the current temperature. Around the temperature bar, there will be two rings indicating the upper and lower level for the optimum range. If the temperature lies outside this range, or if recorded information in the tags indicates that at some time it has been outside the range, an alert will be produced to communicate the problem (for instance, the box will start blinking in a glowing red). Figure 7 shows a product box as will be seen through the augmented-reality glasses used by the operators. The tracking pattern attached to each box (mentioned earlier) will indicate the place where the temperature bar will be displayed, along with the upper and lower levels for the optimal range. Also, the product inside the box will be depicted through pictures and with a name label. We used the ARToolKit [9] for the augmented reality 3D representation of the box.

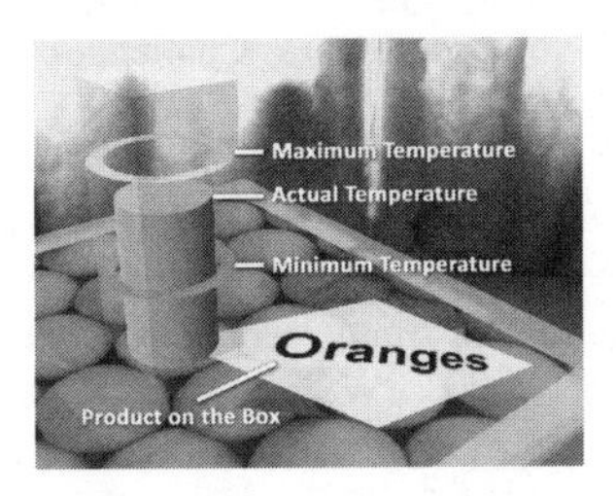

Fig. 7. Close-up of the augmented reality view of a product box in our asset tracking application

5 Concluding Remarks and Future Work

Although there are many infrastructural elements that can allow the creation of ubiquitous computing applications, integrating these elements is still a very difficult task. This is due not only to the heterogeneity present with elements of different nature, but also because application developers have to delve into the low-level details of each of them. In order to cope with this heterogeneity and to provide high levels abstractions for developers, we propose the UbiSOA service-oriented platform. Using UbiSOA developers are not tied to specific programming languages and can reason in terms of application-level details, concentrating on what they want to achieve with their application and not on infrastructural details.

Currently, application scenarios for UbiSOA are those that do not rely on real-time functionality, such as visualization tools or long term monitoring. We are currently working on extending these capabilities to allow access to a greater array of devices, in order provide the basis for the design and implementation of a broader set of ubiquitous computing applications. One important feature in this sense is the inclusion of a data stream management system, which will allow us to support real time capabilities and more reliable and extensible management for the huge amount of incoming data. We are also evaluating the incorporation of standards such as OSGi.

Acknowledgments. We would like to thank to the Mexican Science and Technology Council (CONACyT) for the financial support provided to conduct this research. Also, César Olea-Aguilar should be credited for the development of RSS support in UbiSOA and Rolando Cárdenas-Tamayo for the monitoring network installed in the greenhouse.

References

Weiser, M.: The computer for the twenty-first century. Scientific American 265(3), 94–104 (1991)

Hill, J., Szewczyk, R., Woo, A., Hollar, S., Culler, D., Pister, K.: System architecture directions for networked sensors. In: 9th. International Conference on Architectural Support for Programming Languages and Operating Systems, Cambridge, MA, USA (November 2000)

Davies, N., Gellersen, H.-W.: Beyond prototypes: Challenges in deploying ubiquitous systems. In: IEEE Pervasive Computing, vol. 1(1), pp. 26–35. IEEE Educational Activities Department, Piscataway (2002)

Endres, C., Butz, A., MacWilliams, A.: A Survey of Software Infrastructures and Frameworks for Ubiquitous Computing. Mobile Information Systems Journal 1(1), 41–80 (2005)

Salehi, A., Aberer, K.: GSN, quick and simple sensor network deployment. In: European conference on Wireless Sensor Networks (EWSN), Delft, Netherlands (January 2007)

Sung, J., Lopez, T.S., Kim, D.: The EPC sensor network for RFID and WSN integration infrastructure. In: IEEE International Conference on Pervasive Computing and Communications Workshops, pp. 618–621. IEEE Computer Society, Los Alamitos (2007)

Ubiquitous Web Applications Working Group, http://www.w3.org/2007/uwa/

Aviles, E., Garcia-Macias, J.A.: Providing Service-Oriented Abstractions for the Wireless Sensor Grid. In: Cérin, C., Li, K.-C. (eds.) GPC 2007. LNCS, vol. 4459, pp. 710–715. Springer, Heidelberg (2007)

Kato, H., Billinghurst, M.: Marker tracking and HMD calibration for a video-based augmented reality conferencing system. In: Proceedings of the 2nd International Workshop on Augmented Reality (IWAR 1999), San Francisco, U.S.A (October 1999)

Flexeo: An Architecture for Integrating Wireless Sensor Networks into the Internet of Things

Juan Ignacio Vazquez, Aitor Almeida, Iker Doamo, Xabier Laiseca, and Pablo Orduña

MoreLab-Mobility Research Lab
Tecnológico Fundación Deusto
Avda. de las Universidades 24, 48007 Bilbao
{ivazquez,aalmeida,idoamo,xlaiseca,
porduna}@tecnologico.deusto.es

Abstract. Wireless sensor networks are a hot topic in Ubiquitous Computing for implementing context-awareness scenarios. The connection of sensor nodes to the Internet leads to new ways for remote monitoring of human behavior in real-time. In this paper, we introduce Flexeo: a flexible architecture for implementing monitoring solutions based on wireless sensor networks, with distributed intelligence at different layers. In this way, sensor-populated scenarios may communicate with Internet-based facilities enabling the vision of an Internet of Things.

Keywords: Wireless sensor networks, Internet monitoring, distributed architecture, Internet of Things.

1 Introduction

Context awareness is one of the most important aspects of Ambient Intelligence and Ubiquitous Computing, with a remarkable amount of research that has been carried out during the last years. The most basic form of perceiving context information consists in deploying ubiquitous sensors in the environment, which are able to capture part of the existing data, correlate and synchronize these data, analyze them, and finally, carry out a reactive activity without user intervention.

One of the most popular technologies for creating these sensor-populated scenarios are Wireless Sensor Networks (WSN). The use of an energy-efficient wireless bearer such as IEEE 802.15.4/Zigbee [1], enables the rapid and seamless deployment of sensor nodes in any scenario, with minimum requirements from existing infrastructure. Small size and easy integration of physical sensors into these platforms leverages their flexibility to address different types of applications [2].

On the other hand, there is an existing trend about connecting physical objects to the Internet, acting as first-class citizens both for publishing information or retrieving data that may determine their future behavior. This approach has been dubbed the Internet of Things [3]. On the Internet of Things, real objects may benefit from all the existing knowledge that is available on the Internet for better fulfilling people's needs. For example, an Internet-connected umbrella may access a public weather information website, retrieve the weather forecast in a language such as XML, and switch on

J.M. Corchado, D.I. Tapia, and J. Bravo (Eds.): UCAMI 2008, ASC 51, pp. 219–228, 2009.
springerlink.com

a LED to indicate this situation. In this way, users do not need to access the Internet through a different device (a computer) in order to take the decision whether to use or not a concrete object (the umbrella), achieving a better functional integration.

The other way around is also possible: creating physical objects that capture part of the real world information and publish it on the Internet. This approach enables the possibility of creating communities of Internet-connected devices that share real world information.

As we mentioned, wireless sensor networks are one of the most suitable technologies for capturing real world data. Therefore, connecting WSN to the Internet in order to publish contextual data in standard ways so that they can be shared with other entities, analyzing these data, taking decisions in remote premises, and finally implementing these decisions back in the real world through actuators, is a challenging activity where complementary technologies may be applied.

In this paper, we present Flexeo, a flexible architecture for connecting wireless sensor networks to the Internet, and distribute the intelligence and the decision making process through different layers. Section 2 analyzes previous work in the field of Internet-oriented wireless sensor networks. Section 3 presents the general architecture of Flexeo, while section 4 describes the different layers of the architecture. Section 5 introduces some of the prototypes developed for the system, and finally, section 6 presents some conclusions and future work.

2 Previous Work

Wireless sensor networks have been a hot topic in Ubiquitous Computing research during the last years, facilitating the deployment of Ambient Intelligence scenarios. Different types of platforms have been available such as Berkeley Motes and Smart-Its [4], and more recently Sun SPOT [5] and Intel Mote 2, which promise more powerful and easier application development tools, along with increased computing capabilities.

These wireless sensor platforms have been widely used to prototype a large number of applications, ranging from environmental monitoring [6] to medical monitoring in projects such as CodeBlue [7], Alarm-Net [8] or MyHeart [9]. It is noteworthy how medical applications, which require continuous monitoring of patients' vital signs, are a suitable niche for experimenting with non-intrusive small sensing nodes.

The main problem with all these experiences is that their architecture has been designed in a very *ad hoc* way for the problem being addressed, being difficult to adapt to other scenarios or applications, although the core problem is the same: remote monitoring and operation of dynamic sensor networks. Our previous experiences with WSN-based monitoring also featured this lack of flexibility, forcing us to redesign different aspects of the architecture depending on the concrete target application.

On the other hand, the Internet of Things (IoT) is a newly developed approach, initially promoted by ITU [3], being "*the next step in 'always on' communications, in which new technologies like RFID and smart computing promise a world of networked and interconnected devices that provide relevant content and information whatever the location of the user*". Any device or object can be directly or indirectly connected to the Internet and communicate with other objects, services or people, even forming communities of devices that mimic social behaviour [10].

One of the first experiences on integrating WSN into the Internet is SenseWeb [11], a P2P-like open architecture for sharing sensor data through the Internet. Using the SenseWeb API users can register their own sensors to publish information and create a common repository of sensor data. SensorMap is an example of end-user application consisting on a mash-up that combines sensor streams obtained from SensorWeb and maps from Virtual Earth.

The main limitation of SenseWeb from an architectural point of view is its centralized vision: all the decision process is carried out at a single central point called Coordinator, where all the sensor data are stored and analyzed. That is, all the intelligence in the system is located at a unique place and all the data must me sent through communication gateways to this point in order to analyze them and take the appropriate action.

From our experiences, we found that different decisions levels could be mapped onto different architectural layers, so that part of the analysis of sensor data and the determination of the reactive response could be also done at a local level, where the sensors are deployed. Under this approach, the communication gateway is transformed into an intelligent entity, able to perform reasoning based on existing local data, as well as forwarding the appropriate information (such as reporting taken decisions) to upper levels in the architecture. Of course, a central point of control still exists for global analysis of data and remote monitoring and operation.

In order to solve the two mentioned issues in existing architectures (lack of flexibility and centralized decision taking process) we created a flexible architecture for wireless sensor networks integration into the Internet, that could be customized in different ways, embedding intelligence at different layers, in order to accommodate to the disparate requirements of possible application scenarios with minimum redesign and recoding.

3 General Architecture

Flexeo architecture is divided on three different layers (see Fig. 1): the Sensors and Actuators Layer, the Coordination Layer and the Supervision Layer.

The Sensors and Actuators Layer was formed by the sensors and actuators that interact with the environment. Every sensor/actuator was integrated on wireless nodes (Crossbow Mica2 Motes). These nodes form a mesh network and send the information gathered by the sensors to the Coordination Layer through a special node called the base node. Messages are routed from one node to another until they reach this base node. For instance, sensorized chairs may have two sensors (at the seat and the back, in order to know whether the user is correctly seated or not), while wireless displays have one actuator (an organic screen). Even if these devices have more than one sensor or actuator, only a single wireless node is needed to integrate them.

The Coordination Layer is responsible for the management of the data received from the sensor network and for taking decisions based on this information. This layer

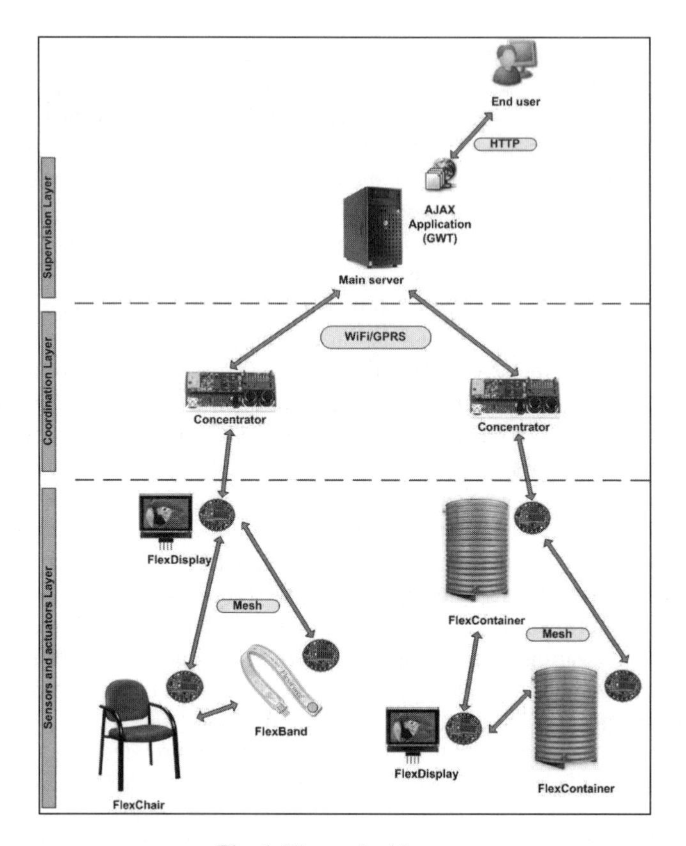

Fig. 1. Flexeo Architecture

will store temporarily the gathered data in a buffer, which is sent to the upper layer once this buffer is full, in order to reduce the number of connections. Since GPRS is used in some scenarios, it is important to minimize the number of connections due to their cost. Preliminary work has been carried out to develop a protocol that uses missed calls to force the coordinating device (called *concentrator*) to establish a new GPRS connection before performing a web request. The use of GPRS is optional, since the concentrator also supports IEEE 802.11b (Wi-Fi). The concentrator is implemented on a Gumstix embedded device [12], powered with an Intel XScale processor and embedded Linux, which has more advanced computational resources compared to the sensor nodes found in the lower layer.

Finally, the Supervision Layer provides the concentrators with a Web server to publish on the Internet the data retrieved from the sensor network they manage. This layer stores the data retrieved from the Sensors and Actuators Layer in a database. The Supervision Layer also offers a Web interface for the end users to manage the sensor data and generate statistics, as well as an API for integrating the monitoring and control functions into any existing business application.

4 Layers and Intelligence Distribution

This section describes how intelligence is distributed among all the layers, along with involved technologies.

4.1 Sensors and Actuators Layer

The selected sensor platform for this layer was Mica2 provided by Crossbow. This platform was selected due to its stability over the other existing model, MicaZ, which was in testing phase when the project was developed. We used this platform in order to design and implement several sensorized objects (see section 5), and the base station that acts as a gateway between the sensor network and the coordination layer.

There is almost no intelligence at this level, due to the computational limitations of the wireless sensor nodes, although some forms of data aggregation strategies were performed in order to provide higher level information to the Coordination Layer.

4.2 Coordination Layer

The Coordination Layer is implemented through some entities called concentrators, which are embedded platforms based on Gumstix, powered with the OSGi framework [13] and different OSGi Bundles, offering a collection of services to access the sensor network. Each concentrator controls one sensor network at a concrete location.

OSGi

In order to provide a modular, loosely-coupled, device-independent framework in the concentrator, each node (sensor/actuator/device) must provide a programming interface with the methods to interact with the node itself. In this way, it is possible to perform high level queries to find and manage the nodes controlled by the concentrator. The OSGi Framework provides this level of abstraction and is available even for some embedded systems, so it was a good candidate for abstracting the sensor layer within the concentrator.

Due to the RAM (64 MB) and flash (16 MB) memory constraints, the selected OSGi implementation must have a small footprint and low platform requirements. In order to select one implementation, existing ones were tested to check how suitable they were for Flexeo. Table 1 shows the results of the tests (the tests were done in the Gumstix platform, using the JamVM Java Virtual Machine and GNU Classpath v0.90 libraries).

Table 1. OSGi Framework comparison

	RAM Memory	Flash Memory
Concierge R3 (without HTTP server)	2864 KB	116.8 KB
Concierge R3 + HTTP Oscar R4	12776 KB	773.4 KB
Concierge R3 + HTTP KF R3	9572 KB	384.8 KB
Concierge R3 + HTTP KF R4	*Invalid classpath*	
Equinox R4 (without HTTP server)	10324 KB	1292 KB
Knopflerfish R4	*Invalid classpath*	
Oscar R4 (without HTTP server)	8080 KB	788 KB

As the table indicates, the Concierge [14] with HTTP Knopflerfish R4 service and the Knopflerfish R4 [16] implementation could not be run in the Gumstix platform due to classpath incompatibility. Equinox R4 [15] and the Oscar R4 [17] implementations were discarded based on their high resource requirements. Finally the Concierge R3 implementation was selected. In order to provide the HTTP Service, Nano HTTPD [18], a tiny embeddable HTTP Server written in Java, is used since the resources it requires are more affordable than the ones required by the Knopflerfish R3 HTTP Service (it only uses 13 KB in Flash Memory and 186 KB in RAM Memory).

Flexeo Concentrator Services

The Flexeo HTTP Service provides a REST-based API to enable the access to the values gathered by the sensors, returning the results based on a given device type or device identifier.

The Network Manager Service deals with the network connections. This service can handle different types of network connections. Although GRPS and Wi-Fi are implemented, other type of connections could be added. In order to minimize the number of connections, the Buffer Management Service stores the values subject to be sent to the upper layer in a buffer. This data is periodically sent when the buffer is full.

Intelligence at the Concentrator

In Flexeo, system intelligence is mostly distributed along two levels in the architecture: the Coordination Layer and the Supervision Layer. The Gumstix-based concentrator is the entity that implements the intelligence at the Coordination Layer.

The strategy for implementing a basic form of intelligence at this level was through a series of domain rule sets that evaluated the current context information, obtained from controlled wireless sensor nodes against some predefined conditions, and triggered a reactive action when those conditions were met. This scheme enabled Flexeo to take rapid local decisions at the concentrator, when all the information required to take those decisions was provided by the directly controlled sensors, while still allowing global intelligence at the Supervision Layer, where these data were integrated with the rest of the information sent by concentrators in other locations.

4.3 Supervision Layer

The Supervision Layer offers a basic graphical interface for monitoring the system, as well as an API for integrating the information into the current business management solution. Using the basic graphical interface (see Fig. 2), users can retrieve the current values of the sensors filtering them by the device type and/or the concentrator ID (each sensor measure has a timestamp associated). Users can also retrieve the alerts generated in one concentrator; these alerts are generated by the domain rules programmed in each concentrator. Finally it is possible to retrieve a histograph with the values of a sensor over a period of time. This graphic can be customized filtering the sensors by concentrator and/or type and configuring the period of time to be represented in it.

The Flexeo Web Interface was developed using Google Web Toolkit (GWT) [19], based on AJAX technology. The applications are developed in the Java programming language and GWT transforms them into HTML and JavaScript code. GWT applications have shorter load times, the generated code is more compact, and applications are supported by Internet Explorer, Firefox, Opera and Safari browsers.

Intelligence at the Supervision Layer

The Supervision Layer has a global view of the whole system, centralizing the data provided by all the concentrators that manage the low-level sensor nodes at the different locations. As already mentioned, we also developed a basic API that enables easy integration of gathered data with any ERP, information system or database. In this way, organizations can benefit from the Flexeo monitoring system completely integrated with their current business monitoring or business intelligence solution.

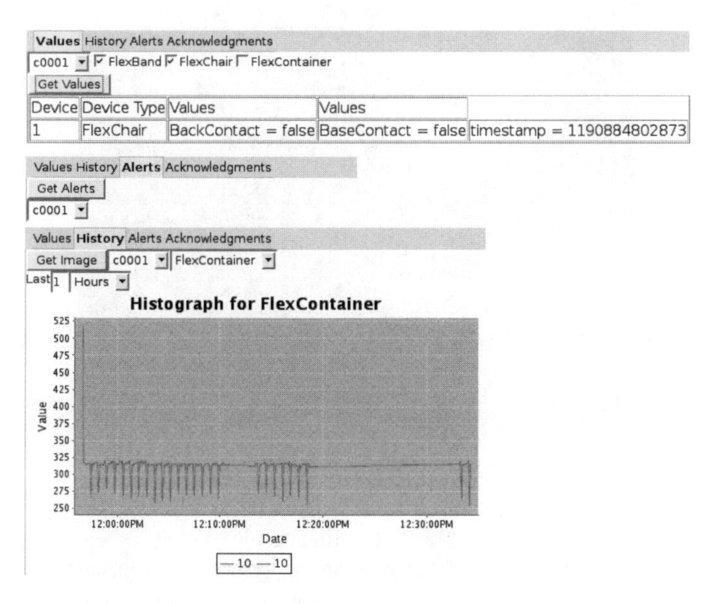

Fig. 2. Flexeo Web Interface at the Supervision Layer

5 Prototype Devices and Scenarios

We have designed several scenarios and devices using the Flexeo architecture, mostly for industrial monitoring and user activity monitoring. These are some of the prototyped devices, integrating wireless sensor nodes and providing data that flowed up all the levels in the architecture:

- *FlexContainer*: this device monitors the state of a chemical fluids container, by detecting the level of liquid. For experimental purposes, we assumed that the contained fluid conducted electricity, so we applied a low intensity electric current to the fluid and detected when the circuit is closed. This sensor node device was based in a MDA-100 device, which also provides temperature and light-intensity values.

- *FlexChair*: this sensorized chair includes a Mica2 mote and two FlexiForce sensors (see Fig. 3). These sensors located at the seat and back of the chair are able to detect the user's position while seated. The FlexiForce sensors provide an intensity based information while the Mica2 wireless node only accepts voltage-based inputs, so in order to solve this problem a transformation function was developed and implemented in the sensor board. This information may be used for tracing the ergonomic behavior of users in a simple way: detecting whether they are leaning back often, or if they remain seated for too long without taking a small break, generating recommendations that may benefit their health.

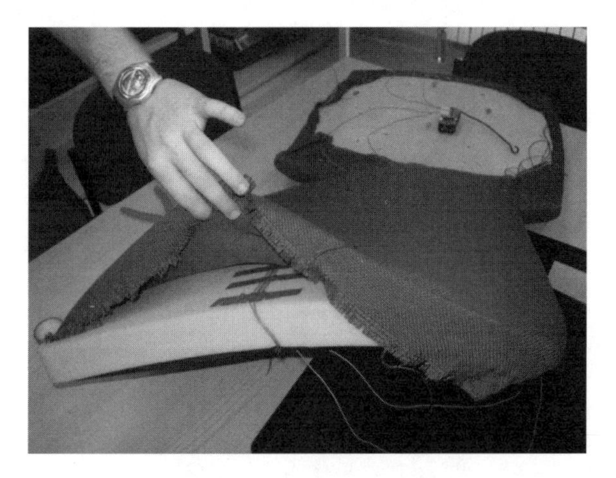

Fig. 3. Prototyping FlexChair

- *FlexDisplay*: this device is a bracelet with a display incorporated and wireless capabilities used to visualize the alarms of the system. It is based on a Mica2Dot mote, the smallest available wireless sensor node from Crossbow. The selected screen for displaying information is the uOLED64, developed by 4D-Systems. This screen has a USART interface compatible with any microcontroller. In order to manage the screen a driver was developed in the Mica2DOT node with three main functions: controlling the synchronization and the start status, controlling the image and text renderization and controlling the display visualization mechanism (through new functions such as DisplayOnOff, FadeToBlack and FadeFromBlack). The FlexDisplay can be used by security personnel or other monitoring staff in order to receive visual alerts about any facility being monitored, as well as further instructions from the monitoring center, via this energy-efficient graphical display.

6 Conclusions and Future Work

Wireless sensor networks are a very popular technology for monitoring user activities and implementing context-awareness scenarios. We introduced an architectural model for Internet monitoring of WSN, based on three layers that enable intelligence distribution and decision taking at different levels. One of the most remarkable aspects of Flexeo is its flexible application-agnostic orientation, which allows the system to be used for industrial purposes, health-at-home monitoring, human activity tracking or even environmental monitoring. We integrated state-of-the-at technologies such as wireless sensor networks, embedded computing platforms (Gumstix) and high-level programming kits (Google Web Toolkit) in a single coherent system.

Our future research will consist on moving more intelligence to lower-levels of the architecture to make it even more independent from the upper layers, particularly enhancing the intelligence at the wireless sensor network nodes by using fuzzy-logic and neural or neuro-fuzzy computational schemes, and semantic annotation of the gathered data.

Acknowledgements. This work has been partially supported by the Biscay Regional Government, Ekinberri program. The authors would like to thank all the people that took part in the development of Flexeo.

References

1. Zigbee Alliance, Zigbee specification v1.0 (2004), http://www.zigbee.org/
2. Sohraby, K., Minoli, D., Znati, T.: Wireless Sensor Networks: Technology, Protocols, and Applications. Wiley-Interscience, Chichester (2007)
3. ITU. ITU Internet Reports 2005: The Internet of Things, ITU (2005)
4. Beigl, M., Gellersen, H.: Smart-Its: An Embedded Platform for Smart Objects. In: Proc. Smart Objects Conference (SOC 2003), Grenoble, France (May 2003)
5. Sun Microsystems. Project Sun SPOT (2008), http://www.sunspotworld.com/
6. Mainwaring, A., Culler, D., Polastre, J., Szewczyk, R., Anderson, J.: Wireless sensor networks for habitat monitoring. In: Proceedings of the 1st ACM international Workshop on Wireless Sensor Networks and Applications. WSNA 2002, pp. 88–97. ACM, New York (2002)
7. Malan, D., Fulford-Jones, T., Welsh, M., Moulton, S.: CodeBlue: An Ad Hoc Sensor Network Infrastructure for Emergency Medical Care. In: International Workshop on Wearable and Implantable Body Sensor Networks (April 2004)
8. Wood, A., Virone, G., Doan, T., Cao, Q., Selavo, L., Wu, Y., Fang, L., He, Z., Lin, S., Stankovic, J.: ALARM-NET: Wireless sensor networks for assisted-living and residential monitoring. Technical Report CS-2006-11, Department of Computer Science, University of Virginia (2006)
9. Luprano, J., Sola, J., Dasen, S., Koller, J.M., Chetelat, O.: Combination of Body Sensor Networks and On-Body Signal Processing Algorithms: the practical case of MyHeart project, pp. 76–79 (2006)

10. Vazquez, J.I., Lopez-de-Ipina, D.: Social Devices: Autonomous Artifacts That Communicate on the Internet. In: Floerkemeier, C., Langheinrich, M., Fleisch, E., Mattern, F., Sarma, S.E. (eds.) IOT 2008. LNCS, vol. 4952. Springer, Heidelberg (2008)
11. Kansal, A., Nath, S., Liu, J., Zhao, F.: SenseWeb: An Infrastructure for Shared Sensing. IEEE MultiMedia 14(4), 8–13 (2007), `http://dx.doi.org/10.1109/MMUL.2007.82`
12. Gumstix Inc. (April 2008), `http://www.gumstix.com/`
13. OSGi Alliance (April 2008), `http://www.osgi.org/`
14. Concierge OSGi – An optimized OSGi R3 implementation for mobile and embedded systems (April 2008), `http://concierge.sourceforge.net/`
15. Equinox OSGi (April 2008), `http://www.eclipse.org/equinox/`
16. Knoplerfish Open Source OSGI (April 2008), `http://www.knoplerfish.org`
17. Oscar OSGi (April 2008), `http://oscar-osgi.sourceforge.net`
18. NanoHTTPD server, `http://elonen.iki.fi/code/nanohttpd/`
19. Google Web Toolkit (April 2008), `http://code.google.com/webtoolkit/`

A Mobile Peer-to-Peer Network of CBR Agents to Provide e-Assistance

Eduardo Rodríguez[1], Daniel A. Rodríguez[1], Juan C. Burguillo[1], and Vicente Romo[2]

[1] Departamento de Telemática
Escola Técnica Superior de Enxeñería de Telecomunicación
Universidade de Vigo
Calle Maxwell s/n, 36203, As Lagos-Marcosende, Vigo, Spain
rodfer@enigma.det.uvigo.es, {darguez,jrial}@det.uvigo.es
[2] Departamento Didáticas Especiais
Facultade de Ciencias da Actividade Física e do Deporte Universidade de Vigo
Campus a Xunqueira s/n, 36005 Pontevedra, Spain
vicente@uvigo.es

Summary. In this paper we combine three recent technologies to provide e-Assistance. On the one hand Ambient Intelligence is an useful paradigm to help people dealing with unknown situations. On the other hand, Case Based Reasoning (CBR) applies knowledge gained through past experiences to solve current problems. Finally, peer-to-peer (P2P) networks let users to share resources among members of a community. A combination of these technologies can be used to build a seamless e-assistance system in order to help people using mobile ad-hoc networks and enabling nodes to establish spontaneous connections among them in order to meet specific needs. Taking this context as a starting point we have developed a CBR multi-agent system (MAS) that exchanges its cases through a P2P mobile ad-hoc network in order to solve daily problems. An example describing an intelligent gym for elder physical training is also provided.

Keywords: CBR, P2P, MANET, e-Assistance, AmI, MAS.

1 Introduction

Ambient Intelligence (AmI) provides a new and natural approach about how people interact with technology. In AmI, devices recognize people and their intentions, as well as objects and changes in the environment. In this digitally equipped environment, populated with embedded systems, intelligence is achieved through the capacity of perceiving the environment through sensors, analyse such environment and adaptively actuate over it to enhance people quality of life.

Elderly or disabled people can and must benefit from the virtues of AmI, especially from e-assistance. But not only them, every person, facing a situation where he does not know how to act, can use pervasive computing and ambient intelligence

J.M. Corchado, D.I. Tapia, and J. Bravo (Eds.): UCAMI 2008, ASC 51, pp. 229–238, 2009.
springerlink.com

in order to achieve e-assistance. To carry out this assistance, collaboration among agents within a multi-agent system for complex troubleshooting can be used.

In this paper, we present an architecture to assist users facing novel or unknown specific situations. This architecture consists in a set of intelligent agents that are able to communicate with each other through an ad-hoc mobile network. We start from the idea that current problems or situations are, at least, similar to the problems or situations that other users had experienced in the past. This means that we can use the previously acquired experiences. In order to achieve those experiences we establish a peer-to-peer network among the agents of the system. Through this network we can exchange the previously acquired experiences, or cases.

There are studies related with using centralized CBR intelligent agents on the one hand; as well as studies about resource sharing through peer-to-peer networks on the other hand. But this paper combines both approaches, modelling the CBR agents as mobile peers and using a P2P network to exchange experiences, problems and their solutions among them.

The remainder of the article is structured as follows. In Section 2 we compile some of the most relevant work related with the multiple disciplines involved in the system. Section 3 exposes the description of the system and explains its architecture. Section 4 presents a case study on a concrete implementation a specific scenario. Finally, Section 5 points out the conclusions of our work as well as future work to extend the research.

2 Related Work

Our system can be seen as a P2P network of intelligent agents working together in order to solve problems. This scheme involves a whole set of technologies. In this section we present a brief introduction those popular technologies involving: Multi-agent Systems (MAS), Case Based Reasoning (CBR), mobile ad-hoc networks (MANETs) and peer-to-peer networks (P2P).

2.1 Multi-Agent Systems (MAS)

Jennings and Wooldridge define agents as entities that enjoy the properties of being reactive, pro-active, autonomous, and have social ability [1]. Agent-oriented techniques are being increasingly considered in a range of telecommunication, commercial and industrial applications, as developers realize about their potential [2]. We can consider a multi-agent system (MAS) as a loosely-coupled network of problem solvers (agents) that work together to solve a given problem [3]. Agents and Multi-agent Systems are a powerful way to model distributed open-ended systems.

2.2 Case Based Reasoning (CBR)

CBR technology takes as premise that new problems usually are similar to previously encountered ones, so that previous solutions are usually valid to solve the

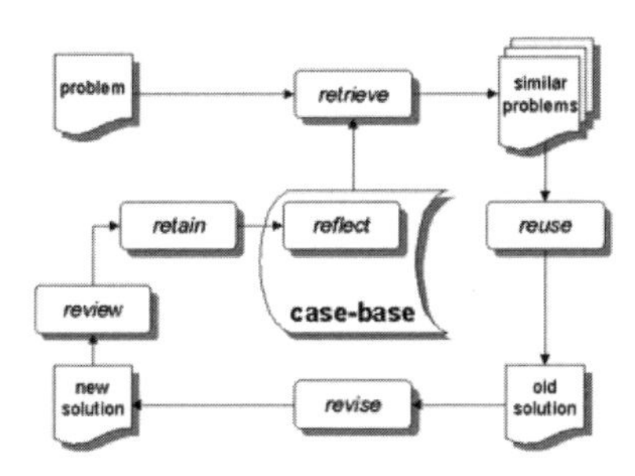

Fig. 1. CBR cycle

current problem. The CBR paradigm considers methods for retrieving, reusing and revising the knowledge from past cases, as well as retaining the new generated knowledge [4]. Cases may be kept as concrete experiences, or as a generalized case, abstracted from a set of similar cases (see Fig. 1).

Depending on the problem, a CBR agent may need to collaborate with other CBR agents to exchange information from their respective case bases. Different approaches are proposed in the literature for managing this situation. Plaza et al. have carried out a research [5] to find possible modes of cooperation among homogeneous agents with learning capabilities. They present two modes of cooperation among agents: Distributed Case-Based Reasoning (DistCBR) and Collective Case-Based Reasoning (ColCBR). In the DistCBR cooperation, the source agent asks to others peer agents to solve the problem. In contrast, in the ColCBR cooperation the originating agent keeps the authority to choose the best solution. It just asks for the experience of other peer agents. They prove that the result of cooperation is always better than no cooperation at all. However, these protocols are domain dependent and are the result of a knowledge modeling process.

Gupta et al. [6] propose a reputation system for decentralized unstructured P2P networks like Gnutella [7] for searching and information sharing. The peer selection strategy is based on the agent reputation score, which is intended to give a general idea of the peers level of participation in the system.

Plaza and Ontañón [8] proposes a strategy to choose the agents that join a committee for solving a problem in the classification tasks. The committee organization improves the classification accuracy with respect to the individual agents. The agent learns to form a committee in a dynamic way and to take decisions such as whether it is better to invite a new member to join a committee, when it must solve a problem individually, and when it is better to ask for a solution to a committee.

2.3 Mobile Ad Hoc Networks (MANET)

Mobile networks allow spontaneous connectivity among multiple wireless devices. Ad-hoc networks enable connections between wireless devices without any infrastructure supporting the communication. Another category of mobile networks include a base station, working as a bridge, to which nodes can be connected in order to establish communications with other nodes, even throught local networks or the Internet. The mixture of both types of nodes results in a hybrid network in which nodes can be connected to mobile nodes, but also with a fixed base station.

In order to maximize the battery of network's devices, it is essential to optimize their consumption and the efficiency of communication protocols. Yang et al [9] proposes a peers discovery service based on multicast requests and multicast responses.

Another aspect that should not be neglected is the routing protocol and the information the nodes have to store in order to forward the packets. Broch et al. give us a comparison of different routing protocols [10].

2.4 Peer-to-Peer Networks (P2P)

Peer-to-peer networks have achieved a great success in multiple applications for exchanging resources among the members of a community. Since its emergence, multiple P2P systems have been developed with different search approaches.

The first peer-to-peer network, Napster [11], consisted of a distributed database with centralized search. Next successful P2P network, Gnutella, used a fully distributed search but it turns out in an unacceptable overloading of the network. Kazaa [12] presented an idea to split the peers between leaders and normal peers. This idea achieved excellent results in the search and acquisition of files. Recently, new P2P models such as eMule [13] and Bittorrent [14] are hybrid systems mixing previous approaches with game theory strategies.

3 System Overview

In this paper we aim to establish the basis of an intelligent mobile e-assistance system. Within this system, users will use an intelligent mobile terminal (usually a commercial smartphone) containing a CBR agent that allows them to interact with the environment devices (sensors and actuators) together with other user terminals. Whenever an unknown situation or problem appears, the intelligent terminal is able to connect with other users terminals or with environment's devices in order to solve that problem or provide e-assistance.

The set of intelligent agents that make up the multi-agent system have two explicit modules. The first one deals with the reasoning process of the system and the second one takes care of the communication process.

3.1 Reasoning Process

The reasoning process is the ability to solve problems. In our system, agents achieve that capacity using Case Based Reasoning. Cases of CBR have a problem part and a solution part. The problem part is defined by a set of attributes that model the case. Attributes refer both to domain independent information like user data or preferences of presentation, and domain dependent information like goals, previous related experience and/or skill level. The solution part consists of a set of guidelines on how to face the problem. These guidelines can be showed with different structures and formats like text or multimedia. Depending on the situation, the solution part can incorporate the results obtained by other user that have already applied this solution.

Whenever a user needs to deal with a given problem P, an initiator agent A_i tries to solve it performing a CBR cycle, which consists of four ordered phases (see Fig. 1): retrieval, reuse, revise and retain phase. In the retrieval phase, the agent A_i searches for similar cases in its local case base. To define the similarity we use a distance weighted k-NN algorithm [15]. The basic idea of the distance weighted k-NN algorithm is to weigh the contribution of each case found according to their distance to the query point, giving greater weight to closer cases [16].

In this sense, the distance between two cases $(C1, C2)$ with a given set of attributes is calculated using the similarity function formula:

$$f_x\left(C_1, C_2\right) = \sum_{a=1}^{n} d_a\left(x, y\right) \tag{1}$$

Where n is the number of attributes used to measure the distance between two cases and d_a is the distance between two values x and y of a given attribute, where: $d_a\left(x, y\right) = 1$ if x or y is unknown, $d_a\left(x, y\right) = overlay\left(x, y\right)$ if x,y are nominal values and $d_a\left(x, y\right) = \sqrt{\left(x, y\right)^2}$ if x,y are real values.

The endorsed solution is considered good enough when the similarity between cases is over a threshold. It sets the maximum distance between the current case and the retrieved one and should be established independently for every scenario. In this situation, A_i continues the CBR cycle without requesting other agents' help. In the contrary, if the endorsed solution is considered not good enough, A_i will disseminate the problem alongside with the threshold. Every available agent of the system receives the problem P and decides whether to participate or not. Everyone of the participating agents apply a CBR cycle to its local case base looking for similar cases. After the initial retrieval phase, they perform the reuse phase.

In the reuse phase, collaborative agents try to solve P using its case base and send back a message to the initiator agent A_i that is either: sorry (if every of its cases have a distance greater than the given threshold) or a set of solutions $< (S_k, C_j^k), P, A_j >$, where the collection of pairs (S_k, C_j^k) mean that the agent A_j has found C_j^k cases in its case base endorsing solution S_k. Agent A_i has to choose the most adequate solution between all the received ones. In order to do this a voting scheme is applied.

The voting scheme defines the mechanism by which an agent reaches an aggregate solution from a collection of solutions coming from other agents. The principle behind the voting scheme is that the agents vote for solutions depending on the number of cases they found endorsing those solutions. Following Plaza [8], we do not want that agents, having a larger number of endorsing cases, may have an unbounded number of votes regardless of the votes of the other agents.

Thus, we define a normalization function so that each agent has one vote that can be for a unique solution or fractionally assigned to a number of solutions depending on the number of endorsing cases. We denote by the set of agents that have submitted their solutions to agent A_i for a problem P. The vote of an agent A_j for solution S_k is:

$$Vote\left(S_k, A_j\right) = \frac{C_j^k}{1 + \sum_{r=1}^{k} C_r^j} \tag{2}$$

Where C_j^k is the number of cases found in the case base endorsing solution C_k. It is easy to see that an agent can cast a fractional vote that is always less than 1. Aggregating the votes from different agents for a solution C_k we have ballot:

$$Ballot\left(S_k, \Delta\right) = \sum_{A_j}^{k} Vote\left(S_k, \Delta\right) \tag{3}$$

Therefore the winning solution is:

$$Sol\left(P, \Delta\right) = argmax_{k=1}^{k} Ballot\left(S_k, \Delta\right) \tag{4}$$

Finally, when A_i has selected a solution S_k, it applies the final steps of the CBR cycle. In the revision phase the voted solution is revised by the initiator agent A_i to remove unsuitable elements and to correct inconsistent attributes. In the retain phase, A_i has to decide if the selected solution is added to the its case base. To make this decision, the agents evaluates how well has worked the proposed solution.

3.2 Pure Distributed Communication Process

The system is deployed as a dynamic, scalable and adaptive P2P network where each agent (located within a user mobile terminal, p.e., a smartphone) of the multi-agent system is a peer of the network. Peers of the network are mobile and everyone contains an agent with accumulated experience. That experience may be useful for another agent so an ad hoc topology where every peer is able to establish a connection with any other peer is useful.

The communication module allows connections among the peers of the network. We have designed a simple source initiated on-demand driven protocol to support that communication. This type of wireless routing protocols fits well with a very highly mobile environment because they create routes only when desired by the source node.

The format of the packets exchanged through this protocol is the following (see Fig. 2). The first field is the direction of the emitting node, then the direction of the destination node, then a time to live (TTL) field, next the direction of the source node and a field with the number of packets generated by that node. These last two fields made the packet can be identified unequivocally. Finally the last field is the payload.

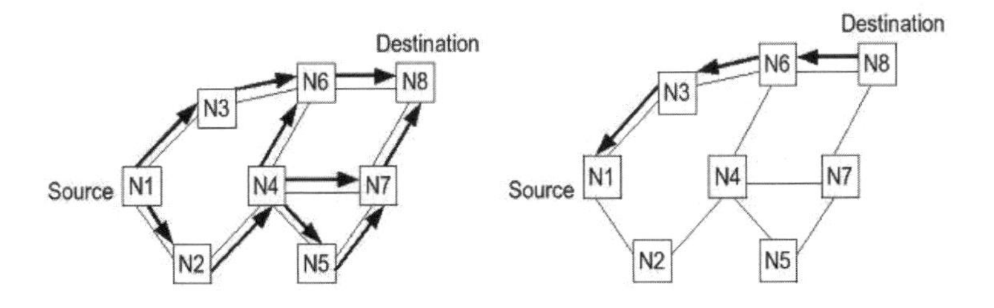

Fig. 2. Request (left) and response (right) processes

The communication process can be summarized as follows. Whenever a node needs to make a request it is because the agent needs help with a specific case. So it disseminates a packet with the request using a broadcast frame sent to its direct neighbors. If they decide to collaborate, as we have seen in the previous section, these one-hop neighbors answer the request.

The broadcast frames should be around only for a limited amount of time so they have a TTL field with the number of hops allowed. Before forwarding the broadcast frame to its own direct neighbors, a node decrease the time to live field. A node only behaves this way the first time it receives the frame. Subsequent times the node discards the frame. As well, when a frame life expires, that is when the time to live field reaches zero, the frame is discarded. This behaviour intends to avoid the packet explosion problem that Gnutella [7] have to deal using a similar protocol.

At the same time a node forwards a frame, it has to save in a table the direction of the node from which it has received the frame, as well as its serial number. In this way, it can forward back to the source node the possible responses coming from other nodes (see Fig. 3).

There are two reasons supporting this decision. First, it makes no sense keeping state information of a highly mobile environment, where every node changes its position constantly. Secondly, as every single node is a potential suitor for the request, it makes sense to broadcast the request among all the direct neighbors of the community.

The protocol needs to define an amount of time while the source node accepts answers. After that time, the source node will assume no other nodes are willing to cooperate. This parameter has to be carefully established. If it is too short, the agent can take a fast decision but, it probably will be taken among a short number of possibilities, which decreases the accuracy of the solution. By contrast, if the time is too long, the agent will have a better solution given the large amount of received solutions, but a bigger wait time to obtain the solution could degrade

Destinat Dir	Emitting Dir	TTL	Source Dir	#	Payload

Fig. 3. Frame format

the performance of the system. Then, we must look for a compromise between speed and accuracy.

3.3 Hybrid Communication Process

We consider here a variation of the pure distributed communication process scenario. The idea is to deploy a hierarchical structure when some nodes have more experience, greater communication power or reasoning capabilities. We call these nodes super-peers [17]. These nodes are intelligent agents too and form an ad-hoc network with the other peers, but they have augmented capabilities in order to increase their functionalities. A super-peer periodically collects cases from other agents what gives it an extensive case base. As well, we added network communication capabilities to this super-peer node. This allows to establish a connection via LAN or the Internet with other super-peer case repositories requesting the help needed.

4 Example: An Intelligent Gym

We have designed a scenario to put into practice our approach: an intelligent gym where users have a mobile intelligent device (smartphone) able to complement trainers' job. These devices know user's long term objectives, like lose weight, rehabilitation or muscle up, and they help users to accomplish their exercises in the gym machines. The machines at the gym have a set of sensors that provide information to the mobile terminals to supervise the right execution and, at the same time, to check how the user performs the exercise.

In this scenario, a problem P is defined by a set of attributes like age, weight, height, user's physical, previous training work, user's objectives and exercise machines at the particular gym. The solution part S includes a suggested training plan, adapted to every user, as well as information about how to carry it out.

The sequence can be like follows. The user enters into the gym and switch on the smartphone, starting the training program. Depending on the day, the training program suggests the user to start doing a particular exercise. Lets assume that the first aim is to run a little bit to warm up. Then, the user gets closer to the running machine (see Fig. 4, right). The smartphone, using Bluetooth, connects and recognize the running machine, sets up the machine for a time and dynamic speed running and tells the user that information in advance. While doing the exercise the smartphone (together with the body devices that (s)he wears) monitors the heart frequency and the arterial pressure and alerts the user if something goes wrong. When finishing the exercise, the smartphone saves the data to keep a record on the user history profile.

Users smartphones run a training program, which uses a CBR agent to suggest the exercises and how to perform them. This may include: information about the correct execution of the exercise, number of series to perform, number of repetitions per serie and execution speed. The case used in the CBR cycle is the present state of the user together with the present state of the gym (number and type of machines, number of users, temperature, humidity, etc.). Smartphones

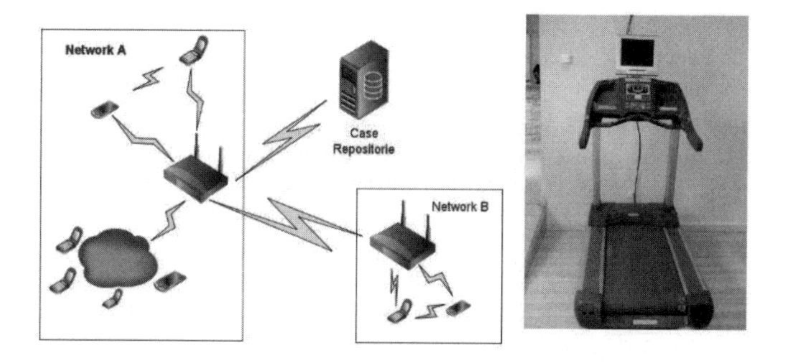

Fig. 4. Topology of the intelligent gym (left) and exercise machine (right)

may interact among them to sequentially organize the occupation of the machines. Besides, when a smartphone does not know how to manage a particular case it may ask other peers (i.e. smartphones) for suggestions about it.

In this scenario we could consider a non mobile node acting as a superpeer (see Fig. 4, left). This object is an intelligent CBR agent too and forms an ad-hoc network with the other peers, but it has bigger computational and communication capabilities (for instance, it can be a normal PC or a laptop). The main application of this superpeer is to collect data from the smartphones to allow the trainer to personally supervise the users training when necessary and, at the same time, to have an updated and complete history profile per user.

At the same time, the superpeer has a global perception of the gym through the interaction with the user smartphones and helps to organize the resources as a global planner. It also collects/provide new cases from/to the smartphones to create a bigger case base. Moreover, superpeers from several gyms could share their cases (see Fig. 4, left) to enhance the system through distributed learning.

5 Conclusions and Future Work

Through this article we presented a network of mobile intelligent agents able to collaborate in order to solve a given problem. Agents of this system base its intelligence on case based reasoning and share their cases to work together. This system can be especially suitable for elder people, due to the nature of their physical training that must be executed in controlled spaces, where all physiologic parameters must be totally under control.

There are several topics to keep this research ongoing. One important aspect is the efficiency of the communication protocol in order to extend the lifetime of the batteries. Another relevant issue would deal with the ability to work in environments with an elevated number of agents. This also involves the computational balancing of communications among the CBR agents and the superpeer itself.

References

1. Wooldridge, M.J., Jennings, N.R.: Intelligent agents: Theory and practice. Knowledge Engineering Review 10(2) (1995)
2. Jennings, N.R.: An agent-based approach for building complex software systems. Communications of the ACM 44(4) (2001)
3. Wang, A., Conradi, R., Lui, C.: A Multi-agent Architecture for Cooperative Software Engineering. In: Proceedings of the Third International Conference on Autonomous Agents (1999)
4. Aamodt, A., Plaza, E.: Case-Based Reasoning: Foundational Issues, Methodological Variations and System Approaches. AI Communications 7(1), 39–59 (1994)
5. Plaza, E., Arcos, L., Martin, F.: Cooperation modes among case based reasoning agents Workshop on Learning in Distributed Artificial Intelligence Systems (1996)
6. Gupta, M., Judge, P., Ammar, M.: A reputation system for peerto-peer networks. In: Proceedings of ACM Networks and Operating. System Support for Digital And Video (2003)
7. Gnutellas' official web site, http://www.gnutella.com/
8. Plaza, E., Ontaon, S.: Cooperative Multiagent Learning. In: Alonso, E., Kudenko, D., Kazakov, D. (eds.) AAMAS 2000 and AAMAS 2002. LNCS (LNAI), vol. 2636, pp. 1–17. Springer, Heidelberg (2003)
9. Yang, Y., Hassanein, H., Mawji, A.: Efficient Service Discovery forWireless Mobile Ad Hoc Networks. In: Proceedings of the IEEE international Conference on Computer Systems and Applications, March 08, 2006, pp. 571–578. AICCSA, IEEE Computer Society, Washington (2006)
10. Broch, J., Maltz, D.A., Johnson, D.B., Hu, Y., Jetcheva, J.: A Performance Comparison of Multi-Hop Wireless Ad Hoc Network Routing Protocols. In: Proceedings of the Fourth Annual International Conference on Mobile Computing and Networking (MobiCom 1998). ACM, Dallas (1998)
11. Napster, Inc. Website, http://free.napster.com/
12. Kazaa's official website, http://www.kazaa.com/
13. eMule project Official site, http://www.emule-project.net/
14. BitTorrent Official Website, http://www.bittorrent.com/
15. Cover, T., Hart, P.: Nearest neighbor (NN) Norms: NN Pattern Classification Techniques. IEEE Computer Society Press, Los Alamitos (1991)
16. Mitchell, T.: Machine Learning. MacGraw-Hill (1997)
17. Beverly Yang, B., Garcia-Molina, H.: Designing a super-peer network. In: 19th International Conference on Data Engineering ICDE (2003)

An Ambient Assisted-Living Architecture Based on Wireless Sensor Networks

Javier Andréu, Jaime Viúdez, and Juan A. Holgado

Software Engineering Department, University of Granada
C/ Periodista Daniel Saucedo Aranda s/n, 18071 – Granada, Spain
javierandreu@ugr.es, jviudez@ugr.es, jholgado@ugr.es

Abstract. Ambient Assisted-Living (AAL) is becoming an important research field in Ambient Intelligence. Many technologies have emerged related with pervasive computing vision, that can give support for AAL. One of the most reliable approaches is based on wireless sensor networks (WSN). Based on this basic assumption we have taken the promising SunSPOT platform as base support, and build upon it a services architecture for our AAL proposal. This architectural model allows us the decoupling of applications in components such as ECG's monitor, position system or location awareness.

Keywords: Wireless Sensor Network, Ambient Assisted-Living, Healthcare.

1 Introduction

Ambient Assisted-Living (AAL) is becoming an active research field on Ambient Intelligence (AmI) due to the demographic change in industrialized countries worldwide, cause a large increment of elderly people which demand more healthcare and assistance services, claiming at the same time a fast reliable access to social, medical and emergency systems [1]. AAL appliances are mainly focused on continuous remote monitoring of health parameters and daily activities for enhancing safety and security of individuals with chronic disease, elderly or disable as well as promoting the independence of these ones inside their living environments.

There are different alternatives approaches to develop AAL appliances. One of the most reliable approaches is based on wireless sensor networks (WSN). During the past five years there has been a great effort to standardize the WSN around the IEEE 802.15.4 [2] and Zigbee standards [3]. The first one specifies the medium access control layer and physical layer of a low rate WPAN (Wireless Personal Area Network), while the second one defines the network layer, security layer and application framework on top of medium access control layer of 802.15.4 to ease the developing of WSN applications. WSN appliances were oriented towards the acquisition of large data information that conveys from any node of WSN system finally into Coordinator node. This one is a centralized mote with more processing and memory resources capable of managing the network, and where the information should be processed and stored. If WSN was connected with other networks such as Internet protocol, then the coordinator node can act as a gateway, deriving the information directly to more powerful computers.

J.M. Corchado, D.I. Tapia, and J. Bravo (Eds.): UCAMI 2008, ASC 51, pp. 239–248, 2009.
springerlink.com

From AAL point of view, WSN are capable of providing several important properties needed on pervasive computing, and extensively to Ambient Intelligence. Different rates of size in motes are producing a lot new little pervasive devices, even wearable devices, which populate the environment providing seamless support for user needs anytime in everywhere.

In this work a new AAL system is presented based on a last generation of WSN. From different options available, we have chosen Sun SPOT platform as base support, and build upon it a services architecture for several reasons. First, it is based on IEEE 802.15.4 for both physical and link layers, and disposes a complete routing mechanism based on AODV, providing a secure platform that can interoperate with other motes. Second, it provides a Java framework, which allows the development of AAL system in Java programming language, easing the implementation, deployment and maintenance of programs. Third, it gives the possibility of build a seamless architecture applicable to different motes depending on its role. The architectural model allows us the decoupling of applications in components that cooperates with each other.

The remainder of this paper is organized as follows. Section 2 examines the characteristic of WSN, its actual status and the most important open issues on this topic. Section 3 presents the TeleCare Ambient Assisted Living Application developed on top of Sun SPOT platform and its main components. In section 4 we discuss the results obtained in this work with related works. At last we present the conclusions of this work and the immediate future works that we are following.

2 Wireless Sensors Networks

WSN consists of large numbers of small-scale nodes with sensing capabilities organized into a cooperative network [4]. WSN are optimized to provide a durable continuous monitoring of environmental conditions aroused from multiple nodes by means of wireless communications, maintaining at the same time a low level of energy consumption of each node. A lot issues might be solved using this technology. First, they are closely coupled to the physical world with all its unpredictable variation, noise, and asynchrony. Second, they involve many energy-constrained, resource-limited devices operating in concert. Third, they must be largely self-organizing, self-maintaining and robust despite significant noise, loss, and failure.

The corresponding core element in a WSN is the motes. Each mote or node of WSN is an interconnectable small, low powered and cheap computing device connected to several sensors (transducer) and a radio transmitter. The main role of a mote is the monitoring of the environment through the sensors connected to motes. These sensors can measure a huge amount of parameters such as light, acceleration, position, stress, pressure, humidity, sound, vibration among others.

The other key ingredient in WSN relies on networking. A sensor network is characterized by the interconnection among motes using a wireless medium. Many wireless network protocols are possible to use in WSN. But the basic applications are oriented towards the acquisition of a huge quantity of data from a self-organized network composed on large quantity of geographically dispersed motes, that, moreover they should continue to operate for years without maintenance. These motes can have both abilities and responsibilities defined in relation to some predetermined roles in the network architecture. On a typical wireless sensor network based on mesh topology we can define the following three roles: (a) end-device mote, (b) router mote, and (c) coordinator mote. An End Device Mote (EDM) contains the minimum processing capability. It senses periodically a physic parameter of the environment and transmits this value to other more powerful mote at specific rate; the rest of time, the mote is asleep. Router Mote (RM) is a more powerful mote capable also to sense the environment, although its main role is passing data among devices. The Coordinator Mote (CM), only one in each network, is the most capable device, and has the role of acting as the root of the network tree and might bridge to other networks.

WSN is still being an active area for research with respect to issues such as routing, MAC layer, time synchronization, security power management, or node localization [5, 6]; thus different approaches may be followed to build a WSN system that can promote some features facing others. An important concern is the network topology. WSN allows topologies such as bus, ring, star, tree and mesh. Although star topology has been widely used in the first generation of WSN to reduce probability to suffer failures, the big disadvantages are the excessive overload of central node and the system fall when a failure is placed in central node. A mesh topology (i.e., a partially connected mesh topology) provides a more robust solution with the interplay of at least two nodes with two or more paths between them to provide redundant paths to be used in case the link providing one of the paths fails. This decentralization is often used to compensate for the single-point-failure disadvantage that is present when using a single device as a central node (e.g., in star and tree networks). The mesh topology is the preferred one in the context of pervasive computing, and it is extensively used in the AAL system presented in this paper.

In table 1 we make overall comparison between some commercial and academic motes. The imote2 is the solution more powerful with a xscale processor of 13 – 416 MHz, although it is clearly penalized over other commercial SPOTs solutions in energy consumption within a interval around 105 – 330 mW in active mode and 1950 μW in passive mode. On the other hand with a 180 MHz processor the SunSPOT mote gives a good performance with regard to energy consumption, being in an interval around 2.5 - 125 mW in active mode and a significant 150 μW in sleep mode. The last one is the selected platform used for our AAL system.

2.1 Wireless Body Area Network

When the interconnectable devices can be wearable or implantable in the human skin it is often used the term WBAN (Wireless Body Area Network) to identify this special kind of network [8]. Such networks are not necessary WSN, and have specific applications in health monitoring and vital parameters through biosensors. Any sensor or device that is integrated into a WBAN, must perform its function on a non-invasive fashion; so neither user himself is aware that the measurement is being made. An AAL system as we proposed should combine the WBAN properties of wearable devices with the overall WSN.

Table 1. Comparison between WSN Motes

	Atlas	MICA2	TMote Sky	MASS	Smart-Its	Sun SPOT	Imote2
Microcontroller	Atmega128L (8 MHz)	Atmega128L (8 MHz)	TI MSP430 (8 MHz)	Cygnal C8051F125 (1-100 MHz)	PIC 18F6720 (20 MHz)	ARM920T (180 MHz)	Marvell PXA271 XScale® Processor (13 – 416MHz)
Flash Memory	128 KB	128 KB	48 KB	128 KB	128 KB	4 MB	32 MB
Communication Protocols Supported	Ethernet, WiFi, Bluetooth, Zigbee	RF, UART	IEEE 802.15.4	RF	RF	IEEE 802.15.4	IEEE 802.15.4
Programming Languages Supported	Java, .NET (ongoing)	C	C	C	C	Java	NesC, C, .NET
Ad-hoc Network Support	No	Yes	Yes	Yes	Yes	Yes	Yes
Modular	Yes	Partial	No	Yes	Partial	Yes	Yes
RAM Memory	4 - 64 KB	4 KB	10 KB	8.25 KB	8 KB	512 KB	32 MB
Power Supply	2.7 - 5.5V	2.7 - 3.3 V (2X AA batt)	2.1 – 3.6 V	2.7 - 3.6 V	2.0 – 5.5 V	3.7V, 750 mAh lithium-ion batt.	3.2 – 4.5 V (USB or 3xAAA batt)
Total Active Power	TBD (> 89 mW)	89 mW	41 mW	0.08 - 125 mW	TBD (<1W)	2.5 - 125 mW	105 – 330 mW
Sleep Power	TBD (> 75 μW)	75 μW	15 μW	TBD	TBD	150 μW	1950 μW
Peripherals	32 inputs (sensors) Some Outputs (actuators)	3 Leds 8 ADCs DIO I2C SPI	8 ADCs, 2 DACs DIO I2C SPI	8 ADC 2 DACs Temp Sensor I2C, SPI 32 I/O pins	SPI I2C 52 I/O pins 12 ADCs	3 axis acceler., temp. sensor, light sensor, 8 tri-color LEDs, 6 ADCs, 2 switches, 5 I/O pins, 4 current output pins, I2C	I2C, 2xSPI, JTAG, Camera Chip Interface, AC97, GPIOs I2S

2.2 SunSPOT Platform

Sun SPOT (Sun Small Programmable Object Technology) [9] is the WSN platform developed by Sun Microsystems to give support to WSN applications. The Sun SPOT device is a small, wireless, battery powered platform. The hardware platform includes a range of built-in sensors as well as the ability to easily interface to external devices. Respect other WSN solutions, SunSPOT platform provides a better I/O connectivity, and support for I2C digital bus to extend the I/O connectivity. SunSPOT motes support IEEE 802.15.4 standard, which includes physical and link layers. Over these layers, SunSPOT supports AODV as routing protocol and implements CSMA/CD. It allows the configuration of different network topologies by customizing several parameters. The base station (coordinator mote) of this platform does not have sensor board. Therefore actuators manipulations or sensors connections are not allowed in this mote, but processing applications can be deployed over this.

An important feature of Sun SPOT is the programming language selected to develop WSN appliances, evidently Java. A Java Virtual Machine, named Squawk, is built on Sun SPOT devices. It is programmed almost entirely in Java to allow regular programmers to customize the WSN stack and improves the performance of applications. Respect to communication, a JavaME datagram packing is used for manipulating radiofrequency communication frames, which simplifies and homogenizes data manipulation from upper layers.

Respect to security, these motes include highly optimized implementations of RSA [10] and Elliptic Curve Cryptography (ECC) [11] that make light processing and secure data transmissions.

3 TeleCare Ambient Assisted Living Application

Assisted living for elderly, disable or individual with chronic disease require in most cases closely monitoring of their activities (Fig. 1), thus limiting their privacy and at the same time giving an excessive workload for caregivers. Using a network of wireless sensors placed at strategic points in the elderly home, as well as objects of everyday use, caregivers can monitor real-time behavior of the elderly, avoiding tedious tasks and focusing on most important aspects such as improving their quality of life.

Fig. 1. Monitoring Elderly Life

A Telecare AAL system is developed upon WSN platform for the monitoring of physiological parameters in addition to different interpretations of events related to healthcare as falls. In this work we explain in detail a part of the system composed of a smart sensor for control the motorized camera, a wearable device for monitoring physiological measurements of assisted-person, and an emergency system that is notified when a risk situation is taken place.

The developed TAAL system has two possible execution scenarios: (a) centralized monitoring scenario and (b) collaborative monitoring scenario. The first scenario represents the simplest use of WSN. The system depends entirely on a coordinator who is responsible for carrying out the data collection to infer the context, and trigger events or system actions. So, the controller mote is responsible of addressing periodically the different sensing tasks such as the monitoring of health-parameters, the localization of assisted-person and the activation of alarm notifications when a patient's fall is taken place. In the collaborative scenario the nodes are able to infer their own information and also have the ability to manage and route information from other sensor nodes. Our system is capable of producing events by themselves without the need for an unavoidable centralized coordinator. In this case the prevention of a health accident is detected by a router node, being responsible for activating a location-awareness service in other router node. In this way we gain capacity of reactivity, what it is really critical in telecare systems.

3.1 Components of System: System Architecture

We have based our system on Sun SPOT mote which provide a more powerful platform that can process data and transmit more processed information instead of simple metric data. In this case motes can infer the state of the environment being sensitive to the context. For this reason our system can generate reactive events with the processed information, communicating the state to a coordinator or calling other services provided by other SPOTs. For example, if there is a fall event of a person carried with a wearable device, the relay information by the mote to other master mote that it can infer the state of patient's health, may perform different actions. First, it can notify the fall to a central coordinator who will transmit the information over Internet and activate a tracking system through a camera and other devices. The system will locate the patient and focus the scene through a web camera and will take place a call via VoIP to a clinical staff (Fig. 1).

Due the distributed nature of the AAL applications, the architectonic model is based on SOA (Service-Oriented Architecture) approach [12]. This model allows us to decouple the functionality of the AAL system in a set of composable loosely-coupled entities, services, which can be distributed on WSN nodes. Each service follows a two-layer architecture: a logic-service layer for manage the life-cycle of services and its communications with outside, and a sensorization/actuation layer responsible of the connection with I/O ports to sensors and actuators.

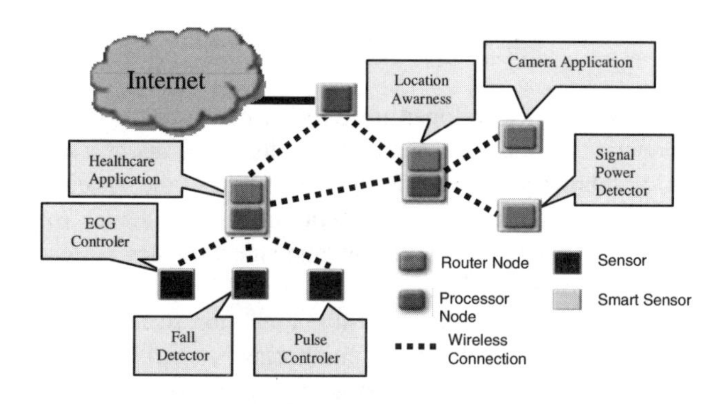

Fig. 2. System architecture with the role assigned to each node

The nodes of TAAL system have assigned roles according to IEEE 802.15.4, EDM for passive *sensing motes*, *RM* for more powerful *processing motes* and *CM* for the *coordination and synchronization* of WSN (Fig. 2). Each node is capable of supporting one or more services, which can be service providers or service consumers depending of the role assigned to them and how they have to interact with other nodes. All the system logic is carried out by a CM node, who also acts as gateway with the outside world via Internet. This coordinator is in charge to drive queries to the responsible nodes from external applications of Internet in order to monitor the health and person's location. Hence it works as a general directory and it is aware of all devices and associated services that compose their network.

The Healthcare node is responsible to have knowledge of the patient state of health. For this reason the healthcare application must arrange the data received from sensors nodes, and the information obtained by the processing of data in order to discover a possible health accident. Moreover, it is able to create an event of localization-awareness if it has detected some health problem directly without need to wait a special signal from a CM node. In this case the healthcare node application acts like a consumer of a location-awareness service, and it also may send an alarm signal in order to activate another services.

The location-awareness node has the responsibility to gather the information referring to settle location as signal power and time delay. With these parameters the node makes calculations about distance and indoor position. So it may send position coordinates to the node which handles the camera movement and other client's node such as CM node and RM node.

3.2 Design and Building of TAAL Components

In this section we brief discuss how the TAAL services are designed and built, specifically the following ones: the fall detector, the camera location-awareness, and the health diagnosis tester. The falls detector creates a distributed vision service for the purpose of remotely monitored to elderly. We have developed an application to monitor falls just as [13] combining the operation of a set of video cameras and an accelerometer inserted on a wearable device. The accelerometer detects possible falls analyzing the changes in accelerations values, while distributed analysis modules in cameras analyze the user's position and location of the head through the comparison of several patterns of image. This information arising from a module of neural networks which makes a final decision depending on what kind of report will be done. Other services module of the TAAL system such as a doctor or an emergency call center can be clients of an emergency VoIP call service in cases where a health disease happens. These services are provided by the coordinator node.

Regarding to services module to monitoring health, we have developed an application for conducting some health parameters with ECG's monitoring as [14], and pulse rate as [15]. But this development use some transducers such as electrodes connected via cable (not really a wireless system at all) using a multiplexer and an amplifier to produce a quantifiable signal. For the development of ECG component we follow a similar power-awareness scheme as [16], which makes portable ECG device in a home care environment combining sending of high data compressed in very short transmission times.

We have tested some ways to measure the ECG with Sun SPOTs. We have tested diverse low-power algorithms which can each individually be wirelessly transmitted onto one ECG service and updated when it is necessary. The ECG monitoring device consists of three main parts: the amplifier, the multiplexing circuit, and the SunSPOT.

We have studied some ways to achieve an efficient position locator services on telecare indoor environments. We are now working with an approach to achieve accurate and reliable position data using methods as hyperbolic positioning using Hybrid TOA/RSS (Time of Arrival) and TDOA/RSS [17].

4 Related Works

There are some related works in the researcher community around the telecare system; many of them use pervasive and ubiquitous nodes integrated in the system. Consequently we will go to contrast our work with several references. The Atlas project from University of Florida [18] suggests a pervasive platform based on services built on its own mote named Atlas. This framework provides an abstraction layer between the physical layer and the services platform by means of a middleware based on SOA architecture. TAAL shares the same architectural vision as Atlas. However, our system is more decentralized than Atlas approach, allowing services collaboration among services placed in different nodes instead of using a central node for handling all services.

Another interesting works is Alarm-Net [19] healthcare system from University of Virginia. In this case the system provides a specific wireless sensor network for Assisted-Living and Residential Monitoring. AlarmNet attends to provide some medical services such as ECG monitor and a service with the purpose to measure the Circadian Activity Rhythm using statistical and predictive algorithm. TAAL shares with Alarm-Net the same goals through a telecare point of view, although ours architectural models is being different, indeed Alarm-Net does not follow a Services-Oriented architecture.

At Harvard University [20] they have developed the plute mote with which have developed a range of medical applications, including pre-hospital and in-hospital emergency care, disaster response, and stroke patient rehabilitation. Interestingly these applications are carried out relaying all the data harvest by plute mote to a centralized processing center. In contrast we have chosen a non ad-hoc hardware platform (SunSPOT) and an architectural model based on services upon a middleware infrastructure, allowing the integration of heterogeneous platforms into the system.

5 Conclusions

On conclusion, we have noticed during the development of our project the importance of using the processing capabilities of smart sensors to define a better architecture whereupon. We improve the ambient intelligence using sensors capable of having reactivity capacity, processing and routing, so that a new coordination of services in ubiquitous spaces may be possible. For that we have developed components to sensorize the patient's environment, such as motorized cameras, detectors location, etc., and also include a wearable unit that allows us to monitor the patient's state of health, and fall situation.

In order to develop the system a wireless networks of sensors based on IEEEE 802.15.4 have been used. Such networks allow data transfer with a low bandwidth, optimal for applications such as AAL monitoring system developed. Although it should be possible to have an interoperable solution among different motes, there are differences over the Mac layer, 802.15.4, as different ways of implementing the network layer, or security frameworks that should be considered. The AAL system developed is built following a SOA approach, which decouples the abstract service level from the WSN stack, allowing adaptations to different WSN stacks.

The processing capacity of motes is a key element that will allow the development of applications for WSN most active within a ubiquitous computing framework, rather than more passive WSN standard in charge of collecting and processing information indirectly into a computer system more powerful, connected to the WSN (coordinator mote).

Finally, the Sun SPOT platform selected for AAL system has demonstrated to be a stable and robust WSN platform, ideal for the development of applications more ubiquitous in which the throughput is shared between the wireless sensor nodes, and thus it may produce more complete and relevant information, making easier the distribution of alarm events. Another advantage is that these devices can be programmed directly in Java, which simplifies the deployment, testing and maintenance of applications. It is remarkable to be aware with the importance that Sun SPOT defines an open standard hardware that can be used for designing new devices based on this mentioned standard.

References

1. Steg, H., Strese, H., Loroff, C., Hull, J., Schmidt, S.: Ambient Assisted Living. Europe Is Facing a Demographic Challenge. Ambient Assisted Living Offers Solutions. European Overview Report, 1–85 (2006)
2. IEEE 802.15 WPAN Task Group 4 (TG4) (Revision dated 2008), `http://www.ieee802.org/15/pub/TG4.html`
3. ZigBee Specification (Revision dated Q4/2007), `http://www.zigbee.org`
4. Hill, J., Horton, M., Kling, R., Krishnamurthy, L.: The Platforms Enabling Wireless Sensor Networks No Journal, 1–6 (2004)
5. Andreu, J., Viudez, J., Holgado, J.: A Survey of Wireless Sensor Networks. In: II Symposium on Software Development, SDS 2008, pp. 271–286 (2008)
6. Stankovic, J.: Wireless Sensor Networks. Handbook of Real-Time and Embedded Systems. Chapman & Hall, Boca Raton (2008)
7. Matheus, K.: Wireless Local Area Networks and Wireless Personal Area Networks (WLANs and WPANs). No Journal, 1–19 (2005)
8. Curtis, S., Biagioni, E., Martha, E., Crosby.: Ad-Hoc Wireless Body Area Network for Augmented Cognition Sensors. In: HCI, pp. 38–46 (2007)
9. SunSPOTWorld, `http://www.sunSPOTworld.com/`
10. Stalling, W.: Cryptography and Network Security: Principles and Practice, 3rd edn. Prentice Hall, Englewood Cliffs (2003)
11. Hankerson, D.R., Vanstone, S.A., Menezes, A.J.: Guide to Elliptic Curve Cryptography. Springer, Heidelberg (2004)
12. Lu, J., Naeem, T., Stav, J.B.: A Distributed Information System for Healthcare Web Services. In: APWeb Workshops, pp. 783–790 (2006)
13. Blasco, R., Casas, R., Marco, Á., Coarasa, V., Garrido, Y., Falcó, J.L.: Fall Detector Based on Neural Networks. BIOSIGNALS 2008, 540–545 (2008)

14. Shen, C., Kao, T., Huang, C., Lee, J.: Wearable Band Using a Fabric-Based Sensor for Exercise ECG Monitoring. In: ISWC, pp. 143–144 (2006)
15. Yamakawa, T., Inoue, T., Harada, M., Tsuneda, A.: Design of a CMOS Heartbeat Spike-Pulse Detection Circuit Integrable in an RFID Tag for Heart Rate Signal Sensing. IEICE Transactions (IEICET) 90-C(6), 1336–1343 (2007)
16. Bumachar, E.M., Andreão, R. V., Pereira, J.G.: A Portable ECG Device in a Home Care Environment Using Burst Transmission. BIODEVICES 2008, 107–110 (2008)
17. Wu, H., Chang, H., You, C., Chu, H., Huang, P.: Modeling and Optimizing Positional Accuracy Based on Hyperbolic Geometry for the Adaptive Radio Interferometric Positioning System. LoCA, 228–244 (2007)
18. King, J., Bose, R., Yang, H., Pickles, S., Helal, A.: Atlas: A Service-Oriented Sensor Platform. In: Proceedings of 31st IEEE Conference on Local Computer Networks, pp. 630–638 (2006)
19. Wood Virone, G., Doan, T., Cao, Q., Selavo, L., Wu, Y., Fang, L., He, Z., Lin, S., Stankovic, J.: ALARM-NET: Wireless Sensor Networks for Assisted-Living and Residential Monitoring. Technical Report CS-2006-13 University of Virginia, 1-14 (2006)
20. Kambourakis, G., Klaoudatou, E., Gritzalis, S.: Securing Medical Sensor Environments: The CodeBlue Framework Case. ARES, 637–643 (2007)

HERMES: A FP7 Funded Project towards Computer-Aided Memory Management Via Intelligent Computations

Jianmin Jiang[1], Arjan Geven[2], and Shaoyan Zhang[1]

[1] University of Bradford, UK
{j.jiang1,s.zhang7}@bradford.ac.uk
[2] Centre for usability research and engineering, Austria
geven@cure.at

Abstract. In this short article, we introduce a new concept in HERMES, the FP7 funded project in Europe, in developing technology innovations towards computer aided memory management via intelligent computation, and help elderly people to overcome their decline in cognitive capabilities. An integrated computer aided memory management system is to be developed from a strong interdisciplinary perspective, which brings together knowledge from gerontology to software and hardware integration. State-of-the-art techniques and algorithms for image, video and speech processing, pattern recognition, semantic summarization and correlation are illustrated, and the general objectives and strategy are described.

Keywords: aging, memory management, semantics, and AI algorithms.

1 Introduction – HERMES' Concept, Services, and Objectives

Aging is often accompanied by various forms of physical and cognitive declines. Cognitive decline in aging brings along reduced capabilities in working memory and information processing, and a reduced ability to encode new memories. Assistive technology in the realm of recording and reminding is very promising in reducing the effects of cognitive decline induced by aging, supporting the older person with day-to-day activities, particularly when combined with a mobile interface. With the increasing amount of recording technology and storage capacity available, research has exploded regarding "life capturing", aimed at recording text, images, and contextual information about places visited and people spoken to. However, the device has to reduce the technology burden placed on the user, instead of making the technology be additional load for the vulnerable memory.

HERMES is an EU FP7 supported project aims to provide the user with useful and usable information at the right place and time, thus supporting people cognitively. HERMES' objective is to use and develop the state-of-art technologies in intelligent audio and visual processing for cognitive care, through advanced, integrated, and assistive technology innovations. Correspondingly, five technical objectives are identified in HERMES, which include: (i) Facilitation of episodic memory through the capture of content in audio and image including additional contextual information,

J.M. Corchado, D.I. Tapia, and J. Bravo (Eds.): UCAMI 2008, ASC 51, pp. 249–253, 2009.
springerlink.com

such as date, time, human emotion, the amount and name of people present, etc; (ii) Cognitive training through games with moments that have been captured previously related to contextual information; (iii) Advanced activity reminding to assist the user's prospective memory in performing independent living; (iv) Conversation support on the grounds of interactive reminiscence based on the recordings of important moments; (v) Mobility support to address the needs of the user outside of the house with cognitive support.

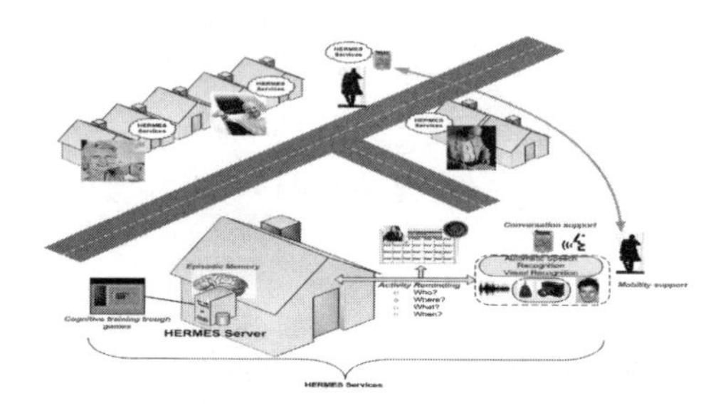

Fig. 1. Illustration of HERMES's concept and services

Fig. 1 illustrates the HERMES concept and services, which show that services provided cover both indoor and outdoor activities. For indoor activities, HERMES provides a server-based system to process all information and host an intelligent computer-aided memory management system to provide elderly with reminders to assist with their daily appointments and other functionalities. For outdoor events, HERMES relies on mobile devices such as PDAs to capture the information and generate multimedia based links for the home-based computer server to provide coordinated and integrated memory management and assistance.

The rest of this paper is organized into two further sections, where Section 2 describes HERMES methodologies and approaches beyond the existing state of art, and Section 3 provides concluding remarks and presents possible future work.

2 HERMES Methodologies and Approaches

A number of ATC (assistive technology for cognition) systems have been developed to address the cognitive disabilities [1, 2]. The arguably most advanced system at the moment is Auto-minder [3], which employs Bayesian inference techniques for scheduling. Its main weakness is that the information about the user's activities is gathered completely manually by tapping, and technologies specifically target older users are limited [4]. Complementary to the ability to enter information manually, HERMES takes a radically different approach and aims at extracting knowledge automatically from the recorded raw materials, including images, videos, speech, and texts. Intelligent processing is to be developed to correlate events, and generate reminders. Fig. 2 illustrates an overview of all technical components inside HERMES.

As seen, HERMES will have a user-friendly interface to bridge the gap between users and the system. Such an interface will provide a range of facilities to enable users to control, influence, and monitoring the system via hierarchical structures and levels. In other words, the interfacing allows the users to manually enter diary information into the system via a simple calendar system by following simple on-screen instructions at basic level. At advanced level, the users can even edit and enter information for metadata records as well as their organization inside the computer-aided memory management system. Such design is essentially to provide a flexible platform enabling users to have the option for the system to work in a manual mode, semi-automatic mode, or fully automatic mode. As a result, the risk level for HERMES can also be reduced to its minimum.

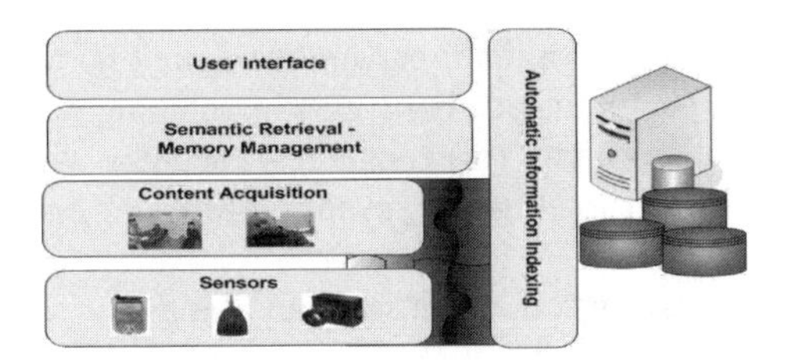

Fig. 2. Overview of HERMES technological components

The next level down the system is the component for semantics retrieval and memory management, where cognitive events and activities are to be organized in terms of metadata that is linked with raw materials. Fig. 3 illustrates the basic structure of the metadata as well as its organization inside HERMES, from which it can be seen that the *ith* metadata essentially have six attributes, providing essential information for cognitive events, paving the way for intelligent computation to process the metadata and establish correlations and associations between events (metadata) and hence generate useful reminders to provide cognitive assistance.

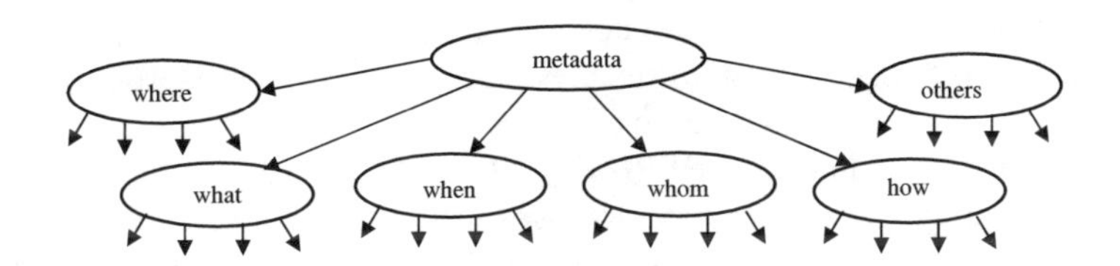

Fig. 3. Illustration of metadata and its organization

Metadata processing not only provides parameters for events correlation and memory association, but also provides a platform for semantics retrieval from raw material to index the database of raw materials for further interpretation and analysis. This is illustrated in Fig. 2 on the right, where server spaces are allocated to hold the database for all information access and processing.

Further down the system as illustrated in Fig. 2 is the content capture module, which includes a range of technology innovation platforms and scenarios. Typical examples are through sensors, cameras, digital recorders, and pre-arranged indoor content capture facilities. All the build-up for content and information capture provides a rich source of information for HERMES consortium to carry out intensive research and innovations across the boundaries of computer vision, image processing, video processing, speech processing, AI, and information systems.

In summary, HERMES methodologies include three stages of innovation and developments: (i) combination of manual and semi-automatic entry of metadata to characterize daily events, meetings and appointments; (ii) automatic metadata extraction via multimedia and sensor-network technologies, which is to be carried out by indoor activity capture and outdoor activity capture. While outdoor activity capture relies on speech recognition technology and manual operation of portable devices (PDA), the indoor activity capture will rely on domain knowledge together with video analysis. This is primarily represented by a multi-camera monitored home, in which a range of domain knowledge can be automatically captured, including face recognition, timing of events, venue of dialogues etc. (iii) reminder generation via intelligent metadata processing through artificial intelligence, event correlation, pattern recognition, and machine learning approaches.

In comparison with existing efforts in relevant communities, HERMES will provide the following innovations beyond the existing state of the art: (i) In terms of Assistive Technology for Cognition, HERMES will provide a holistic approach in which training and support are functionally integrated, including the capture, retrieval, and reminding, where automated reminders through intelligent reasoning remains the key target for HERMES; (ii)In terms of speech recognition technology, HERMES will develop a person adaptive ASR system [5] that learns about its main user, gains experience and improves its proficiency over time. Such a system will have additional strength with combined word-based and phone-based search [6] using n-best lists and confidence information, and thus making the system more robust and accurate in speech recognition and its indexing; (iii) In terms of visual information processing, HERMES will introduce a range of intelligent computation technologies, which can be highlighted by: (1) while existing visual content analysis intelligence is limited to rule based approaches, HERMES will develop pattern recognition and machine learning approaches to extract semantics features, such as case-based-reasoning, artificial intelligence and neural networks to analyse the video content in consistent with human content understanding; (2) HERMES will select a range of video processing tools on competitive basis from partners' previous research for temporal segmentation, semantic object segmentation, to enable video content processing be carried out in terms of consistent video sections and semantic objects rather than video sequences and image regions; (3) Extracting MPEG-7 content descriptors directly in compressed domain [7, 8] without decompression to reduce the computing costand speed up the processing; (iv) In terms of content capture, HERMES's approach is characterized by fusion of multiple algorithms to improve accuracy and deal with the unconstrained nature of HERMES recordings; (v) Finally, HERMES will deliver a computer aided memory management system via integration of all the above technological innovations and modules, where captured multimedia content will be analysed, interpreted, and processed via means of intelligent information processing and computations to generate reminders. As a result, elderly users can overcome their cognitive decline through such reminders, providing a solid technology platform for further research, development and innovations.

3 Conclusions

In this paper, we described HERMES, the FP7 project with a range of technological concepts towards establishment of a computer aided memory management system to assist with elderly people in fighting cognitive declines. HERMES consortium contains industrial companies with their technology development capabilities, academic partners with their research expertise, and user partners with their valuable contributions in terms of validating the project concept, evaluating the innovations, and assessing the overall system. In addition, HERMES also provides an enormous platform with huge potential for further research, which can be identified as: (i) intelligent audio and visual content understanding, interpretation, and recognition like humans; (ii) intelligent computer-aided memory to enhance human strength and coverage in their life management and recordings.

Acknowledgments. The authors wish to acknowledge the financial support from the European Commission for HERMES project under Framework-7 Programme under the contract: 216709, and contributions from all other partners in the consortium.

References

[1] LoPresti, E., Mihailidis, A., Kirsch, N.: Assistive technology for cognitive rehabilitation: State of the art. Neuropsychological Rehabilitation 14, 5–39 (2004)

[2] Pollack, M., Brown, L., et al.: Autominder: an intelligent cognitive orthotic system for people with memory impairment. Robotics and Autonomous Systems 44, 273–282 (2003)

[3] Vemuri, S., Schmandt, C., Bender, W.: Remember: a personal, long-term memory prosthesis. In: CARPE 2006: Proceedings of the 3rd ACM workshop on Continuous archival and retrieval of personal experiences, pp. 65–74 (2006)

[4] Byrne, W., Doermann, D., et al.: Automatic recognition of spontaneous speech for access to multilingual oral history archives. IEEE Transactions on Speech and Audio Processing, Special Issue on Spontaneous Speech Processing 12, 420–435 (2004)

[5] Chelba, Acero, A.: Position specific posterior lattices for indexing speech. In: Proceedings of the 43rd Annual Conference of the Association for Computational Linguistics (ACL 2005), Ann Arbor, MI (2005)

[6] Seide, Yu, P., Ma, C., Chang, E.: Vocabulary-independent search in spontaneous speech. In: Proceedings IEEE ICASSP (2004)

[7] Jiang, J., Weng, Y.: Video Extraction for Fast Content Access to MPEG Compressed Videos. IEEE Transactions on Circuits, Systems and Video Technology 14(5), 595–605 (2004)

[8] Jiang, J., Weng, Y., Li, P.: Dominant colour extraction in compressed domain. Image & Vision Computing Journal 24, 1269–1277 (2006)

Reinforcement Learning of Context Models for a Ubiquitous Personal Assistant

Sofia Zaidenberg[1], Patrick Reignier[1], and James L. Crowley[1]

Laboratoire LIG, 681 rue de la Passerelle - 38402 St-Martin d'Hères, France
{Zaidenberg, Reignier, Crowley}@inrialpes.fr

Summary. Ubiquitous environments may become a reality in a foreseeable future and research is aimed on making them more and more adapted and comfortable for users. Our work consists on applying reinforcement learning techniques in order to adapt services provided by a ubiquitous assistant to the user. The learning produces a context model, associating actions to perceived situations of the user. Associations are based on feedback given by the user as a reaction to the behavior of the assistant. Our method brings a solution to some of the problems encountered when applying reinforcement learning to systems where the user is in the loop. For instance, the behavior of the system is completely incoherent at the be-ginning and needs time to converge. The user does not accept to wait that long to train the system. The user's habits may change over time and the assistant needs to integrate these changes quickly. We study methods to accelerate the reinforced learning process.

1 Introduction

New technologies bring a multiplicity of new possibilities for users to work with computers. Not only are spaces growingly equipped with computers or notebooks, but more and more users carry mobile devices (smart phones, PDAs, etc.). Ubiquitous computing takes advantage of this observation. Its aim is to create smart environments where devices are dynamically linked and provide new services and new human-machine interaction possibilities. *The most profound technologies are those that disappear. They weave themselves into the fabric of everyday life until they are indistinguishable from it* [18]. This network of devices must perceive the context in order to understand and anticipate the user's needs. Devices should be able to execute actions helping the user to fulfill his goal or simply accommodating him. Actions depend on the user's situation. Possible situations and the associated actions reflect the user's work habits. Therefore, they should be specified by the user him-self. However, this is a complex and fastidious task.

The objective of this work is to construct automatically a context model by applying reinforcement learning techniques. Rewards are given by the user when expressing his satisfaction of the system's actions. A default context model assures a consistent initial behavior. This model is then adapted to each particular user in a way that maximizes the user's satisfaction.

J.M. Corchado, D.I. Tapia, and J. Bravo (Eds.): UCAMI 2008, ASC 51, pp. 254–264, 2009.
springerlink.com

In the remainder of this paper, we present our research problem and objectives before evaluating the state of the art. Afterwards, we explain the accomplished and future work. Finally, we present our first results.

2 Research Problem

A major difficulty when applying RL (Reinforcement Learning) techniques to real world problems is their slow convergence. We need to accelerate the learning and to obtain results with few examples. The user will not give patiently rewards while the system is exploring the state and action space. In addition, user rewards maybe inconsistent or not given all the time (Sect. 5.1).

Furthermore, we deal with a very large state space and it is necessary to reduce it. For this purpose we need to generalize our states at first, and then apply techniques to split states when it is relevant (Sect. 5.1).

Lastly, working in a ubiquitous environment adds difficulties. Detecting the next state after an action is not obvious because another event may occur meanwhile. The environment is non-stationary because it includes the user.

In this research context, our goal is to create a ubiquitous personal assistant. Devices of our ubiquitous environment and mobile devices provide information about the user's context [4]. Knowing this, we offer relevant, *context-aware* services to the user. Examples are task migration or forwarding a reminder when the user is away. Most of current work on pervasive computing pre-defines services and fires them in the correct situation [17, 12]. Our assistant starts with this pre-defined set of actions and adapts it progressively to its particular user. The default behavior makes the system ready-to-use and the learning is a life-long process. At first, the assistant is only acceptable but with time it gives more and more satisfying results.

3 State of the Art

Our work relates to two primary areas. 1) *Context-aware applications* which use context to provide relevant information and services to users. 2) *Reinforcement learning* where an agent learns to behave from feedback on its actions.

Context is recognized a key concept for ubiquitous applications [7, 1, 4]. Dey defines context as *any information that can be used to characterize the situation of an entity, where an entity can be a person, place, or physical or computational object.* In [4], a context is represented by a network of situations. A situation is a configuration of entities, roles and relations. Entities play roles and relations are semantic predicate functions on entities. An example of ambient systems is the Gaia Operating System [13] which manages the resources and services of an active space. First-order logic is used to model context and define rules, but no learning components are included. Christensen states that it is not easy to build context-aware systems because *the gap between what technology can "understand" as context and how people understand context is significant* [3]. He believes that it might be an error to build completely autonomous systems removing humans from the loop. Our learning depends on user rewards. He can specify initial preferences, get explanations of automatic actions and we keep the option to ask him questions when necessary.

Personal learning agents were studied in particular by [14] on providing context-specific assistance while optimizing interruption. Schiaffino builds user interaction profiles using association rules, learned and incrementally updated with new experience. Our goal is for the assistant to work without needing an initial amount of experience. Additionally, rules provide a behavior only for observed experience, not new situations.

RL was applied to interface agents for instance by [10, 11], where RL is completed by memory-based learning. Their agent assists users in scheduling group meetings and sorting email. A similar project [6] uses different machine learning techniques such as neural networks. Our constraint is to keep the model understandable. We believe explaining the works of the assistant to the user is fundamental to gain his trust. Neural networks do not meet this requirement. Furthermore, we took a greater interest in accelerating the learning process. We were inspired by indirect RL techniques first introduced by [15] and implemented for instance by [5].

4 Reinforcement Learning

Reinforcement learning is a computational approach to learning whereby an agent tries to maximize the total amount of reward it receives when interacting with a complex, uncertain environment [16]. A learning agent is modeled as a Markov decision process defined by $\langle \mathcal{S}, \mathcal{A}, \mathcal{R}, \mathcal{P} \rangle$. $\mathcal{S}$ and $\mathcal{A}$ are finite sets of states and actions; $\mathcal{R} : \mathcal{S} \times \mathcal{A} \rightarrow \mathbb{R}$ is the immediate reward function and $\mathcal{P} : \mathcal{S} \times \mathcal{A} \times \mathcal{S} \rightarrow [0,\ 1]$ is the stochastic Markovian transition function.The agent constructs an optimal Markovian policy π: $\mathcal{S} \rightarrow \mathcal{A}$ that maximizes the expected sum of future discounted rewards over an infinite horizon. We define $Q^{\pi}(s,\ a)$, the value of taking action a in state s under a policy. The Q-learning algorithm allows computing an approximation of Q^*, independently of the policy being followed, if $\mathcal{R}$ and $\mathcal{P}$ are known.

4.1 Indirect Reinforcement Learning

In our case, the transition ($\mathcal{P}$) and reward ($\mathcal{R}$) functions are unknown. Indirect RL techniques enable the learning of these functions by trial-and-error and the computation of a policy by planning methods. This approach, described in [15], is implemented by the DYNA architecture. The DYNA-Q algorithm (Fig. 1) is an instantiation of DYNA using Q-Learning to approximate V^*.

In steps 2a-2c, the agent interacts with the world by following an ϵ-*greedy* exploration [16] based on its current knowledge. Step 2e is the supervised learning of $\mathcal{P}$ and $\mathcal{R}$. Step 2f is the planning phase where $\mathcal{P}$ and $\mathcal{R}$ are exploited to update the Q-table that is used for interaction with the real world.

Input: $\emptyset$, Output: π

1. Initialize $Q(s,\ a)$ and $\mathcal{P}$ arbitrarily.
2. At each step:
   a) $s \leftarrow$ current state (non terminal).
   b) $a \leftarrow \epsilon\text{-}greedy(s,\ Q))$.
   c) Send the action a to the world and observe the resultant next state s' and reward r.
   d) Apply a reinforcement learning method to the experience $\langle s,\ s',\ a,\ r\rangle$: $Q(s,a) \leftarrow Q(s,a) + \alpha(r + \gamma \max_{a'} Q(s',a') - Q(s,a))$
   e) Update the world model $\mathcal{P}$ and $\mathcal{R}$ based on the experience $\langle s,\ s',\ a,\ r\rangle$.
   f) Repeat the following steps k times:
      i. $s \leftarrow$ a hypothetical state that has already been observed.
      ii. $a \leftarrow$ a hypothetical action that has already been taken in state s.
      iii. Send s and a to the world model, obtain predictions of next state s' and reward r: $s' \leftarrow max_{s' \in S}\mathcal{P}(s'|s,\ a)$, $r \leftarrow \mathcal{R}(s,\ a)$
      iv. Apply a reinforcement learning method to the hypothetical experience $\langle s,\ s',\ a,\ r\rangle$: $Q(s,a) \leftarrow Q(s,a) + \alpha(r + \gamma \max_{a'} Q(s',a') - Q(s,a))$

Fig. 1. The DYNA-Q algorithm

This algorithm accelerates the convergence of Q-values by repeating real examples virtually. Examples are better and quicker integrated into the Q-table, providing a satisfactory behavior faster.

5 The Ubiquitous Assistant

Figure 2 sums up the works of the assistant. When interacting with the user, the assistant uses its current policy to choose actions. It also gathers data about the environment to update the world model. Then it uses the current world model to learn a new policy through offline Q-learning.

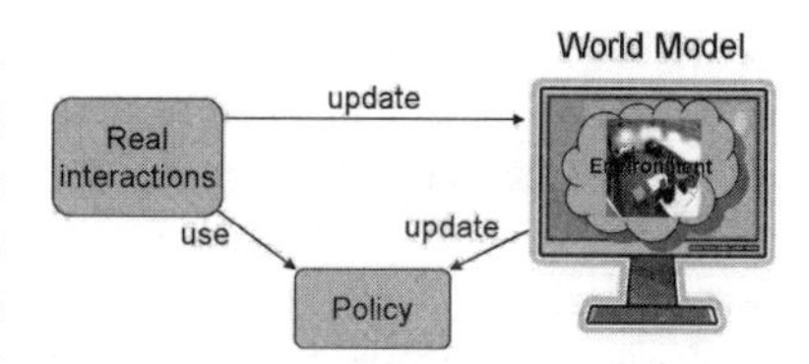

Fig. 2. Overview of the assistant

5.1 Application of Reinforcement Learning and DYNA-Q

The Components of the Reinforcement Learning Agent

The State Space. Our assistant must be able to provide explanations to the end user. State representation must not be a black box. Therefore, we use predicates. Each predicate represents a relevant part of the environment. Predicates are defined with arguments. A state is a particular assignment of argument values, which may be null. These predicates are described below.
alarm(title, hour, minute) A reminder fired by the user's agenda.
xActivity(machine, isActive) The activity of the X server of a machine.
inOffice(user, office) Indicates the office that a user is in, if known, null otherwise.

absent(user) States that a user is currently absent from his office.
hasUnreadMail(from, to, subject, body) The latest new email received by the user.
entrance(isAlone, friendlyName, btAddress) Expresses that a bluetooth device just entered the user's office. *isAlone* tells if the user was alone or not before the event.
exit(isAlone, friendlyName, btAddress) Someone just left the user's office.
task(taskName) The task that the user is currently working on.
user(login), userOffice(office, login), userMachine(machine, login) The main user of the assistant, his office and main personal computer (not meant to be modified).
computerState(machine, isScreenLocked, isMusicPaused) Describes the state of the user's computer regarding the screen saver and the music.

Each predicate is endowed with a timestamp accounting for the number of steps since the last value change. Among other things, this is used to maintain integrity of states, *e.g.* the predicate `alarm` can keep a value only for one step and only one of `inOffice` and `absent` can have non-null values.

Our states contain free values. Therefore, our state space is very large; the Q-table would be too large. This exact information is not always relevant to choose an action. The user might wish for the music to stop when anyone enters the office, but to be informed of emails only from his boss. As soon as we observe the state "Mr. Smith entered the office", we have an estimated behavior for the state "someone entered the office", which is more satisfying for the user. We generalize states in the Q-table by replacing values with wildcards: "`<+>`" means any value but "`<null>`" and "`<*>`" means any value.

The Action Space. Possible actions are: display a written message or read it, send an email, lock or unlock the screen of a computer, pause or unpause the music. "Do nothing" is an action as well.

Reward. Since the user is the target of the assistant's services, the user is the one to give rewards to the assistant. But, as pointed out by [9], user rewards are often inconsistent and can drift over time. He will not always give a reward and when he does, it may concern not only the last immediate action, but the last few actions.

We gather reward from explicit and implicit sources. For instance, if we inform the user of a new email and he views the message, then he probably was satisfied with the action. However, such implicit reward is numerically rather weak.

Model of the Environment

Applying the DYNA-Q algorithm (Fig. 1) requires modeling the environment, through the transition and reward functions $\mathcal{P}$ and $\mathcal{R}$.

The transition model. We use common sense to initialize $\mathcal{P}$, and at regular time intervals we apply supervised learning on examples to complete the model. To do so, during interactions with the user we register the previous state, the last taken action and the current state which is thus the next state of the tuple.

The transition model is a set of transformations from a state to the next, given an action. A transformation is composed of starting predicates, an action, modified predicates and a probability of being applied. Starting predicates define required values, possibly using wildcards. The transformation can be applied to every state matching them, when the given action has just been taken. If several transformations match a state, one is chosen randomly based on their probabilities. The next state is a copy of the previous state on which we apply the given modifications. A modification operates on an argument and can be to erase the value, set a given value, set the value of another predicate's argument or reset the timestamp.

The reward model. Likewise, we define initial obvious rules and learn $\mathcal{R}$ using examples observed during real interaction. The reward model is a list of triplets $\langle s, a, r\rangle$, the reward r earned when taking action a in state s. s can be defined with wildcards so an entry can be used when the states match and with action a.

Supervised learning of the models. The two algorithms are given Fig. 3. It makes sense to run these algorithms rather often at first and to space out the runs as the models are complete enough, *i.e.* when new transformations are rarely created because then the model has already seen most of the environment. From the assistant's point of view the user is part of the environment thus this model is non stationary and we can not stop updating it.

In the second algorithm (Fig. 3), entries are added with exact states, without generalizing values since we can not know which ones are important. We can apply an offline treatment to possibly merge entries that express the same piece of information (see below). Furthermore, we need to define a merging function *mix* that translates the weight of the new example against the previous value. Currently $mix(r, r_e) = 0.7r_e + 0.3r$. This choice needs to be validated empirically, or changed.

Input: A set of examples $\{s, a, s'\}$, Output: $\mathcal{P}$

- For each example $\{s, a, s'\}$ do
  - If a transformation t that obtains s' from s with the action a, can be found, then
    - Increase the probability of t.
  - Else
    - Create a transformation starting with s, having the action a and ending in s', with a low probability.
    - Decrease the probability of any other transformation t' that matches the starting state s and the action a but whose ending state is different from s'.
  - End if.
- Done.

Input: A set of examples $\{s, a, r\}$, Output: $\mathcal{R}$

- For each example $\{s, a, r\}$ do
  - If an entry $e = \{s_e, a_e, r_e\}$ such as s matches s_e and $a = a_e$, can be found, then
    - Update e, set $r_e = mix(r, r_e)$, where *mix* is a merging function.
  - Else
    - Add a new entry $e = \{s, a, r\}$ to the reward model.
  - End if.
- Done.

Fig. 3. The supervised learning of the transition (left) and reward (right) models

---

Input: Initial transition and reward models, Output: the user's context model.

1. Run an episode (algorithm Fig. 5).
2. At each step i:
   a) Receive the new state s_i.
   b) Store the example to the database: $\{s_{i-1}, a_{i-1}, s_i\}$.
   c) Choose an action using the current policy $a_i = \pi(s_i)$.
   d) Display to the user s_i and a_i.
   e) If the user gives a reward then store it to the database: $\{s_i, a_i, r_i\}$.
   f) If i is a multiple of n then
      i. Run the supervised learning of the transition model (algorithm Fig. 3).
      ii. Run the supervised learning of the reward model (algorithm Fig. 3).
3. In parallel, at regular time intervals, run an episode (algorithm Fig. 5).

Fig. 4. The global learning algorithm of the RL agent

Global Learning Algorithm

At this point we have defined all the elements of our learning algorithm, let us formulate their interweaving (Fig. 4). At the beginning of the assistant's life, the only knowledge of the RL agent is the initially predefined environment model. Firstly, the RL agent performs several episodes, with random initial states, to initialize the Q-table. This provides a consistent initial behavior.

Events are sent to the RL agent as a state change. We add this as an example for the transition model. The RL agent uses its current policy to choose an action and sends it to the assistant, in charge of executing it. These state and action are displayed to the user (in a nondisruptive manner) and his reward, if he gives one, is stored for the reward model. We choose not to perform a step of Q-learning (Fig. 1/2d) here in order to modify the behavior not too frequently and avoid surprising the user.

The supervised learning of the models is performed every n steps, when enough new experience has been acquired. The planning step (Fig. 1/2f) consists in running an episode of RL and is shown Fig. 5. At the beginning, we should perform frequent, short episodes in order to quickly integrate everything that happens into the Q-table. Later on, the assistant can run longer episodes less often, for instance once a day. An episode consists of executing k steps of RL. At each step, a state change leads to the choice of an action and to the update of a Q-value, using the transition and reward models. A state change is triggered by an event, which we generate. We can only replay previously seen events (DYNA-Q) or we can generate random events. The first option makes the most of past experience while the second emphasizes exploration. It is a means to have an estimate for a Q-value even if the situation never happened yet. This makes sense when the transition and reward models are somehow complete. Both methods, plus a mixture of them (starting with the first and as the models evolve, add progressively the second) need testing.

Input: $\mathcal{P}$, $\mathcal{R}$, Output: π

1. Repeat the following steps k times:
   a) Choose a state s.
   b) Choose an action $a = \pi(s)$.
   c) Send s and a to the world model and obtain predictions of next state s' and reward r: $s' \leftarrow max_{s' \in S}\mathcal{P}(s'|s, a)$, $r \leftarrow \mathcal{R}(s, a)$
   d) Apply a reinforcement learning method to the hypothetical experience $\langle s, s', a, r\rangle$: $Q(s,a) \leftarrow Q(s,a) + \alpha(r + \gamma \max_{a'} Q(s',a') - Q(s,a))$

Fig. 5. An episode of Q-learning used for planning by the RL agent

Split and Merge

As mentioned above, our state space is extremely large and we need to reduce it by generalizing states. We replace actual values by wildcards.

Then, we reveal cases where actual values matter. This way we can make the assistant inform the user of an email from his boss but not from "newsletter@nytimes.com". We intend to accomplish this through an offline treatment inspired by [2]. The idea is to detect conflicting rewards given by the user for similar events, corresponding to a merge of states. We split these states and learn different Q-values for each of them.

Finally, new entries of the reward model (algorithm Fig. 3) are added with exact states, without generalization. The offline treatment would reveal similar entries with similar reward values for merging. In the history of given rewards, it would pick out tuples with different rewards used to update one entry and split the entry.

Further Improvements

The transition model will be split into two: one model for the next state after an action (the current model) and one for the next state after an event. If the last event is a consequence of the agent's actions, it saves an example of the old state, *action* and next state: $\{s, a, s'\}$. If not, it saves an example of the old state, *event* and next state: $\{s, e, s'\}$. This example is used to learn the event transition model, the same way we learn the action transition model (Fig. 3). This is a better way of computing the next state of an event.

6 Preliminary Results

We implemented modules to interact with the environment using a framework well adapted for ubiquitous systems [19], based on a combination of the middleware OMiSCID [8] and OSGi (`www.osgi.org`). But to begin, we use an experimental platform to test our algorithms. The world, including the user, is simulated. The simulator plays input scenarios by sending events as sensors would do. It answers commands as effectors would do. It behaves exactly like the real environment. The aim of the test is to bring the system to a desired state. For evaluation we measure 1) the distance to the expected state (the number of transitions in the graph defined by the transition model). 2) How fast we learned this behavior. Deterministic scenarios facilitate experimentations: we can replay them with variations of our algorithm. Much more tests need to be done; this is the very first result.

For this test we used a very simple scenario in three events: 0. "Sofia is in the office" – 1. "Sofia leaves" – 2. "Sofia enters". The 9 possible actions concern the screen saver and the music: they are combinations of (nothingAboutMusic || pause || unpause) and (nothingAboutScreen || lock || unlock).

The desired behavior is (lock && pause) when the user leaves, and (unlock && unpause) when the user enters. The reward when this happens is 75, it is -50 when both are wrong and -25 when one is wrong. We start with a good environment model. At the beginning, the Q-table is empty and we run 100 episodes of 10 iterations where we choose randomly initial states and events from the database. Further experiments will use random initial states and events both from the database and random to cover more of the environment. We skip an event when the transition model does not know the next state to avoid modifying a Q-value when not necessary. Later on we run an episode every minute.

We observed the Q-values and which one the best action in the states corresponding to our test scenario was. After ∼195 episodes, the desired behavior was reached: as episodes went on, Q-values corrected themselves and the best ones corresponded to the actions we wanted the system to execute. This proves that we need to run a lot of episodes focusing on exploration. We remind the reader that episodes are the off-line learning phase. The user is not at all involved in this experiment. His input is only needed to build the environment models, which are learned by a supervised algorithm, which does not need a huge amount of examples.

The convergence of the Q-values is encouraging: Fig. 6 shows the mean update value of the Q-table during Q-learning steps. At each step, several Q-values are modified in order to spread a change. The Y value is the mean amount of differences between old and new Q-values. As it is clear Fig. 6, this difference decreases drastically. This can be used to stop an episode prematurely because the Q-table has converged. We are fully aware that this experiment is far from complete.

It is only the beginning. The next tests will start with minimal models and learn them online. Episodes will mix saved and random events and start in random states. Of course, we will perform tests with the real environment. Everything is in place, we need only a precise validation method.

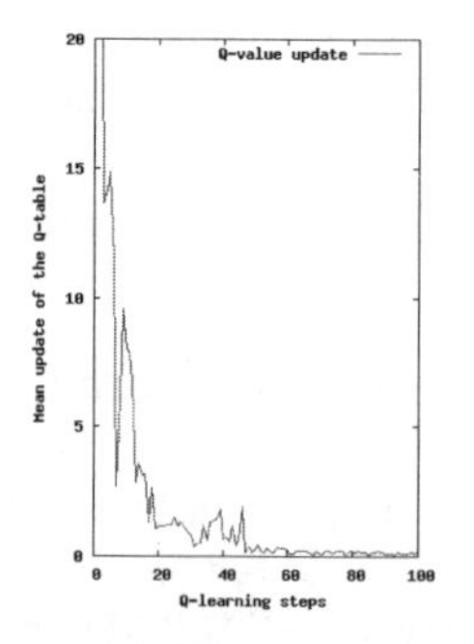

Fig. 6. The mean Q-value update/step

7 Conclusion and Expected Outcome

The aim of this research is to investigate the learning of a context model in the frame of ubiquitous environments. A context model defines the observable situations and what actions should be executed in each situation in order to provide a useful service to the user. We achieve this goal by applying a RL algorithm and then following the resulting policy. We use techniques to initialize and accelerate the learning process in order to bother the user with as few undesirable actions as possible. To do so, firstly we use common sense to build an initial behavior. Secondly we perform offline virtual learning steps to simulate real interaction. The system learns quicker, with potentially inconsistent and retroactive rewards. Our assistant is deployed into a ubiquitous environment equipped with video cameras, bluetooth sensors, microphones, speakers and mobile devices. These devices gather information about the user's context and activity, and provide him services. To complete this work we need to implement the split and merge algorithms, perform evaluations, compare techniques and carry out tests with real users.

References

1. Bardram, J.E.: The Java Context Awareness Framework (JCAF) - A Service Infrastructure and Programming Framework for Context-Aware Applications. In: Pervasive Computing (2005)
2. Brdiczka, O., Reignier, P., Crowley, J.L.: Automatic development of an abstract context model for an intelligent environment. In: PerCom (2005)
3. Christensen, J., Sussman, J., Levy, S., Bennett, W.E., Wolf, T.V., Kellogg, W.A.: Too much information. ACM Queue (2006)
4. Crowley, J.L., Coutaz, J., Rey, G., Reigner, P.: Perceptual components for context awareness. In: International conference on ubiquitous computing (2002)
5. Degris, T., Sigaud, O., Wuillemin, P.-H.: Learning the structure of factored markov decision processes in reinforcement learning problems. In: ICML (2006)
6. Dent, L., Boticario, J., Mitchell, T., Sabowski, D., McDermott, J.: A personal learning apprentice. In: AAAI (1992)
7. Dey, A.K., Abowd, G.D.: The context toolkit: Aiding the development of context-aware applications. In: SEWPC (2000)
8. Emonet, R., Vaufreydaz, D., Reignier, P., Letessier, J.: O3MiSCID: an Object Oriented Opensource Middleware for Service Connection, Introspection and Discovery. In: SIPE 2006 (2006)
9. Isbell, C., Shelton, C.R., Kearns, M., Singh, S., Stone, P.: A social reinforcement learning agent. In: AGENTS (2001)
10. Kozierok, R., Maes, P.: A learning interface agent for scheduling meetings. In: IUI (1993)
11. Maes, P.: Agents that reduce work and information overload. ACM, New York (1994)
12. Ricquebourg, V., Menga, D., Durand, D., Marhic, B., Delahoche, L., Log, C.: The smart home concept: our immediate future. In: ICELIE (2006)
13. Roman, M., Hess, C.K., Cerqueira, R., Ranganathan, A., Campbell, R.H., Nahrstedt, K.: Gaia: A middleware infrastructure to enable active spaces. In: IEEE Pervasive Computing (2002)

14. Schiaffino, S., Amandi, A.: Polite personal agent. Intelligent Systems (2006)
15. Sutton, R.S.: Integrated architectures for learning, planning, and reacting based on approximating dynamic programming. In: ICML (1990)
16. Sutton, R.S., Barto, A.G.: Reinforcement Learning: An Introduction (1998)
17. Vallée, M., Ramparany, F., Vercouter, L.: Dynamic service composition in ambient intelligence environments: a multi-agent approach. In: YRSOC (2005)
18. Weiser, M.: The computer for the 21st century. Scientific American (1991)
19. Zaidenberg, S., Reignier, P., Crowley, J.L.: An architecture for ubiquitous applications. In: IWUC (2007)

An Approach to Dynamic Knowledge Extension and Semantic Reasoning in Highly-Mutable Environments

Aitor Almeida[1], Diego López-de-Ipiña[2], Unai Aguilera[1], Iker Larizgoitia[1], Xabier Laiseca[1], Pablo Orduña[1], and Ander Barbier[2]

[1] Tecnológico Fundación Deusto
{aalmeida,uaguiler,ilarizgo,xlaiseca, porduna}@tecnologico.deusto.es
[2] Universidad de Deusto. Avda. de las Universidades, 24 - 48007, Bilbao, Spain
{dipina,barbier}@eside.deusto.es

Abstract. AmI environments are dynamic. They change rapidly and continuously due to the appearance and disappearance of devices, people and changes in their situation. These changes need to be reflected in the context information which is collected and maintained by the Ambient Intelligence applications. In this work we present a semantic infrastructure whose context information can be dynamically enriched and extended by the dynamically discovered objects in the environment, and which enables to reason over it.

1 Introduction

The management of environment and user context is one of the key aspects of Ambient Intelligence (AmI). Without context knowledge, an intelligent environment cannot adapt itself to user needs. This explains why management of context should be one of the main activities in any Ambient Intelligence system. The Smartlab project[1] has produced a semantic middleware platform that aims to facilitate this task while making explicit the previously hidden context. Semantic technologies are used to represent context and to infer from it reactions to be taken. The use of a context ontology allows Smartlab to integrate information coming from heterogeneous sources and to share it easily.

The most remarkable feature of the SmartLab middleware is its ability to dynamically extent the context ontology with new context-modelling concepts and rules provided from non-previously available devices within an environment. Dynamic discovery and installation of previously non-available semantic devices is automatically performed by this middleware. Furthermore, its context ontology is enriched taking into account the semantic knowledge and behavioural information in the form of rules provided by these newly discovered devices, which were not previously considered in the design phase.

The following sections describe the context managing and semantic reasoning mechanisms used by this middleware. In section 2 previous work on semantic context management is discussed. Section 3 explains the system architecture and the dynamic

[1] http://www.tecnologico.deusto.es/projects/smartlab

J.M. Corchado, D.I. Tapia, and J. Bravo (Eds.): UCAMI 2008, ASC 51, pp. 265–273, 2009.
springerlink.com

context enrichment mechanisms it offers. Section 4 describes the Smartlab Context Ontology created to model the context. In section 5 the semantic reasoning capabilities of this middleware over the context are explained. Finally in section 6 conclusions are given and possible future work is suggested.

2 Related Work

In recent years several projects [1][2][3] have created ontologies that model the context. SOUPA [4] is a set of ontologies oriented to ubiquitous and pervasive applications used by the COBRA project[1]. It is composed by two sub-sets SOUPA Core and SOUPA Extensions. The SOUPA Core defines the elements that are present in any ubiquitous application while SOUPA Extensions gives support to more specific applications. CONON [5] is used by the SOCAM [2] project to model the context of pervasive computing applications. It is also divided in two sets, one with the general information shared between all the applications and the other one domain specific. CODONT [6] is used by the CODAMOS [3] project and its main aim is to create an adaptable and flexible infrastructure for AmI applications. These three ontologies have in common some similar elements (see Table 1) like information about location, time, people/users, actions/activities and devices. There are also some elements unique for each system, like the information about the environmental conditions, events, policies and services.

Table 1. Context modeling ontologies

	SOUPA	CONON	CODONT
Similar elements	Person, Agent, BDI, Policy, Event, Action, Time, Space	Location, Person, Activity, Computing, Entity	User, Service Platform, Location, Time, Environmental condition

While all these ontologies are easily extensible, this process takes place offline. In Smartlab this extension of the ontology is done dynamically, on the fly, when a new device is discovered in the environment. The device is able to add new concepts to the ontology creating new classes that are able to model new context information. The consistency of this new concepts is checked automatically before adding them to the ontology. Moreover new domain rules are also added dynamically, adapting the system reactions to the new devices. This mechanism enables the system to be more flexible, adaptable and future-proof.

Another distinguishing feature of Smartlab is the decoupling of the devices from the context managing service. This is achieved with the use of OSGi [7] and an event-based architecture. Events generated in the ontology are translated to OSGi events and propagated to the devices. Other projects have also used OSGi to develop their context-aware middleware platform. Gu et al.[2] do not use the OSGi event service, making our services more decoupled from the infrastructure. Helal et al. [14] do not model the context semantically, not taking advantage from the capabilities for an easy extension that the ontologies offer.

3 The Smartlab Architecture

The system has a multilayer architecture composed by four layers: Sensing and Actuation Layer, Service Abstraction Layer, Semantic Context and Service Management Layer and Programming, Management and Interaction Layer.

1. **Sensing and Actuation Layer:** This layer consists on the different devices that exist in the system environment. Representatives of these devices exist on Layer 3. Currently a wide range of devices have been implemented: EIB devices, IP Cameras, VoIP devices, SMS gateways, sensor netwoks, Indoor Location Systems and some custom-make devices like sensorized chairs and wristbands with displays and wireless capabilities.
2. **Service Abstraction Layer:** This layer encapsulates the devices functionality into OSGi bundles. Devices are discovered dynamically by the third layer and their information added to the ontology.
3. **Semantic Context and Service Management Layer:** In this layer resides all the context management and reasoning functionality of the system. It is divided in three modules:
   - Continuous and Dynamic Environment Service Management Module: This module constantly monitors the environment searching for new devices and manages addition of these new devices to the system.
   - Semantic Context Management Module: This module manages the context information stored in the ontology and provides an inference mechanism over it. When a new device is discovered all its metadata is stored in the ontology and new concepts are added if they did not exists previously.
   - Web Gateway Module: Offers a Web gateway that allows managing the devices of the system.
4. **Management and Interaction Layer:** In this final layer reside the applications that take advantage of the registered services and the ontology. One example application is the Environment Controller, which uses the Web Gateway Module to offer a Web Page where all devices in the system can be controlled.

3.1 The Dynamic Ontology Enrichment in Smartlab

Dynamic Ontology enrichment allows to seamlessly incorporate into the system not previously considered devices, knowledge and behavior. This facilitates the building of a more adaptative system able to react to changes in the environment. To enable this every device service in Smartlab must implement the ISmartlabService interface. This interface exposes the methods that allows the ContextManager to query the bundle about the its metadata, the new ontology concepts and the new domain rules related to that device. The interface methods are the following:

- *getOntology*: returns the new ontology concepts in RDF/OWL to be added. When new concepts are added to the ontology the consistency of the resulting ontology is checked automatically by the Semantic Context Manager. If the new ontology is inconsistent the new additions are discarded, the service is uninstalled and its rules are ignored.

- *getIndividual*: returns the service instance with the device metadata to be added to the ontology in RDF/OWL. As with concepts the consistency of the new individuals is also checked before adding them to the ontology.
- *getRules*: services can also add new rules to the system to enrich its behavior.
- *getEventsToRegister*: returns an String array with all the events the service is interested in.
- *startUpdatingContext*: Once all the concepts, individuals and rules have been added to the system the service can start updating the ontology with its measures. Measure consistency is also checked when it is inserted in the ontology.
- *stopUpdatingContext*: This method is called to stop the updating of the context.
- *getInterface*: returns the bundle interface name.

Currently Smartlab supports two types of devices: physical devices and logical devices. The concepts and rules provided by the physical devices (lights, windows, doors…) are not domain specific. These devices are not required to know what other devices and rules exist in the knowledge base. Domain specific concept and rules are supplied by logical devices (for example a MeetingManager) that are familiar with the other existing devices and can model the domain behavior

4 The Smartlab Context Ontology

The context in Smartlab is stored in an ontology using RDF/OWL. There are several advantages in this approach:

- The knowledge is modelled using standard languages (XML/RDF[8]/OWL[9]) and knowledge can be easily shared between applications.
- Knowledge can be easily extended adding new entities. Smartlab take advantage of this feature to dynamically enrich the ontology with new concepts.
- The semantic relations of the ontology allow inferring new context making explicit the implicit context.
- Several ontologies can be united providing they share some common concepts.

The ontology is based in the systems previously discussed in the Related Work section. There are four key elements in those ontologies: who, what, when and where. For this reason the main entities in the ontology are the space, time, events and actors (devices and people):

- *TimeItem:* This class models the time and chronological relations in the ontology. Time can be expressed in two ways, as an instant or as a time period composed by two instants. Chronological relations express how different TimeItems relate to each other: before, during, after…
- *SpatialItem:* This class models the location and spatial relations in the ontology. SpatialItems can be points, two dimensional areas, rooms or buildings. Spatial relations express how SpatialItems and LocableItems relate to each other: subsumedBy, locatedIn, contains, limits, isAdjacent …
- *LocableItem:* This class models the actors in the ontology. All the actors in the Smartlab Context Ontology have a spatial location. LocableItem have three

subclases: DeviceItems which models the devices, PersonItem which models the people and ValueItem which models the measures taken by the devices. Each measure is taken in a given time and has a location associated.

- *Event:* An event is a reaction to a change in the context. These events are created by the domain rules of the system and are translated to OSGi events. The device bundles can subscribe to these events using OSGi Event Admin [10].

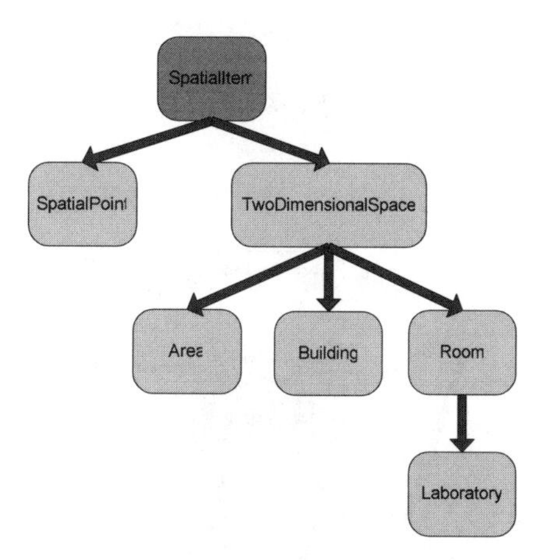

Fig. 1. SpatialItem hierarchy

These four elements are the basic building blocks of the knowledge stored in the ontology. Based on these elements the new devices can add domain specific knowledge to the ontology.

5 Semantic Context Manager

This module located in Layer 3 of the system architecture is the element in charge of managing the knowledge of the system. It can be subdivided in three other sub-modules (see Figure 2):

- *Context Manager:* The measures taken by the devices do not have per se any semantic meaning. This sub-module "semantizes" this data giving meaning to the measures and inserting them into the ontology. An example of this would be the location data given by the Indoor Location System. This system provides the X and Y coordinates of a LocableItem. The Context Manager interprets this data adding semantic information, identifying the object to which those values belong and their reference point.
- *Context Reasoner:* This module infers new context information based on the measures provided by the devices. This is done making explicit the implicit context and using domain specific rules. Continuing with the Location System example the Context Reasoner can infer that the LocableItem is inside a room, collocated with other LocableItems.

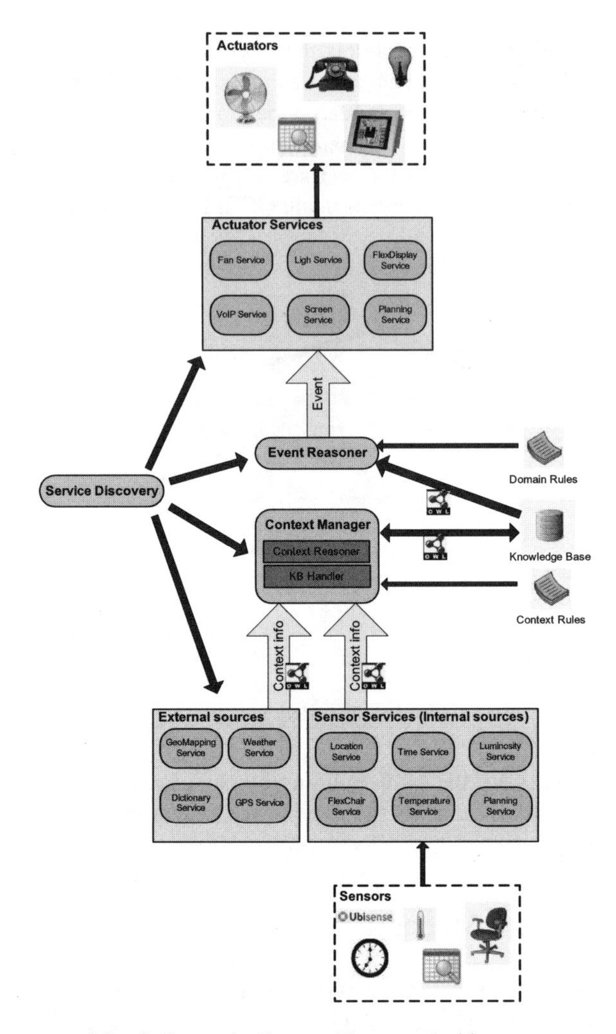

Fig. 2. Semantic Context Manager Architecture

- *Event Reasoner:* This module periodically checks via SPARQL [11] if a new event has been generated in the ontology responding to a change in the context. These Event entities are translated to non-semantic OSGi events and propagated using the OSGi Event Manager.

Typical context reasoning in Smartlab follows these steps:

1. A device provides new measures that are analyzed to check their consistency and added to the ontology by the Context Manager.
2. This knowledge is processed by the Context Reasoner and new knowledge is generated by the Knowledge Eliciting rules (see Section 5.1).
3. All the knowledge is processed using the Domain Rules (see Section 5.2) and events are created in the ontology.

4. The Event Reasoner discovers those events in the ontology and extracts them from the ontology. These events are translated to OSGi Events and propagated by the OSGi Event Manager.
5. The devices subscribed to those events receive them and act accordingly.

Two types of rules can be found in the system, the knowledge eliciting rules and the domain rules (see Fig 2).

5.1 Knowledge Eliciting Rules

In this category are placed the rules that expand the ontology making explicit the hidden implicit knowledge in the ontology. There are two subcategories of knowledge eliciting rules: the semantic rules and the heuristic rules. The semantic rules implement partially the RDF Model Theory [12] and the OWL Model Theory [13]:

```
OWL-InverseOf:
       (?x ?prop1 ?y),
       (?prop1 owl:inverseOf ?prop2)
       ->
       (?y ?prop2 ?x)
]
```

The heuristic rules infer new spatial and temporal knowledge. One example would be to infer the relation between two time instants, reasoning which one take place before:

```
[TEMPORAL1-eventInstantBefore:
       (?ev1 rdf:type smart:Event),
       (?ev1 smart:time ?ins1),
       (?ins1 smart:value ?v1),
       (?ev2 rdf:type smart:Event),
       (?ev2 smart:time ?ins2),
       (?ins2 smart:value ?v2),
       lessThan(?v1,?v2)
       ->
       (?ev1 smart:before ?ev2)
]
```

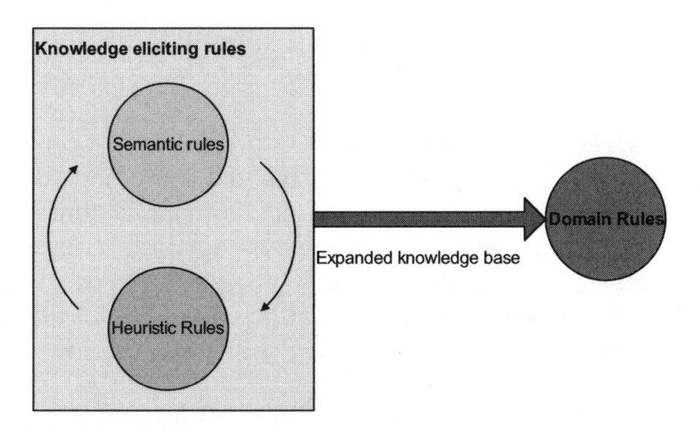

Fig. 3. System rules

5.2 Domain Rules

Domain rules define the behavior of the system, modeling the reactions to the changes of the context. These rules are domain specific and change from application to application. We have defined a prototype for an intelligent laboratory. “A new bundle is added to the system, the “MeetingBundle”. This bundle extends the knowledge of the system adding new domain rules to manage the meetings in the laboratory. Unai, Pablo and Aitor are sited in the sensorized chairs in a meeting area. The system infers that a meeting is in progress if three or more people are seated in a meeting area or if four or more people are present. A MeetingEvent is generated and the MeetingEvent receives it. Automatically the presentation planned for the meeting is displayed in the proyector in the room. More domain rules are fired buy the MeetingEvent. The lights on the meeting area are dimmed so the presentation can be seen correctly. Using the location data and thanks to the information stored in the ontology an SMS is send to the people not present in the meeting area alerting them about the ongoing meeting. The people in the area receive an alert in their SmartDisplays (a bracelet with an integrated display and wireless capabilities) telling them to turn of their mobile phone”. An example of a domain rule would be:

```
[EVENT-SendMeetingSMS:
        (?meetingEvent rdf:type smart:LocableMeetingEvent),
        (?meetingEvent smart:meetingAreaName ?nameArea),
        (?p1 smart:isLocatedIn ?area2),
        (?area2 smart:name ?name2),
        notEqual(?nameArea, ?name2),
        (?p1 smart:mobile ?mobile),
        makeTemp(?smsEvent)
        ->
        (?smsEvent rdf:type smart:SMSEvent),
        (?smsEvent smart:areaName ?nameArea),
        (?smsEvent smart:mobile ?mobile),
        (?smsEvent smart:smsText 'You are late'),
]
```

A SMS is send to the people not present in the area alerting them about the meeting:

6 Conclusions and Future Work

This work has presented various mechanisms to store and extend context and use it to infer reactions. Environments are highly mutable and can change drastically over a short period of time explaining why any attempt to model a complex environment must tackle this issue. The solution proposed in this paper is to dynamically enrich the ontology that models the environment and the system rules to adapt on the fly to all this changes. Using this mechanism the applications react dynamically to the changes on the environment brought forward whenever a previously unknown device is added to the system.

Further work will try to develop a mechanism to identify patterns in user’s behaviour and the changes in the context that take place when those patterns occur to create rules dynamically and to predict future behaviour. These actions/reactions will be used to create on the fly OWL-S workflows. To do this the capabilities of the services will be semantized to enable the use of automatic service composition and so dynamically create complex responses to context changes.

Acknowledgments. Thanks to the Industry, Commerce and Tourism Department of Basque Government for sponsoring this work through grant S-PE06FD02 of the SAIOTEK 2006 program.

References

1. Chen, H.: An Intelligent Broker Architecture for Pervasive Context-Aware Systems. PhD thesis, University of Maryland, Baltimore County (2004)
2. Gu, T., Pung, H.K., Zhang, D.Q.: Toward an OSGi-Based Infrastructure for Context-Aware Applications. In: Pervasive Computing (2004)
3. The CoDAMoS Project: Context-Driven Adaptation of Mobile Services, `http://www.cs.kuleuven.ac.be/distrinet/projects/CoDAMoS/`
4. Chen, H., Perich, F., Finin, T., Joshi, A.: SOUPA: Standard Ontology for Ubiquitous and Pervasive Applications. In: Proceedings of the First Annual International Conference on Mobile and Ubiquitous Systems: Networking and Services (Mobiquitous 2004), Boston, MA, August 22-26 (2004)
5. Wang, X.H., Zhang, D.Q., Gu, T., Pung, H.K.: Ontology Based Context Modeling and Reasoning using OWL. In: Proceedings of the Second IEEE Annual Conference on Pervasive Computing and Communications Workshops (2004)
6. Preuveneers, D., Van den Bergh, J., Wagelaar, D., Georges, A., Rigole, P., Clerckx, T., Berbers, Y., Coninx, K., Jonckers, V., De Bosschere, K.: Towards an Extensible Context Ontology for Ambient Intelligence. In: Proceedings of Ambient Intelligence: Second European Symposium, EUSAI 2004 (2004)
7. OSGi Alliance, OSGi Alliance Home Site (April 2008), `http://www.osgi.org/Main/HomePage`
8. RDF, Resource Description Framework (April 2008), `http://www.w3.org/RDF/`
9. OWL Web Ontology Language Overview (April 2008), `http://www.w3.org/TR/owl-features/`
10. OSGi Event Admin (April 2008), `http://www2.osgi.org/javadoc/r4/org/osgi/service/event/EventAdmin.html`
11. SPARQL Query Language for RDF (April 2008), `http://www.w3.org/TR/rdf-sparql-query/`
12. RDF Model Theory (April 2008), `http://www.w3.org/TR/2001/WD-rdf-mt-0010925/`
13. OWL Model Theory (April 2008), `http://www.w3.org/TR/2002/WD-owl-semantics-20021108/`
14. Helal, S., Mann, W., El-Zabadani, H., King, J., Kaddoura, Y., Jansen, E.: The Gator Tech Smart House: A Programmable Pervasive Space. IEEE Computer Magazine, 50–60 (March 2005)

Learning Accurate Temporal Relations from User Actions in Intelligent Environments

Asier Aztiria[1], Juan C. Augusto[2], Alberto Izaguirre[1], and Diane Cook[3]

[1] University of Mondragon, Mondragon, Spain
{aaztiria;aizaguirre}@eps.mondragon.edu
[2] University of Ulster, Jordanstown, United Kigndom
jc.augusto@ulster.ac.uk
[3] Washington State University, Pullman, Washington, U.S.A
cook@eecs.wsu.edu

Summary. Ambient Intelligence environments depend on their capability to learn user's preferences and typical behavior. In this paper we present an algorithm that taking as starting point information collected by sensors finds out accurate temporal relations among actions carried out by the user.

Keywords: Ambient Intelligence, Context Aware Computing, Learning behavioral patterns, Temporal relations.

1 Introduction

Ambient Intelligence (AmI) [2] [7] [14] can be understood as 'a digital environment that proactively, but sensibly, supports people in their daily lives' [3]. Such systems can improve the life of users in many ways: for example, by making an environment safer, more comfortable and more energy efficient. These environments should achieve such goals without creating any extra burden to the users so as to maximize users' acceptance. In order to accomplish aims like adjusting the temperature automatically, turning off the lights, or issuing an alarm when it detects an unsafe situation, the environment will need information about user preferences and habits (i.e., their 'normal' behaviour).

This paper makes a contribution to the problem of discovering patterns of activity. The problem is inherent to any intelligent environment. Here we will focus on a Smart Home scenario but the results can be easily extrapolated to similar environments. The next section summarizes previous related work. Section 3 explains the nature of the collected data and how patterns are represented. Then in Section 4 we explain how those patterns are obtained using the algorithm $\mathcal{A}_{PUBS}$ which is part of PUBS (Patterns of User Behaviour System). Section 5 shows the results of the validation experiments. Section 6 explains our plans for current and future work in this system and finally we provide our conclusions in Section 7.

J.M. Corchado, D.I. Tapia, and J. Bravo (Eds.): UCAMI 2008, ASC 51, pp. 274–283, 2009.
springerlink.com

2 Related Work

Learning is an essential feature in any AmI system. However, given the diversity of elements in AmI systems, learning has not been devoted as much attention in the literature as it may require. Even so, some notable exceptions have been. Artificial Neural Networks were the first technique used to infer rules for smart homes, and a survey of those works can be found in [6]. A first attempt was made in the MavHome project to predict the next smart home inhabitant action using pattern discovery and Markov model techniques [8]. Jakkula and Cook [11] extend this work to predict actions using temporal relations, defined by means of Allen's temporal logic relations [1].

Other techniques, such as Fuzzy-Logic in iDorm [10], Case-Based Reasoning in MyCampus [13] or Decision Trees in SmartOffice [9] have been used. Taking into account the characteristics of each problem we can state that each problem favours the use of a certain technique, but as Muller pointed out [12] 'the overall dilemma remains: there does not seem to be a system that learns quickly, is highly accurate, is nearly domain independent, does this from few examples with literally no bias, and delivers a user model that is understandable and contains breaking news about the user's characteristics'.

3 Data Collected, Required Patterns and Their Formalization

Our approach aims at obtaining user patterns from sensor data. Different sensors provide different types of information; therefore the learning process has to consider each accordingly. Three main different groups of sensors are considered (we are aware there are more types of sensors, e.g., alarm pendants, RFID, etc. but these are the relevant ones for the examples listed in this paper).

- (type O) Sensors installed in objects (devices, furniture, domestic appliances, etc). They provide direct information about user actions. For example, a sensor installed in the bedroom lamp may indicate when that lamp was switched on and off.
- (type C) Context sensors. These sensors provide information about context and not about user actions directly. Temperature, light and smoke sensors are examples of type C sensors.
- (type M) Motion sensors. These sensors can be used to infer where the user is (in the bedroom, outside the house, etc.).

Let us consider the event sequence illustrated in Figure 1. As the figure shows, sensors installed in the objects provide clues about users' actions that can be used to define patterns such as the one defined in (1):

'Bedroom Lamp is turned on 5 seconds after Motion Bedroom is turned on If and When bLight is <10' (1)

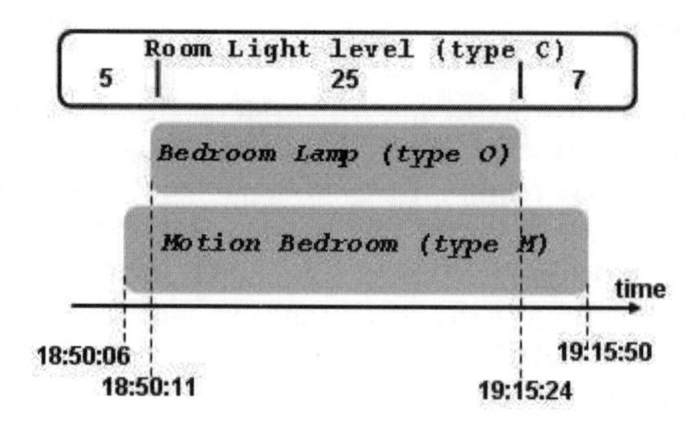

Fig. 1. Event sequence and context information

In order to use a clear and non ambiguous representation which can facilitate future automation, patterns will be described as ECA (Event-Condition-Action) rules [4]. They will capture situations where an Action that is detected through an activated sensor, called main sensor triggering (*mainSeT*) (e.g., Bedroom Lamp), is related (in terms of time, e.g., 5 seconds) to an Event involving another activated sensor, called the associated sensor triggering (*associatedSeT*) (e.g., MotionBedroom), if some Conditions (e.g., Bedroom light <10) are true. For more details see [5].

4 Learning Temporal Patterns with $\mathcal{A}_{PUBS}$

The process to identify patterns in data collected by sensors is summarized in the following algorithm:

$\mathcal{A}_{PUBS}$ Algorithm (for learning patterns)

---

for each sensor of type O (consider it as *mainSeT*)
    Identify the *associatedSeT* of type O or M (See Section 4.2.1)
    for each *associatedSeT*
        Identify possible time relations (See Section 4.2.2)
        if there exists a time relation **then** make it more accurate using context information, i.e., by using C-type (See Section 4.2.3)

Notice the emphasis on sensors type O as *mainSeT* given they are those more closely linked with user's explicit and intended actions.

4.1 Identifying Associated Sensor Triggering

The aim of this first step is to get a list of possible related sensors (associated) in order to minimize the complexity of the learning process. For the purpose of discovering possible *associatedSeT*, we search for previous events of other sensors that happened before each event related to the *mainSeT*. If $\mathcal{A}_{PUBS}$ discovers that before an event instance of *mainSeT* there are frequent occurrences of an event of another sensor triggering, then the later will be considered as *associatedSeT*. Finding an *associatedSeT* does not mean definitively there will be a pattern that describes a relation *associatedSeT* - *mainSeT*, but indicates there could potentially be one.

A list of possible *associatedSeT* is obtained with a similar approach to the Apriori method for mining association rules with only two differences.

- Limit possible associations to the object we are analyzing (*mainSeT*).
- The result does not consider a pair (*mainSeT*, *associatedSeT*) as a sequence, but only as sensors that can be potentially related in a meaningful way.

A modified Apriori algorithm (adding the aforementioned constraints) has been used in this step. As in every association mining process, minimum coverage, support and window size values must be provided. Let us consider the *mainSeT* 'Bedroom Lamp' again; the result of this first step will be a list of *associatedSeT*s.

$$\textit{BedroomLamp} => [\textit{MotionReception, MotionBedroom, LuxoLamp}]$$
$$\textit{where } n < \textit{TotalSensorNumber}$$

4.2 Identifying Time Relations

Once we know what other actions triggering sensors could be related to actions triggering *mainSeT*, the next step is to discover if there are possible meaningful relations. Those relations could be either quantitative (2) or qualitative (3).

'BedroomLamp is turned on 5 sec. after MotionBedroom is turned on' (2)

'BedroomLamp is turned on after MotionBedroom is turned on' (3)

Quantitative relations include a specific measurement (e.g., 5 seconds) whereas qualitative relations emphasize the relative occurrence of the actions/events (e.g., after). Numerical relations carry extra valuable information with regards to qualitative ones which are useful for automation. For example, knowing the user likes to listen to the radio news at a particular time or to have the temperature of the house within 20°C-23°C at a particular month of the year. Qualitative relations like 'after dinner and before going to bed' are not constrained enough to schedule a favorite TV program. Our work has focused on numerical relations, as we know other reports in the literature had covered qualitative relations.

Grouping Instances

The starting point for discovering numerical time relations will be to generate a table listing time distances between occurrences of *mainSeT* events and previous appearances of *associatedSeT*. Consider the scenario in Figure 2 where the relations between Bedroom Lamp and *associatedSeT*s are depicted.

Once the temporal relations are collected, the next step is to find out if there is any interesting time distance that is repeated frequently enough (see subsection 'extracting patterns' for more details). In order to discover these patterns, groups are made taking into account similarities among them. In our

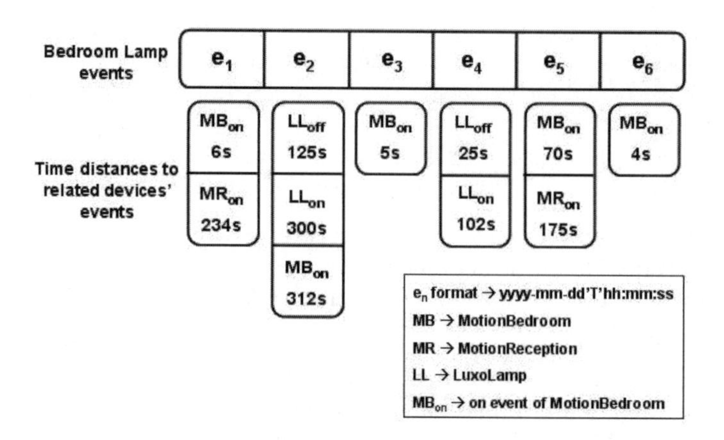

Fig. 2. Time distances between *mainSeT* and *associatedSeT*

case, considering the *associatedSeT* MBon (MotionBedroom's 'on'), the distances would be ({e1,6s} {e2,312s} {e3,5s} {e4,-} {e5,70s} {e6,4s}) and groups among (6,312,5,70,4) must be done. The technique to make groups could be as complex as we can imagine and different techniques could be suggested to accomplish this task. In this case the technique used is based on joining values that are within a range established by:

$$[min, max] = \overline{x} \pm (\overline{x} * tolerance) \qquad where \qquad \overline{x} = \frac{\sum_{i=1}^{n} a_i}{n} \tag{1}$$

with: tolerance = tolerated deviation from $\overline{x}$ (%); a_i = time distance of a element; and n = number of elements

If a value does not fulfill the requirements to join any group, a new group is created using that value as the group mean. Every time a new value is added to a group the mean value of that group is recalculated. Considering the possible relation between Bedroom Lamp and MBon, the process of making groups will be (considering 50% as tolerance percentage in all cases).

(e_1,6s); There is no group, create(group0, $\overline{x}$ (6s), [3,9])

$$\text{Tolerance} = 6 \pm (6 * 0.5)$$

(e_2,312s)$\neq$[3,9], create(group1, $\overline{x}$ (312s), [156,468])

$$\text{Tolerance} = 312 \pm (312 * 0.5)$$

(e_3,5s)=[3,9], join(group0, $\overline{x}$ (5.5s), [2.75,8.25])

$$\text{Tolerance} = 5.5 \pm (5.5 * 0.5)$$

(e_4,70s)$\neq$[3,9] and 70$\neq$[156,468] create(group2, $\overline{x}$ (70s), [35,105])

$$\text{Tolerance} = 70 \pm (70 * 0.5)$$

$(e_6$,4s)=[2.75,8.25], join(group0, $\overline{x}$ (5s), [2.5,7.5])

$$\text{Tolerance} = 5 \pm (5 * 0.5)$$

Extracting Patterns

Once groups are made, the following step is to evaluate what information groups do and do not reveal about a pattern. As in the first step (identify *associatedSeT*) we consider interesting patterns to be those that cover more instances than established by the minimum confidence level. If the minimum confidence is, say, 25% the only group considered as interesting in our case would be group 0, which covers 3 instances out of 6, creating a pattern:

'BedroomLamp is turned on 5s after MotionBedroom is turned on' (4)

4.3 Identifying Appropriate Conditions of Pattern Occurrence

Every pattern has a confidence level that indicates how many *mainSeT* instances are covered by the pattern. The pattern (4) has a confidence level of 50% because it covers the instances e_1, e_3 and e_6 out of a total of 6. It means that only in 3, out of 6 times, the Bedroom Lamp was switched on when there was a motion detection at bedroom 5 seconds before. Having such a confidence level (rarely a confidence in this point will be close to 100%) does not mean that a pattern is not useful, but it means that it is not as accurate as it would be desirable. Finding out (if possible) under what conditions a pattern appears or not will be the last step to increase accuracy in patterns detection.

The possible conditions are given by: (a) Time specifications (e.g., such as: time of day, day of week, season, etc.) and (b) Context sensors (type C sensors such as those to measure temperature, light, humidity, etc.).

Adding Conditions

In order to discover the conditions, two tables, *covered* and *non-covered* tables, are generated. In the *covered* table there will be instances classified well by the pattern together with the context information collected when they happened. The *non-covered* table will contain instances where the pattern fails. Dividing both tables, using the information they contain helps to provide more accurate patterns. Some example of these conditions are 'when time is between 18hs and 20hs', 'when day of the week is Saturday', 'when temperature is less than 25°C' or 'when light level is less than 10'.

Let us consider the tables in Figure 3 show the context information collected when instances of our example happened. Separating both tables will allow us to know when our pattern defines properly the relation between Bedroom Lamp and Motion Bedroom. In this case the easiest way to separate *covered* and *non-covered* tables seems to be by using the sensor bLight that indicates the light level in the bedroom when action happened. Adding that condition, our pattern will be:

non-covered

	e_2	e_4	e_5
time of day	17:30:28	16:05:37	16:17:22
day of week	monday	tuesday	wednesday
bTemp Sensor	26	27	24
bLight Sensor	12	15	13

covered

	e_1	e_3	e_6
time of day	18:50:12	17:15:30	19:05:10
day of week	monday	tuesday	thursday
bTemp Sensor	25	26	22
bLight Sensor	5	6	5

Fig. 3. *Covered* and *non-covered* tables

'BedroomLamp is turned on 5s after MotionBedroom is turned on When BedroomLight level is <10' (5)

Adding these conditions do not increase the instances the pattern can classify well, (still classifyies the same instances, 3/6), however we make sure that it does not include instances that do not have that pattern.

Patterns achieved up to now explain when and under what conditions instances of *mainSeT* happen. That will allow us to understand the user's behavior, but this is insufficient if the objective is to automate a house. For example, we may know that every time the bedroom lamp was turned on and the bedroom light level was less than 10 units, there was a motion sensor triggering in the bedroom 5s before. But this does not inform us about the converse relation, i.e. we cannot make sure that every time there was a motion sensor triggering in the bedroom and the bedroom light level was less than 10 units, 5s after that there was a bedroom lamp sensor activation. Therefore, it is necessary to analyze the pattern from the opposite point of view and find out the conditions to be able to automate. The way of getting these conditions is the same as before. The only difference is that to generate the tables, *associatedSeT* instances are considered instead of *mainSeT* instances and an *associatedSeT* instance is considered as covered if there is an instance of *mainSeT* just as the pattern indicates.

Classification Technique

In order to separate the tables and get the conditions, we have used a modified JRip algorithm (JRip is a rule learning algorithm provided by Weka) [15]. The unique modification done in the algorithm has been made to get always conditions related to the *covered* table, i.e., JRip algorithm provides rules without taking into account if they are related to one class or another, its unique aim is to separate classes. Because in our case separation gives us the conditions under which the pattern is useful, it is desirable that the classification technique always provides rules about the *covered* class.

5 Results

$\mathcal{A}_{PUBS}$ has been validated by applying it to artificial data generated by the authors and then to a real dataset collected from MavPad, a smart apartment that was created to test algorithms as part of the MavHome project [16]. The dataset we used was collected in three different time periods: Trial1 (spanning 15 days), Trial2 (spanning almost 2 months) and Trial3 (spanning 3 months). The types of sensors installed are: 26 sensors on objects such as lamps, lights or outlets, 53 context sensors such as light, temperature or humidity, and 37 motion sensors distributed in all the rooms.

Figure 4 shows the number of patterns discovered by $\mathcal{A}_{PUBS}$ in different trials, considering different minimum confidence levels (25%, 50%, 75% and 100%). A 25% confidence level means that, at least, a quarter of the instances must be covered by the pattern to consider it as interesting. On the other hand, the number of accurate patterns indicates the number of patterns out of total where it has been possible to define conditions to know when they can be used.

Some conclusions can be extracted from these results. First of all, it seems clear that it is almost impossible to define patterns associated to a specific object based on only one relation (in fact, there is no pattern with 100% confidence level) so the importance of defining conditions is clear. Moreover figures show in most of the patterns (92 out of 121) it has been possible to define conditions of occurrence in order to obtain patterns with relatively high levels of accuracy. The system can be run successively with different requests of accuracy to find the optimal balance in between a high accuracy in the detection and achieving patterns with minimum level of confidence. Table 1 shows the runtime needed by $\mathcal{A}_{PUBS}$ in each trial.

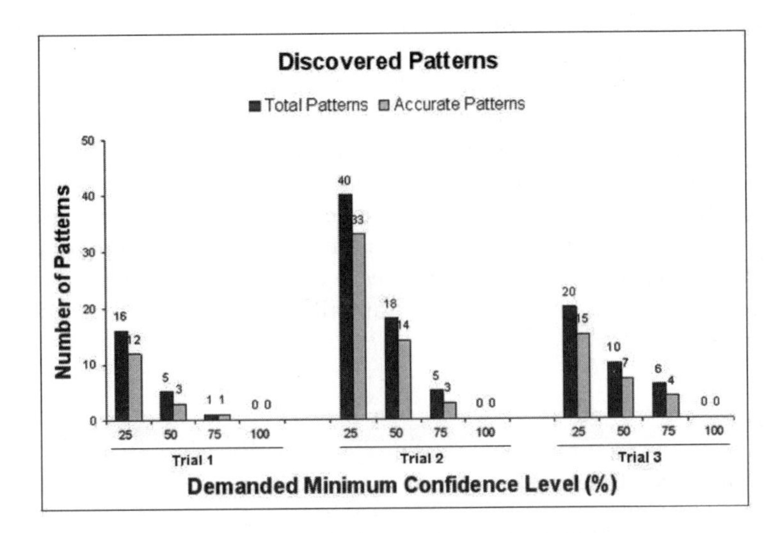

Fig. 4. Number of patterns and accurate patterns obtained in different trials considering different confidence levels

Table 1. Experiments' runtime considering different confidence levels

Confidence level (%)	Trial 1	Trial 2	Trial 3
25	45.7s	336.2s	137.5s
50	30.1s	227.4s	101.7s
75	26.8s	194.2s	88.3s

6 Future Work

Once a methodology to learn patterns is clear, the way of getting the objectives of each part can vary and different techniques or approaches can be suggested. Our short-term efforts will be aimed to discover patterns with qualitative relations, getting complex patterns where more than two activated sensors would be related and experimenting with some other datasets. Further future work will also include the possibility of incorporating user preferences and user feedback which will provide a useful heuristic to achieve much more personalized and user-adapted patterns.

7 Conclusions

Learning in AmI systems is a task that must be carried out as unobtrusively as possible, for example, by considering the information collected from sensors as the sole (or at least primary) source of information for user's actions. Discovering patterns on how objects are used, the places visited and the time associated with those events is informative on user's behaviour and preferences. This in turns helps an environment to accomplish the task of providing a service tailored to the inhabitants of the environment.

A three step algorithm, $\mathcal{A}_{PUBS}$, to find out temporal relations has been described, where the first two steps are focused on finding out relations among activated sensors and defining what type of relation these are, and the third step attempts to define when or in what conditions they must be used.

Results obtained after validating $\mathcal{A}_{PUBS}$ on artificially generated dataset and datasets collected by colleagues in a Smart Home indicate the importance of including conditions as part of a pattern specification and also showed the efficacy of $\mathcal{A}_{PUBS}$ to find such patterns in substantially large datasets with acceptable performance.

References

1. Allen, J.: Towards a general theory of action and time. Artificial Intelligence 23, 123–154 (1984)
2. Augusto, J.C.: Ambient Intelligence: the Confluence of Ubiquitous/Pervasive Computing and Artificial Intelligence. In: Intelligent Computing Everywhere, pp. 213–234. Springer, London (2007)

3. Augusto, J.C., Cook, D.J.: Ambient intelligence: applications in society and opportunities for ai. In: 20th International Joint Conference on Artificial Intelligence (IJCAI 2007) (2007)
4. Augusto, J.C., Nugent, C.D.: The use of temporal reasoning and management of complex events in smart homes (2004)
5. Aztiria, A., Augusto, J.C., Izaguirre, A.: Spatial and temporal aspects for pattern representation and discovery in intelligent environments. In: Workshop on Spatial and Temporal Reasoning at 18th European Conference on Artificial Intelligence (ECAI 2008) (to published, 2008)
6. Begg, R., Hassan, R.: Artificial neural networks in smart homes. In: Designing Smart Homes. The Role of Artificial Intelligence, pp. 146–164. Springer, Heidelberg (2006)
7. Cook, D.J., Das, S.K.: Smart Environments: Technology, Protocols and Applications. Wiley-Interscience, Chichester (2005)
8. Cook, D.J., Huber, M., Gopalratnam, K., Youngblood, M.: Learning to control a smart home environment. In: Innovative Applications of Artificial Intelligence (2003)
9. Le Gal, C., Martin, J., Lux, A., Crowley, J.L.: Smartoffice: Design of an intelligent environment. IEEE Intelligent Systems 16(4), 60–66 (2001)
10. Hagras, H., Callaghan, V., Colley, M., Clarke, G., Pounds-Cornish, A., Duman, H.: Creating an ambient-intelligence environment using embedded agents. IEEE Intelligent Systems 19(6), 12–20 (2004)
11. Jakkula, V.R., Cook, D.J.: Using temporal relations in smart environment data for activity prediction. In: Proceedings of the 24th International Conference on Machine Learning (2007)
12. Muller, M.E.: Can user models be learned at all? inherent problems in machine learning for user modelling. In: Knowledge Engineering Review, vol. 19, pp. 61–88. Cambridge University Press, Cambridge (2004)
13. Sadeh, N.M., Gandom, F.L., Kwon, O.B.: Ambient intelligence: The mycampus experience. Technical Report CMU-ISRI-05-123, ISRI (2005)
14. Weiser, M.: The computer for the 21st century. Scientific American 265(3), 94–104 (1991)
15. Witten, I.H., Frank, E.: Data Mining: Practical Machine Learning Tools and Techniques, 2nd edn. Elsevier, Amsterdam (2005)
16. Youngblood, G.M., Cook, D.J., Holder, L.B.: Managing adaptive versatile environments. In: IEEE International Conference on Pervasive Computing and Communications (2005)

Advanced Position Based Services to Improve Accessibility

Celia Gutierrez

Departamento Ingeniería de Software e Inteligencia Artificial
Facultad de Informática
Universidad Complutense de Madrid
Ciudad Universitaria, 28040 Madrid, Spain
cegutier@fdi.ucm.es

Abstract. Accessibility is a growing interesting area for research as elderly or disabled population is increasing in developed countries. INREDIS is a project to make research in emergent technologies to provide solutions to different disabilities in several environments. Objectives of the project have been achieved by the analysis of the state of art of the current technologies. At this stage, context-aware technologies will play an important role. Position services are part of these technologies, so a full description of it, their applications in several contexts and the accessibility solutions they provide are explained in this paper. Special interest is focused on positional commerce.

Keywords: accessibility, INREDIS, state of the art, context-aware, positional services, positional commerce.

1 Introduction

Technology research has not taken into consideration end users until recently. The reasons why there is a growing interest in making technologies accessible are the increasing elderly and disabled population and the social and political concern to give equal opportunities to them. This population is estimated to be 10% out of the total in the developing countries, and it is estimated to be larger in the following years as the age expectation is getting higher.

INREDIS[1] (**IN**terfaces for **Re**lationships between environment and **DIS**abled people) is a research project to create new channels to improve accessibility by the use and

[1] Inredis Project, approved in July 2007, is a CENIT project (National Strategic Consortium of Technical Research) which belongs to the INGENIO 2010 Spanish government initiative and is managed by the CDTI (Industrial Technological Development Center) and financed in part by e-laCaixa. The aim of this initiative is to increase the technological transfer between research organizations and enterprises within the R+D+I frame. Project duration estimation is from 2007 to 2010. Leadership project is undertaken by Technosite, which is the Fundosa Group's technological enterprise, depended by ONCE Fundation. Technosite performs the project integral management and the relationship management between INREDIS Consortium with CDTI and other external agents. Technological surveillance leadership during the life of the project is undertaken by e-laCaixa.

J.M. Corchado, D.I. Tapia, and J. Bravo (Eds.): UCAMI 2008, ASC 51, pp. 284–292, 2009.
springerlink.com

interaction of emergent Information and Communication Technologies [1].The starting point of the project is the analysis of the state of art and its application in accessibility context in ten environments: mobile communications, communication media, tele-assistance and personal surveillance, domestic environment, urban environment, banking, product and service purchasing, education, work, and transversal technologies.

Although there are some initiatives related to accessibility in shape of consortiums, conferences, funded programs, etc., like eInclusion@eu by the European Commission [2] and WAI - Web Accessibility Initiative [3], there are few published works that make an exhaustive analysis of the applications of technologies for accessible purposes. The difficulties of the disabled collectives to enter the Information Society - the Internet, mobile and TV services - focusing on the audiovisual services and the technologies and adaptations to ensure this inclusion are analyzed in [4]. Stephanidis and Savidis [5] give an example of universal access application, showing how adaption can be used to accommodate the requirements of different user categories and contexts of use. This application is then used as a vehicle for discussing a new engineering paradigm appropriate for the development of adaptation-based user interfaces. In summary, most of the publications offer a solution to accessibility problems of some kind, but do not make an exhaustive analysis of technology.

Context-aware technologies play an important role to improve accessibility in several of these environments. Context aware systems are concerned with the acquisition of context (e.g. using sensors to perceive a situation), the abstraction and understanding of context (e.g. matching a perceived sensory stimulus to a context), and application behavior based on the recognized context (e.g. triggering actions based on context). Context awareness is regarded as an enabling technology for ubiquitous computing systems, [6] and [7] define human factors and physical environment as two important aspects related to computer science. Human factors related context is structured into three categories: information on the user, the user's social environment and the user's tasks. Likewise, context related to physical environment is structured into three categories: location, infrastructure and physical conditions. Services based on the user location or position are suitable for several of the environments in Inredis and have been proven to be relevant because of the recent patents, funded projects and publications in this field. For these reasons they are the objective of this article under the point of view of the integration with other technologies.

Section 2 is related to location based services and their evolution towards modern systems that integrate with other technologies; in section 3 there is a full description of one of most emergent location based technologies (positional commerce) and its architectures. This section also contains privacy, security and authentication concerns about this services; in section 4 there is an accessibility analysis of the emergent location based services; sections 5 contains the conclusions and finally there are acknowledgements and bibliography.

2 Location Based Services

Location based services (LBS) are information services accessible with mobile devices through the mobile network and utilizing the ability to make use of the location of the mobile device.

Depending on the environment there are different techniques to get a location:

- Inside a building and depending on the service: ZigBee, RFID, Bluetooth, UWB.
- Outside: Global Position Systems (GPS) is a Global Navigation Satellite System (GNSS), that allows determine the position of anything in real time, with a centimeter precision by using differential GPS.

Assisted GPS (AGPS) is a new technology created to overcome some disadvantages of traditional GPS. The latter were neither capable to reach the satellite signal in indoor areas, nor able to transfer GPS data at a reasonable speed. Newer phones and PDAs typically have AGPS receiver built into the phone, so they connect to an assistance server within the mobile network. This service has three objectives: they can reach a mobile device signal, they can reach the satellite signal and have power to make position calculations, it gives more precise location coordinates because it is no affected by other errors affecting GPS signal.

In figure 1, extracted from [8], there are the two performances: a standalone GPS and the GPS assisted by a network. AGPS architectures increase the capability of a stand-alone receiver to conserve battery power, acquire and track more satellites, thereby improving observation geometry, and increase sensitivity over a conventional GPS architecture.

In figure 2, extracted from [8], there is a schema of the integration of the elements that compound the AGPS service. Assisted-GPS requires a worldwide tracking network

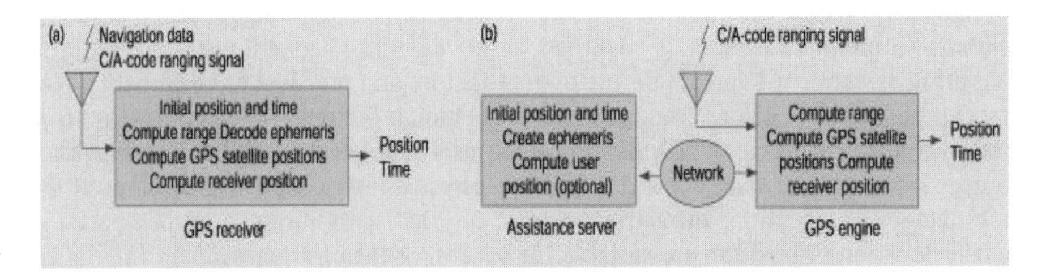

Fig. 1. Stand-alone GPS and GPS assisted by a network

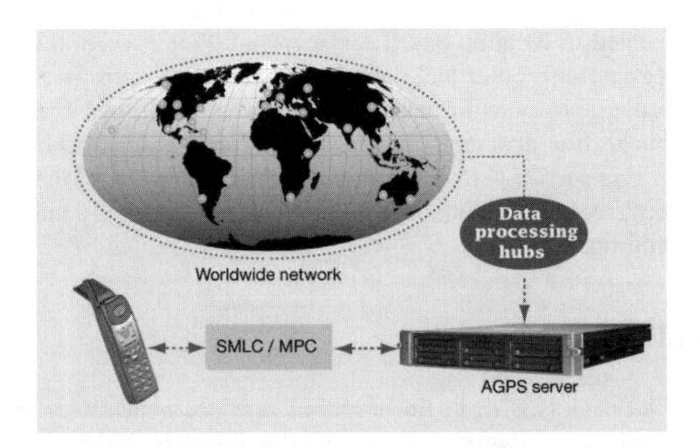

Fig. 2. Assisted GPS tracking network

for obtaining data from all satellites and data processing hubs along with a server which feeds data to a Serving Mobile Location Center (SMLC) or Mobile Position Center (MPC) operated by a network service provider. Data is sent to individual cell phones using Hypertext Transfer Protocol (HTTP) and the Short Messaging Service (SMS).

Although in the past only some PDA have had integrated GPS service, there are some mobile phones that provide AGPS service: Motorola E1000 [9], Siemens SXG75 [10], Nokia 6110 [11]. AGPS is fuelled via E-911 in North America by the Federal Communications Commission (FCC) and via E-112 in Europe, [8].

2.1 AGPS Applications

AGPS applications can be found in several of the Inredis environments, all them involve real-time features:

- In urban environment:
  - They can be applied to traffic regulation: AGPS may provide a feedback of the real-time traffic or car accidents, so it can be organized using traffic lights in a more suitable manner.
  - For eTourism, people may get real time touristic information brought in a multi-modal way and adapted to their preferences, so that they can make personalized electronic guided visits.
  - As a guidance and routing technology, to know the route provided on a map to get to a place.
  - As a localization and tracking system, when one wants to get the coordinates of a person and keep track of him.
- In mobile technology environment: As part of the technology, the use of Location based Services may be used in this environment, depending on the advances on middleware, protocols, interfaces, contextualization, navigation that are achieved in next generation of mobile.
- In teleassistance: they can be used to get the assistance center location or to get the patient location.
- In service and good purchasing: involves any kind of sale, advertisement or purchasement made through location based services. This technology is called positional commerce and is going to be studied in the next section.

3 Positional Commerce

Positional or Location based Commerce (p-commerce ó l-commerce) is one of the natural evolutions of the m-commerce or mobile commerce, which aim is to provide location based information with business purposes. This information may be provided in any format (voice, text, images) and can be related to sales, purchases or announcements. This technology takes into account the user context, so it is considered as a context-aware computation technology. In fact it is regarded as a mixture of m-commerce and context-aware computation, though sometimes is wrongly considered as plain m-commerce.

A scenario of p-commerce is a person with a mobile phone who wants to know about the nearest restaurants which menus are within a range of price. The service will provide the information based on his position. Another example is a person with a mobile phone walking in certain area. An announcement of a car promotion from a car marketplace nearby is sent to his device. In the first case the initiation of conversation is performed by the user; in the second case it is initiated by the establishment.

Although it is a technology that follows the tendency of flexible and personalized consume, there are some legal restrictions imposed by the privacy concerning information about the user localization in the sales and marketing field (see section 3.3).

Apart from location based commerce, there are other synonyms used in literature: pervasive commerce, public commerce, pocket commerce, personalized commerce, though each of them refers to a functionality of the positional commerce [12].

3.1 Description of Positional Commerce

In e-commerce context, positional commerce shares the meaning of these terms:

- Business-to-consumer (B2C), as it is information that goes from a business to the consumer.
- Mobile peer-to-peer (P2P), as the exchanged information is between two nodes from different kinds.
- Business-to-business (B2B), as it facilitates the virtual union between two small commerces and so it can increase its presence in the marketplace.[13]

  Positional commerce has several functionalities related to the synonyms, [12]:
- Positional commerce: Mobile device access to information about entertainment and commerce services based on user preferences and localization. The applications of this kind may be make a booking in a nearby restaurant and receive information about its localization on a map or send a mail to all car sellers nearby asking for a car model.
- Pocket commerce: based on the idea of a Universal Product Code (UPC), real-time information about a certain product may be provided just by giving its UPC.
- Personalized commerce: by the use of personal profiles, more suitable content and announcement are supplied. Three types of personalization are distinguished: service personalization (users can refine service structure based on requirements and functionality), content personalization (users may configure preferences to filter contents and content providers may use them to develop strategies to catch customers), dynamic personalization (users may reveal their profile during the service use so that this information is caught to improve their profiles).
- Pervasive commerce: it is related to the transition from traditional HTML browsers to multi-modal interfaces, so that these interfaces must adapt to new devices.
- Public commerce: integration of different types of information with geographical data , and manage different profiles related to preferences, security, privacy, transaction management,… [13]

In summary, three system qualities, ie.: completeness, real time and intelligence, present the grand challenge to the study of p-commerce, [14].

3.2 P-Commerce Architectures

To provide information based on a user position, the system must have a mobile positioning system. This device captures the consumer position.

Computational architectures for p-commerce vary upon the scenarios where they are involved. All of them will have a LBS and a mobile device. In [12], there is a multilayered architecture to manage different client profiles. These may be customers, companies, information providers. The LBS will be part of the external environment.

In [14] another architecture is presented which center is a P-Mall center, which manages the store servers belonging to the same shop chain. These ones are communicated through the P-Mall server. There are sensors in the store that detects the presence of a person and communicate with the PAM system. In figure 4 it is shown the overall architecture and in figure 5 it is shown the PAM architecture.

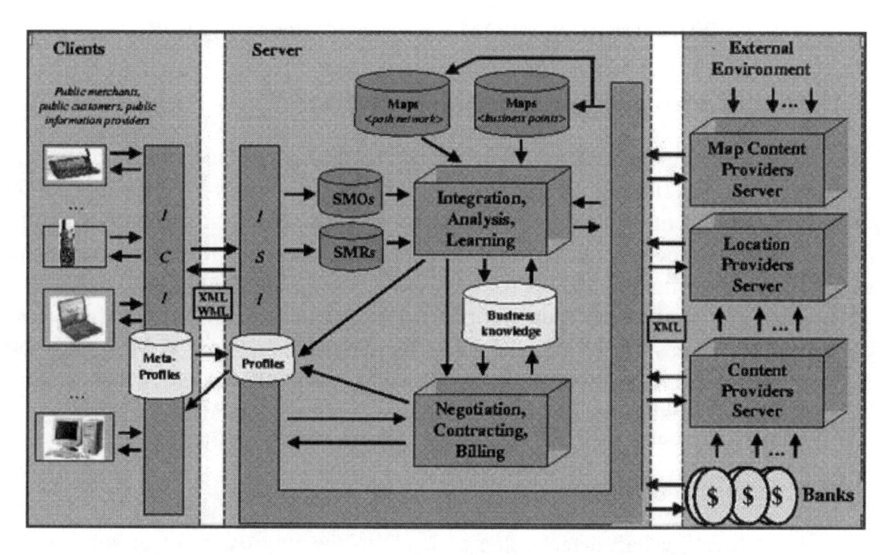

Fig. 3. Multilayered architecture manages different client profiles, based on [12]

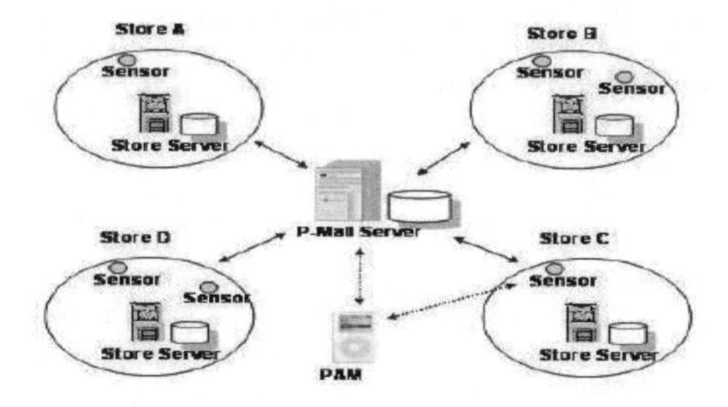

Fig. 4. Design of the overall architecture described in [14]

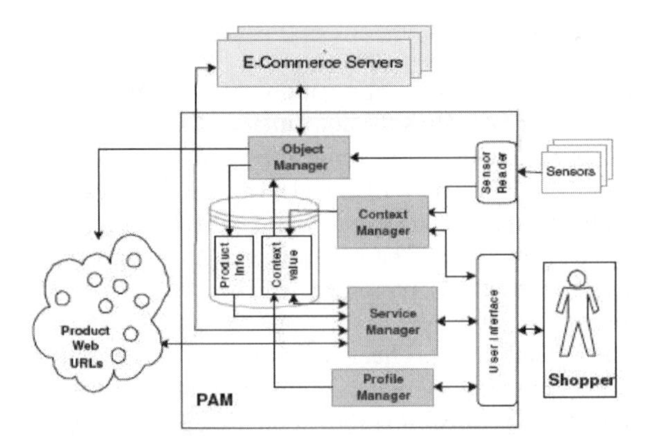

Fig. 5. Design of the PAM architecture described in [14]

3.3 Privacy, Security and Authentication Concerns

In any context-aware technology, there are privacy, security and authentication requirements when developing the middleware. Privacy laws deal with the user's control upon the contextual information and therefore who can access it. When designing a context-aware architecture, there must be a balance between the system efficiency based on user's information and his desire of privacy. Another legal aspect is the security: l-commerce transactions must not be interfere by third parts. With respect to authentication, it is necessary an authentication service in l-commerce WAP. In this way a new protocol, WTLS, fuelled by WAP Forum in year 2000, specifies that a mobile device must be adhered to an individual, telephone number or localization [15].

4 Accessibility Analysis

The study of location based technologies leads to the improvement in accessibility under two points of view:

1. The nature of the technology provides universal access because:
   - AGPS actions are performed with the exact location obtained by using the GPS receiver with assistance and network positioning in those areas which are not covered by stand-alone GPS. Advances against traditional GPS are just an advance.
   - Personalized interfaces may be provided to this technology.
   - Calculations are more precise in AGPS than in conventional GPS.

2. There are special features or uses that make AGPS accessible to disabled:
   - People with mobile disability may use the AGPS to get provided with route guidance. The use of mobile devices integrated with GPS provides an improvement in guidance systems.
   - For eHealth services AGPS may provide location based service, like where the nearest health center service is. Or even it can be use to locate people suffering from madness who are in emergency.

- For security in urban areas, they can be used to alert people when they enter a dangerous area. This may be applies to a wide range of disabilities like cognitive and visual ones.
- For eTourism, disabled people may get real time touristic information brought in a multi-modal way and adapted to their preferences, so that they can make personalized electronic guided visits.
- In general, location based services which aim is to improve accessibility to visually impaired people are with audible interfaces.

5 Conclusions

Convergence of technologies makes possible a wide range of applications to improve accessibility. For advanced location based services, the tendency is to fusion location based services, mobile technology and multimodal interfaces. These convergence needs middleware services so that all elements are correctly interconnected. Although some of the accessible applications have not been implemented yet, they constitute a promising starting point for this technology.

However they still depend on how well and soon progress on new generation mobiles are made. Legacy and authentication features must be also taken into consideration, as it is a tendency imposed by social demands. Although there are numerous advantages in the application of advanced location based services for accessible purposes, there is one disadvantage in the current price, though in the future it is expected to get lower the same way GPS are also getting cheaper.

Acknowledgments. The research described in this article is a result of the INREDIS project (CEN-2007-2011) [1], which is part of the CENIT program (National Strategic Consortium of Technical Research), supported by CDTI (Industrial Technological Development Center) and e-laCaixa within the INGENIO 2010 framework. This work has been also supported by research project TIN2005-08501-C03-01, funded by the Spanish Council for Science and Technology.

References

1. Inredis (Accessed 23 April 2008), `http://www.inredis.es/`
2. WAI. Web Accessibility Iniciative (Accessed 23 April 2008), `http://www.w3.org/WAI/`
3. eInclusion@EU (Accessed 23 April 2008), `http://www.einclusion-eu.org/`
4. Alvarez, F., Cisneros, G., Lopez, R., Menendez, J.M.: Towards an information society for all. The Journal of the Communications Network 4(3), 64–69 (2005)
5. Stephanidis, C., Savidis, A.: Universal Access in the Information Society. Methods, Tools, and Interaction Technologies 1(1), 40–55 (2001)
6. Schmidt, A.: Ubiquitous Computing-Computing in Context. PhD dissertation. Lancaster University (2003)
7. Schmidt, A., Beigl, M., Gellersen, H.W.: There is more to Context than Location. Computers & Graphics Journal 23(6), 893–902 (1999)

8. LaMance, J., Jarvinen, J., DeSalas, J.: Assisted GPS: A Low-Infrastructure Approach. GPS World, March 1 (2002)
9. Motorola 1000 (Accessed 23 April 2008), http://www.motorola.com/motoinfo/product/details.jsp?globalObjectId=43
10. Siemens SXG75 (Accessed 23 April 2008), http://www.infosyncworld.com/news/n/5840.html
11. Nokia 6110 (Accessed 23 April 2008), http://www.engadgetmobile.com/2007/07/20/nokia-to-add-assisted-gps-to-all-new-gps-devices/
12. Terziyan, V.: Architecture for Mobile P-Commerce: Multilevel Profiling Framework. In: Workshop Notes for IJCAI 2001 Workshop on E-business & the Intelligent Web (2001)
13. Mena, A.: Un cliente en movimiento, un posible comprador. Revista ESIDE 5, 39–41 (2004)
14. Lin, K.J., Yu, T., Shih, C.Y.: The Design of a Personal and Intelligent Pervasive-Commerce System Architecture. In: Proceedings of the 2005 Second IEEE International Workshop on Mobile Commerce and Services (WMCS 2005) (2005)
15. WAP Forum (2000) (Accessed 23 April 2008), http://www.wapforum.org/what/technical_1_2_1.htm

PIViTa: Taxonomy for Displaying Information in Pervasive and Collaborative Environments

Ramón Hervás[1], Salvador W. Nava[2], Gabriel Chavira[2], Vladimir Villarreal[3], and José Bravo[1]

[1] Information and Technologies Systems Department, Castilla-La Mancha University, Paseo de la Universidad 4, 13700, Ciudad Real, Spain
ramon.hlucas@uclm.es, jose.bravo@uclm.es
[2] Faculty of Engineering A.N.S. Autonomous University of Tamaulipas, Tampico-Madero 89138 Tampico, México
snava@uat.edu.mx, gchavira@uat.edu.mx
[3] Faculty of Computational Systems, Technological University of Panama, Chiriquí, Lassonde, David, Panamá
vladimir.villarreal@utp.ac.pa

Abstract. In the years ahead, there are increasing demands for ubiquitous and continuous access to information and interactive devices embedded into a physical context are proliferated. Users need support for getting required information anywhere and anytime, therefore, we cannot study advances in Ambient Intelligence and Information Visualization separately, it is necessary to consider the relevant features for displaying information into intelligent environments. A taxonomical approach can make another step towards understand the design space of information visualization in intelligent environments by extracting crucial characteristics.

Keywords: Ambient Intelligence, Information Visualization, Pervasive Displays, Context-Awareness, Taxonomy.

1 Introduction

Substantial advances in ubiquitous computing and communications, natural interfaces and pervasive displays have enable intelligent environments that are responsive and sensitive to the presence of people, adaptative to their needs, habits and emotions. In general, Ambient Intelligence (AmI) is the vision in which technology becomes invisible, embedded, present whenever we need it, enabled by simple interactions, attuned to all our senses and adaptive to users and contexts [1].

Visualization provides a powerful mechanism for assisting immersed users in Ambient Intelligence. Context-sensitive environments can respond to changes in visualization by accommodating techniques to change the interface in displays (varying from cell-phones, desktop computers to high-resolution displays) at run-time [2]. However, it is not possible a direct transition from the desktop screen to this wide range of display devices. This transformation brings up numerous research questions in the space of user interfaces, so an attempt to apply user interfaces designed for the desktop to screens embedded in the environment leads to problems [3].

The key goal of this paper is to offer a taxonomy in order to consider the relevant characteristics for displaying information in intelligent environments, distinguishing

J.M. Corchado, D.I. Tapia, and J. Bravo (Eds.): UCAMI 2008, ASC 51, pp. 293–301, 2009.
springerlink.com

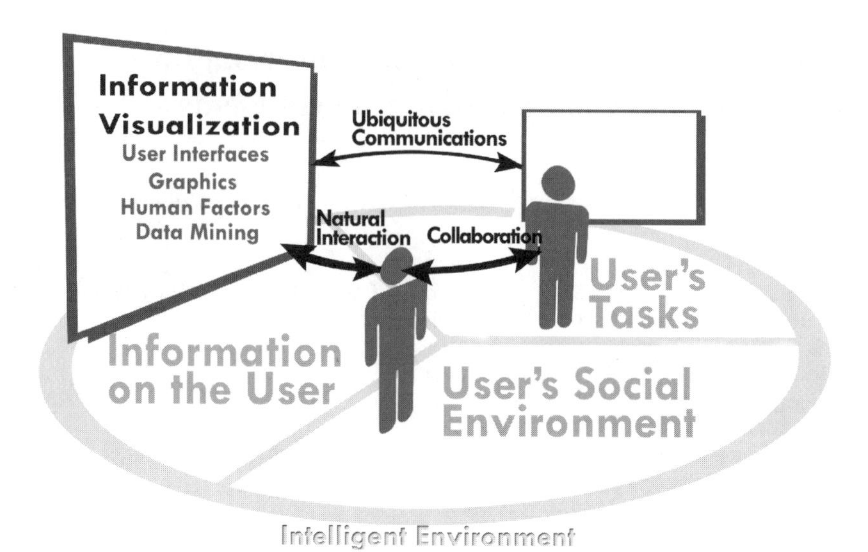

Fig. 1. Relationship between information visualization and user integrated into a context-aware and intelligent environment

information visualization details and context-awareness attributes (Figure 1). Recognition of these issues has given rise to work in automated visualization in Ambient Intelligence.

Section 2 deals with information visualization issues, specially focused on general aspects concerned to visualization in pervasive environments. If information is displaying in an intelligent and context-awareness environment, we can identify the environment characteristics to improve the information visualization. In section 3 we analyze these characteristics, i. e. which attributes, related to the environment around the displays, are significant to pervasive visualization. Next section presents the proposal of the taxonomy PIViTa (Pervasive Information Visualization Taxonomy).

2 Information Visualization

In everyday life, people have to access and analyze a variety of personal information such as agendas, news, emails, and digital documents. In the office, usually we need to share information between partners and access digital repositories. In our free time, we continuously manage information, for example, we make decisions among a variety of products, models, prices and characteristics when we go to shopping. In general, many decisions and actions are based on information that is gathered from a diversity of sources, in fact, information is becoming a ubiquitous and everyday task. Consequently, advances in information visualization can enhance Ambient Intelligence by analyzing high level design issues.

Regarding the purpose of the visualization, it is possible to recognize the role of the displays by identifying frequent situations with similar characteristics. A display role is a visual structure with a specific functionality in a usual situation. In [4] three key roles in pervasive and collaborative environments are proposed: (a) presentation: The visualization has the main function of displaying formatted information that has been previously prepared. (b) News: the objective is to show general information that is of interest to users near the display and (c) group coordination: several users who have common goals and collaborate frequently. Some important issues in visualization of information are related with the role. Each visualization role may be supported by specific schemas of interaction. The classic WIMP (Windows, Icon, Menu, Pointer) paradigm requires reconsiderations if visualization are to operate in a coherent fashion, because this paradigm is deeply oriented around processing a single stream of user input and the actions staged on a relative small and personal scream [5]. Innovative interaction schemas and paradigms are necessary, e. g. implicit interaction [6], touching approaches [7], gesture recognition [8], etc. Not only the schema of interaction is significant but also, the interaction flow is important too. In [9] a classification of interaction regarding the control-flow is proposed. The authors distinguish between three types: one-way interaction, i. e applications only need to be able to receive content from users; two-way, this category requires that data can be sent to the display landscape from users and vice versa; and high degree of interactions, applications require a permanent interaction between display landscape and users in both directions. Other classifications have been proposed. For example, the proposals in [10] also focused on the tasks that users wish to perform during visualization of information, organizing them in a higher-level way (e.g. prepare, plan, explore, present, overlay, re-orient).

Data source of information presents challenges mainly because of their diversity, volume, dynamic nature and ambiguity. Understanding the nature of the data, we can provide mechanisms that help the visualization process. Regarding the data source, we considerer some data types: text, databases, images, video and contextual data (typically obtained from sensors, but it may be inferred). Another core concept is the scalability. Usually, filter methods are necessary to scale the data, reducing the amount, defining the latency policy or adapting the complexity. These concepts are input variables; we also can analyze the scalability as an output variable that determinates the capability of visualization representation and visualization tools to effectively display massive data sets (called visual scalability [11]). There are some factors directly related with visual scalability: display characteristics such as resolution, communication capabilities, and size; visual metaphors, cognitive capabilities, and interaction techniques [12].

3 Pervasive Displays

Traditionally whiteboards in collaborative environments play an important coordinative role. The boards are typically situated in places where users often come by. The information displayed may offer an overview of the flow of work, reveal its status, enable communication between users and manage contingencies and unexpected situations. While whiteboards and desktop PC screens used to be the norm, the range of available display devices has exploded in the past years. Modern displaying systems bring us core benefits: bigger screens, new interaction schemas, more coordination and information resources, and so on.

Rui José [13] defines Pervasive Display Systems (PDS) as embedding infrastructures into the environment and composing many multi-purpose and public displays. A PDS should carry out three key principles (a) It should be made up of public displays, i.e displays that are not under the control of a single user, the main difference between PDS and Distributed Displays Environment, (b) it has to be multi-purpose and (c) a PDS is assumed to support coordination between multiple displays.

It is possible to recognize the spaces where display devices are placed, by analyzing the usual situations. In [14] four types of spaces are defined: transient spaces (e.g. walk-ways), social (e.g. coffee shops, bars), public/open (e.g. park) or informative spaces (e.g. notice boards). It is important to include a semi-public space: group rooms (e.g. conference hall, classroom). Group rooms are non-public spaces that may change into semi-public spaces depending upon the situation. For example, a classroom has a specific purpose during the class-time, however during the break it has a purpose similar to a social or an informative space. In this sense the pervasive display systems should change their services. Therefore, each display should handle extra information in order to change its function, according to the space type and the rest of the context information.

The type of space and the privacy aspect are high-level contextual issues. In a general vision, understanding the world around the applications, they can be developed improving the visualization of information. For example, context-aware applications can know how each situation causes different configurations and adaptations of the visualization. Furthermore, the actions of a user could be predicted by an analysis of his or her situation and these actions and others contextual features (e.g. number of users, roles, profiles, temperature, brightness…) are valuable input-characteristics in order to improve the visualization.

4 Pervasive Information Visualization Taxonomy

In previous sections we have analyzed the visualization in pervasive environments from two perspectives: information visualization and Ambient Intelligence issues. The first one is focused specially on general aspects concerned to perceptive and cognitive issues, graphic attributes, and data aspects. The second one is about the identification of environment characteristics to improve the information visualization.

4.1 Input-Output Classification

The identified features may directly influence on the transformation of information into visual representation. In this case we can define then as input-elements. In the other side, there are characteristics taken out from the final visual representation, influencing indirectly by the feedback of the system. For example, the luminosity in a room is an input-attribute that influence on the brightness of the visualization as an output-attribute.

Figure 2 shows input and output attributes into the visualization pipeline, describing the step-wise process of creating visual representation of data. It includes two key processes in the conversion of information into visualization: *visual mapping*, selecting the data portions to be visualized (usually user-centered or, even, context-centered) obtaining the visual form; and *view transform* that creates the final visualization. Input attributes contribute to improve both processes. The output attributes can be taken out of the generated view and are feed off input attributes in order to generate new views.

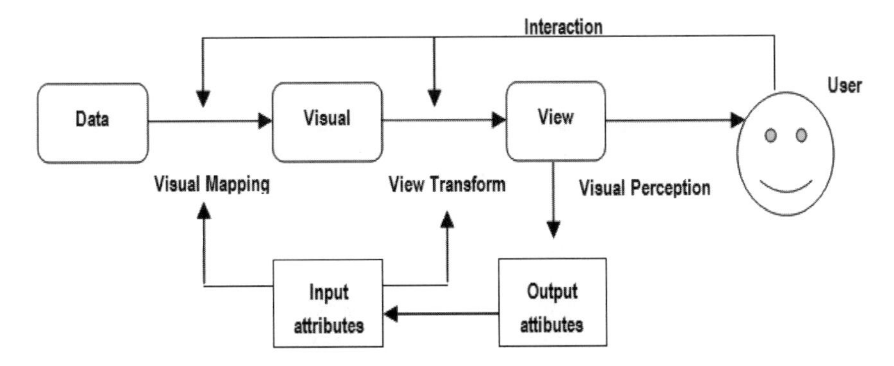

Fig. 2. Input and output attributes in the process of converting information into interactive visual representation (adapted from [15])

4.2 Type of Characteristics Classification

Another classification is expected, we have identified 5 core attribute types: Human Factors, Data, Interaction, Context-Aware and Services:

- *Human factors*: physical or cognitive properties of an individual or social behavior which is specific to humans and influences functioning of visualization systems. How people perceive and interact with a visualization system can strongly influence their understanding of the data and therefore, ergonomic issues should be input attributes in the visualization pipeline.
- *Data*: They present challenges mainly because of the diversity of the data sources, the huge growth rate of data and their dynamic and ambiguous nature. It is necessary mechanisms that help understanding the characteristics of the data being evaluated.
- *Interaction*: It allows the user to implicitly form mental models of the correlations and relationships in the data, through recognition of patterns, marking or focusing on those patterns, forming mental hypotheses and testing them, and so on. Interaction in information visualization has many facets to be studied. R. Kosara [16] presents thoughts on the nature of interaction and its importance in visualization, following this by descriptions of interaction methods that are used in Information Visualization (e.g. Focus+Context or Linking+Brushing). The classic interaction taxonomy [17] contained a mix of both tasks and datatypes, by offering a task-by-datatype taxonomy. It consists of seven tasks: Overview, Zoom, Filter, Details-on-Demand, Relate, History, and Extract. Both methods and task are interaction factors related to information visualization. However, interaction is also strong related to the environment, in fact, interaction schemas and paradigms depend on environment information and context-awareness system capabilities.
- *Context-Awareness*: “Context is any information that can be used to characterize the situation of an entity. An entity is a person, place, or object that is considered relevant to the interaction between a user and an application, including the user and application themselves.” [18]. Context-Awareness attributes can provide visualization services, integrated into an intelligent environment where the contents are

adapted to the situation (context) all the time. These services can be offered through an implicit process of the user's identification combining location and time awareness. All of these concepts allow us to provide task awareness and to offer attendance integrated into the user activity

- *Services*: Visualization services are intrinsically unassociated units of functionality. They typically implement functionalities most humans would recognize as a service, such as viewing an online bank statement or a route-planning facility on a map. Analyzing the service nature we can approach a better visualization.

4.3 Taxonomy

Table 1 shows the taxonomized information visualization elements and environment characteristics.

Table 1. Taxonomy for Displaying Information in Pervasive and Collaborative Environments

	Input	Input/Output	Output	
InfoVis	Capabilities of the Human Cognitive System. Motor functions and skills Social and cultural factors		Visual Metaphor	**Human Factors**
InfoVis	Source Volume Veracity Data Structure	Scalability(Filter) Scalability(Latency) Scalability(Complexity) Relationship between data Data functions Preattentive proprieties	Density Size Visual Scale View Structure	**Data**
InfoVis / AmI	Frequency Time response	Interaction Methods Interaction Paradigms Flow Control Navigation Graph	Change-awareness Interaction elements	**Interaction**
AmI	Display Features (Size, Resolution, Communication, Energy) Types of spaces Users (number, profile, situation, motivation, expectations, etc.)	Display Role Available Devices Environment attributes (Lighting, temperature, humidity, etc.) Time and schedule	Users (number, profile)	**Context-Awareness**

Table 1. *(continued)*

	Types of tasks Priority level (primary, secondary or peripheral)	Feedback (services used, type of use, time, level of interaction)	Types of services	**Services**

5 Conclusions and Future Work

This paper is the result of the analysis about characteristic features in pervasive information visualization. These features have not been described into meticulous detail, but this work can help to designers to improve user interfaces and can approach automatic and adaptative information visualization. Below, detailed contributions and future works are presented:

5.1 RDF/OWL Ontology

In previous works [19, 20], we have proposed a context model that is represented by an ontology describing parts of the real world encompassing visualization services. The context information is obtained from sensors embedded into the environment (tags and RFID/NFC devices) and from the static system information (stored in Data Bases) and it is continuously inferred and selected according to the ontology rules. The ontological model is mainly based on user profile and user situation. Formalizing the taxonomy in OWL (Ontology Web Language) classes and attributes and defining the relationship between taxonomy elements and previous ontological concepts we can obtain a form of knowledge representation about information visualization in Ambient Intelligence.

5.2 Design Patterns and Guidelenes

This paper shows that PIViTa not only helps researches understand the information visualization in intelligent environments but also, helps software developers understand how context-awareness and visualization techniques can be applied together. User interface designers use guidelines and patterns to unify applications developers and research advances. PIViTa may form part of useful guidelines and should enable developers to capitalize on pervasive displays and enhance Ambient Intelligence. Future work will include taking the taxonomy and making more meta-analysis of the similarities and differences between the elements in different cases of study, in order to obtain a repository of design patterns. Many of the situations share similar taxonomy elements that can easily be reused by means general visualization patterns.

5.3 AmI Usability and Evaluation

Integrating visualization taxonomy into Ambient Intelligent services development is possible to identify feature requirements and interaction plans based upon the user's goals. The benefits associated integrating these design principles include:

- Improving the overall usability of the system
- Incorporating context-awareness while catering to the user
- Capturing user experience.
- Approaching Universal Access, improving accessibility in design phase.

Acknowledgments. The authors would like to acknowledge the Spanish Science and Education Ministry and the Spanish National Plan I+D+I 2005-2007 the support for this paper as the project TSI2005-08225-C07-07 "Mosaic Learning: Mobile and Electronic Learning of open code, based on standards, secure, contextual, personalized and collaborative".

References

1. ISTAG, Scenarios for Ambient Intelligence in 2010 (February 2001), http://cordis.europa.eu/ist/istag-reports.htm
2. Alimohideen, J., Renambot, L., Leigh, J., Johnson, A., Grossman, R.L., Sabala, M.: PAVIS - Pervasive Adaptive Visualization and Interaction Service. In: Information Visualization and Interaction Techniques for Collaboration across Multiple Displays. Workshop associated with CHI 2006, Montreal, Canada (2006)
3. Baudisch, P.: Interacting with Wall-Size Screens. In: Information Visualization and Interaction Techniques for Collaboration across Multiple Displays. Workshop associated with CHI 2006, Montreal, Canada (2006)
4. Hervás, R., Nava, S.W., Chavira, G., Sánchez, C., Bravo, J.: Towards implicit interaction in ambient intelligence through information mosaic roles. In: Engineering the User Interface: From Research to Practice - Invited papers from Interaccion 2006. Springer, Heidelberg (2006)
5. Dempski, K., Harvey, B.: Multi-User Display Walls: Lessons Learned. In: Information Visualization and Interaction Techniques for Collaboration across Multiple Displays. Workshop associated with CHI 2006, Montreal, Canada (2006)
6. Schmidt, A.: Implicit Human Computer Interaction through Context. Personal Technologies 4(2&3), 191–199 (2000)
7. Bravo, J., Hervás, R., Nava, S.W., Chavira, G., Sánchez, C.: Towards Natural Interaction by Enabling Technologies, A Near Field Communication Approach. In: The 1st Workshop on Human Aspect in Ambient Intelligence (European Conference on Ambient Intelligence), Frankfurt (2007)
8. Dempski, K., Harvey, B.: Multi-User Display Walls: Lessons Learned. In: Information Visualization and Interaction Techniques for Collaboration across Multiple Displays. Workshop associated with CHI 2006, Montreal, Canada (2006)
9. Vincent, V.J., Francis, K.: Interactivity of Information & Communication on Large Screen Displays in Public Spaces through Gestures. In: Information Visualization and Interaction Techniques for Collaboration across Multiple Displays. Workshop associated with CHI 2006, Montreal, Canada (2006)
10. Hibino, S.L.: Task analysis for information visualization. In: Visual Information and Information Systems (VISUAL 1999), pp. 139–146 (1999)
11. Eick, S., Karr, A.: Visual Scalability. Journal of computational and Graphical Statistics 11(1), 22–43 (2002)

12. NVAC, National Visualization and Analytics Center. Illuminating the Path: The Research and Development Agenda for Visual Analytics (2004), http://nvac.pnl.gov/agenda.stm
13. José, R.: Beyond Application-Led Research in Pervasive Display Systems. In: Workshop on Pervasive Display Infrastructures, Interfaces and Applications (at the Pervasive 2006), Dublin (2006)
14. Race, M.K.: Oi! Capturing User Attention within Pervasive Display Environments. In: Workshop on Pervasive Display Infrastructures, Interfaces and Applications (at the Pervasive 2006), Dublin (2006)
15. Card, S.K., Mackinlay, J.D., Shneiderman, B. (eds.): Readings in Information Visualization: Using Vision to Think. Morgan Kaufmann, San Francisco (1999)
16. Kosara, R., Hauser, H.: An Interaction View on Information Visualization. In: Proceedings of the EUROGRAPHIC 2003, Granada, Spain (2003)
17. Shneiderman, B.: The eyes have it: A task by data type taxonomy for information visualizations. In: Proceedings of the IEEE Symposium on Visual Languages, pp. 336–343. IEEE Computer Society Press, Los Alamitos (1996)
18. Dey, A.K.: Understanding and Using Context. Personal and Ubiquitous Computing 5(1), 4–7 (2001)
19. Bravo, J., Hervas, R., Chavira, G.: Ubiquitous computing at classroom: An approach through identification process. Journal of Universal Computer. Special Issue on Computers and Education: Research and Experiences in eLearning Technology 11(9), 1494–1504 (2005)
20. Bravo, J., Hervás, R., Chavira, G., Nava, S.: Mosaics of Visualization: An Approach to Embedded Interaction through Identification Process. In: Luo, Y. (ed.) CDVE 2006. LNCS, vol. 4101, pp. 41–48. Springer, Heidelberg (2006)

Data Management in the Ubiquitous Meteorological Data Service of the America's Cup

Eduardo Aldaz[1], Jose A. Mocholi[2], Glyn Davies[1], and Javier Jaen[2]

[1] America's Cup Management S.A., Port America's Cup, Valencia, Spain
{eduardo.aldaz,glyn.davies}@acregatta.org
[2] Departamento de Sistemas Informáticos y Computación, Universidad Politécnica de Valencia, Camino de Vera s/n, 46022, Valencia, Spain
{jmocholi,fjaen}@dsic.upv.es

Abstract. In this paper we present an overview of the system designed to provide the 32nd America's Cup teams with the most accurate and up-to-date meteorological data available of the race course, paying special attention to how the information is gathered in a Grid of meteorological sensors, massively managed and robustly delivered to participating teams.

Keywords: Data Fusion, Sensor Grids, Ubiquitous Data Management.

1 Introduction

The America's Cup (AC) is a challenge-based competition where the Yacht Club of the previous winning boat makes the rules and hosts the event [1]. It was born after the yacht called America, representing the New York Yacht Club (NYYC) [2], won the 100 Guinea Cup of the Royal Yacht Squadron (RYS) [3] in a race held in 1851 in the Isle of Wight, England. Then, in 1857 the cup was presented to the NYYC as perpetual challenge trophy for friendly competition between nations. However, the first race for the America's Cup did not take place until 1870. Therefore, the America's Cup is often called the oldest trophy in international sport.

Sailing is a sport in which a better knowledge of the climate conditions over the course area may have a great impact on the performance of ships. It allows making better weather predictions and therefore better strategies for the race: which sail would be better to choose, which side of the course will have better winds, etc. This was for years the most important handicap challengers (teams that challenge the cup holder) had, since the defenders (winner team of the previous AC edition) always choose locations where their better knowledge and greater experience might play in favor of their victory. In order to counteract this advantage, in previous editions of the AC challengers with the greatest budgets started their own meteorological programs using several boats to collect weather data from the race course. In 2003 each one of the largest teams had at least 7 boats for this task. However, boats from the different teams ended up collecting data from the same areas. This was an expensive way of collecting weather data and the America's Cup Management company (ACM) estimated that each team would need 1.3 millions of euros to carry out their own meteorological program. To avoid these expenses and to equalize teams'

J.M. Corchado, D.I. Tapia, and J. Bravo (Eds.): UCAMI 2008, ASC 51, pp. 302–311, 2009.
springerlink.com

opportunities at the meteorological level, ACM created the Meteorological Data Service (MDS) program.

The MDS system does not provide predictions, it is just a meteorological service that provides weather and localization data when requested and with a certain format. It is the same service for all the teams and all of them receive the same data. One of the main goals of the MDS service is to design and build the necessary systems and devices in such a way that they could provide a constant amount of accurate weather data from the courses 24 hours a day, 7 days a week and could be reused in next AC editions. For this reason the system of the MDS for the 32nd America's Cup was designed to benefit from the *sensor fusion*, defined as "the combining of sensory data or data derived from sensory data such that the resulting information is in some sense better than would be possible when these sources were used individually" [4]. This way weather analysis and predictions (or other applications) over the course area can improve their accuracy [5]. However, as the rationale of the MDS system is to provide weather data as it is (no analysis or predictions are made), it follows the sensor fusion approach in a shallow manner.

The organization of this paper is as follows: in section 2 we describe the devices designed and built by the MDS staff and their deployment by the cost of Valencia. Section 3 presents how weather data is collected and stored for later use. Section 4 describes the mechanisms that allow the MDS staff to have a better control over the devices and the data AC teams receive from them. Section 5 defines how the gathered data is managed to obtain a great performance on the delivery to teams while maintaining a suitable structure for further analysis of device behavior by the MDS staff. In section 6 we describe the characteristics of the MDS system core and present some data that verifies the goodness of the design decisions made in order to achieve a good performance. Finally, in section 7 we present the conclusions of this paper.

2 MDS Sensor Grid

In the 32nd edition of the America's Cup, the MDS has provided the sailing teams with a distributed meteorological system that consisted of a grid of approximately 50 wireless weather sensing units in the form of buoys, boats and land stations. These buoys and boats are mostly deployed off the coast of Valencia, on the race course itself, and thus they are a regular feature on the TV coverage of the event.

The buoys (see Fig. 1) have been custom designed for high performance sailing events and use the most advanced technologies available to provide the best possible weather data. The electronics basically consist of a microprocessor that collects data from the different sensors (wind direction, wind speed, air pressure, air temperature, humidity and location coordinates), a GPRS modem (see Fig. 3) to send the collected data and the necessary lightning protection. The data they provide could be used for design purposes, for real time racing predictions and to ensure the safety of the event.

The core of the electronics is called the VIPER box and was designed, developed and programmed by MDS staff from a basic motherboard. It runs on Linux, has serial ports, USB ports, an Ethernet connection and a Compact Flash memory slot for data storage as well as short circuit detection and lightning protection.

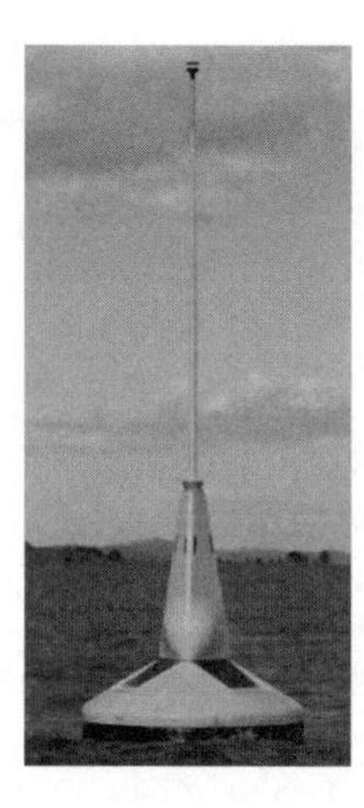

Fig. 1. Depiction of one of the MDS buoys deployed off the coast of Valencia

Fig. 2. Sonic anemometers use ultrasonic transducers to estimate direction and speed of wind

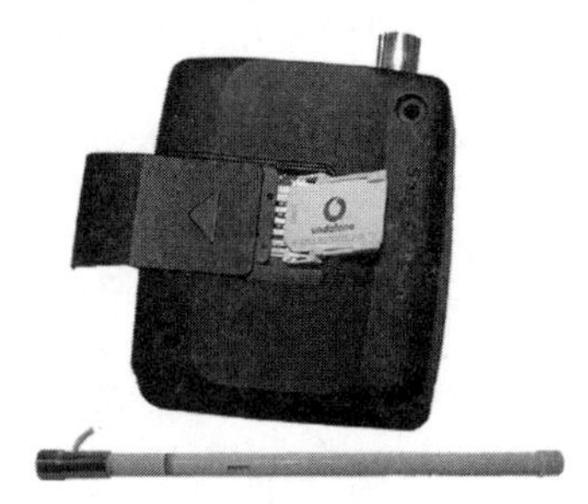

Fig. 3. GPRS modem and high performance sea antenna

To collect true wind direction and speed with accuracy, the buoys have been equipped with a sonic anemometer (Fig. 2). They are used in conjunction with an electronic compass.

Design recommendation 1. Always check if there are new ways to solve old problems. Sonic anemometers have several advantages over paddle wheels: higher accuracy, no movable parts, no need for calibration, and dust and salt do not affect them.

Once collected the weather data is sent via a GPRS modem (see Fig. 3) that is also installed inside the Viper box. The modem is connected to a high performance sea antenna installed in the buoy mast.

For safety reasons buoys are equipped with a GPS device. In case a buoy detects that it has drifted more than 30 meters away from its position, it will immediately send an alarm message to the MDS staff.

Fig. 4. Watertight box where buoy electronics are located

The buoys were designed to be in operation 24 hours a day, 7 days a week, so the electronics box (Fig. 4) is placed inside a watertight compartment, and the box itself is water tight to reduce the risk of corrosion to the elements. The electronics box was also designed to be able to work installed on a boat, thus boats can be easily used to create a temporal course race or to check the correctness of any buoy. For these reason, boats have also an accurate GPS device to correctly identify where the collected weather data come from.

Design recommendation 2. A portable design of (the core of) your sensor has its benefits: it allows you to perform faster replacements, to assess easily the correct operation of (new) deployed sensors, and to deploy temporary sensors quickly.

Devices on the buoys receive the power supply from 3 gel batteries that ensure a minimum autonomy of 2 months. They also supply energy to a LED based beacon light that can be remotely programmed. The batteries are recharged by 3 solar panels installed around the base of the buoys.

Design recommendation 3. An ideal world is not the real world. In other terms, try to design your system for the worst conditions of the environment where it will be placed in order to tackle with power shortage, weather conditions, safety and security.

Regarding land stations, they were also designed and built by the MDS staff and, as they are fixed to a location, they have a simple but resistant design that consists of an 11.5 meters long mast, one solar panel and a sonic anemometer on top of the mast (Fig. 5).

In Fig. 6 we can observe a map with the localization of each buoy and land station deployed off the coast of Valencia for the 32nd edition of the America's Cup. In the current edition there were 2 courses available to hold the races. The North course (blue buoys) were placed in two concentric arcs, and not in a circle like in the South course (red buoys), to allow the start line to be placed as close as possible to the beach.

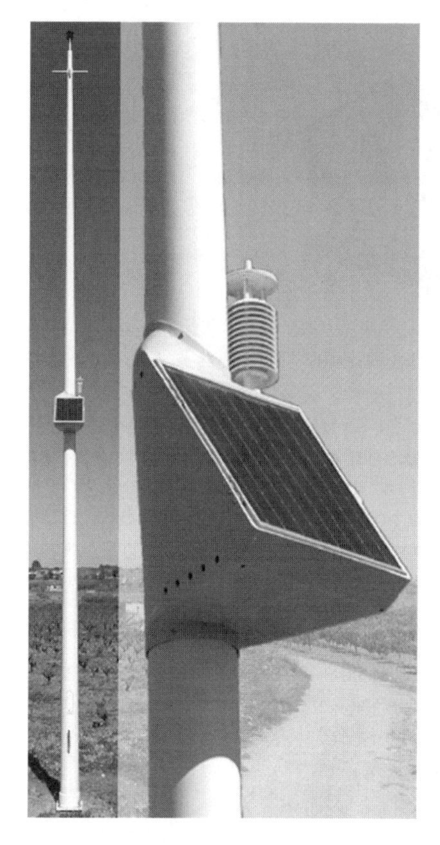

Fig. 5. Depiction of land station and detail of the solar panel

Fig. 6. Map of buoys and land stations deployment

3 Information Gathering

Every certain amount of time, depending on the device type and the moment of the day (for example, every 15 seconds from 9AM to 5PM and every 5 minutes from 5PM to 9AM for buoys sending wind data), buoys and land stations (and any other MDS boat in the sea) send the meteorological data collected to the MDS base.

Design recommendation 4. Carefully estimate the amount of data that covers the needs of your system. Receiving too much data may overkill your system (waste energy on sensors, saturate communication links, need more storage capacity). Too few data may make impossible to achieve the goal of the system.

Prior the sending, VIPER averages the weather data of the last 15 seconds (or 5 minutes), and the result is what is sent via the GPRS modem (Fig. 7). Data is sent as a string of comma separated values that basically contains the type of the sender device,

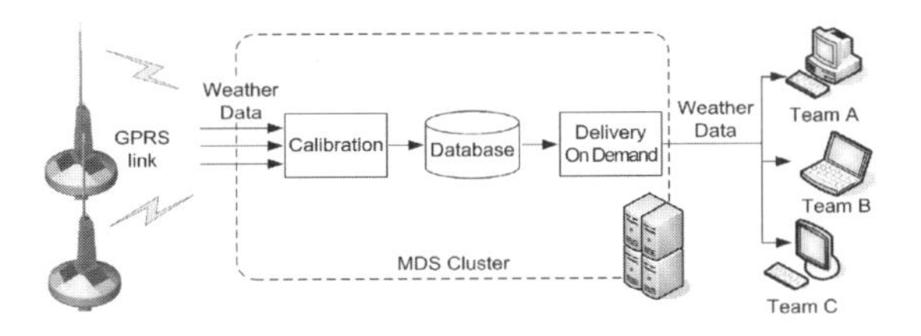

Fig. 7. Complete weather data management overview

the identifier of the station where the device is installed, the timestamp of the moment when data was sent, and the measured values themselves.

Upon reception through the GPRS link, an application at the MDS computer cluster checks the timestamp of the data received against the timestamp of the last stored data of that device to avoid duplicates and out of sequence data. If the data is sent by boats acting as race marks, the application will store the data only if any of the following 2 criteria is true:

- More than T seconds have passed since the last data from that device was sent.
- The boat has moved more than D meters with respect to the average of the N previous values.

Distance D, time interval T and the number of average values N are configurable parameters of the application. This way, different levels of data fusion granularities are allowed. Then it prepares the data received to be stored by applying the specified calibration (if there is any) for the sender device. For example, if MDS staff detects that the pressure sensor of a buoy is not sending accurate data, they may measure the necessary offset and gain to apply to reach the desired level of accuracy.

Once applied any necessary calibration, the string of data is parsed to store each value separately, to allow any possible further analysis by the MDS staff in order to detect misconfigured or malfunctioning devices. After this step, any team connected to the system may receive the gathered meteorological data.

4 Information Delivery

Once the meteorological data collected by buoys, boats and land stations have been stored, any team may retrieve them provided that the MDS staff has not marked data sent by a device as disabled. There are several reasons to disable the delivery of data from a specific device, for example: the device may have suffered a breakdown or a misconfiguration which could cause it to send completely inaccurate data, or may be the MDS staff is testing the device and they want to avoid teams getting its likely inaccurate data. For this task the system allows marking the delivery of the data of a certain device as *All* (any user can retrieve the data, it is the default value), and *None* (nobody will receive data from that device). Then, since that instant and until the MDS staff changes its disable value, data sent by that device will be stored but no team will be able to get it.

Another possible reason why teams may not receive data form a certain device is that it may have suffered a breakdown or a power shortage. To detect those situations the system permits to assign a *watchdog* (an alarm signal) per device. All devices have a watchdog assigned and if data is not received for a specified time for a device type, the watchdog is activated and a warning mail is immediately sent to the MDS staff.

Design recommendation 5. Make your system work for you or, at least, ease your work: in any case, you should be warned that something is not working properly (directly by the malfunctioning sensor or by other means), and you should be able to fix most errors remotely.

4.1 Delivery Mechanisms

Whenever teams want to retrieve meteorological data of the course race to prepare their strategies for the race, they can use the 2 mechanisms the system provides them to do this: either they can retrieve weather data in real-time or by downloading historical data files.

Real-time Delivery
When a team wants to know the current weather conditions, they have to visit a web page through a secure connection and, after successfully identifying themselves with their login and password, the system will provide them with the data received from the last 60 minutes then continually update the user with the real-time data from the system. The data is supplied in plain text with no HTML decoration. To avoid flooding teams with data from past days when they have not previously logged on, the system is configured to send only the weather data received in the last hour. If they want to retrieve data older than one hour, they can download a historical data file. It is the responsibility of the team, or any application they may be using to get the data, to regularly request the latest weather data in order to keep alive the session on the system. If more than half a minute passes from the last request made, the team will have to log in again to get weather data in real-time.

Historical Data Files
If a team has not been connected for a long time and wishes to recuperate the data of that period, they will have to log into a specific web page where they will be able to specify the day (or the date interval) they wish to download. If the team specifies a past single day, they will be given a link to a compressed ZIP file that will contain all the weather data that was sent the selected day. Otherwise, if the team specifies the current day or a date interval, the MDS system will automatically create the ZIP file with the historical data. In any case, the compressed file will contain one folder for each day with weather data, and each folder will contain one plain text file per device that sent data.

In any case, meteorological data supplied by both mechanisms is exactly the same: teams receive the weather data as strings of comma separated values, just as devices sent it. They also take into account the periods of time in which data from a device has been disabled when retrieving it from the database.

5 Information Management

One of the most important goals while designing the MDS system was to build a system with a high performance to be able to manage the requests of all the teams, i.e., it should tolerate both real-time and historical data requests. With this goal in mind, the following decisions were taken:

- To reduce up to some extent the load of the system when historical data is requested, the MDS system dumps (after midnight) all the data received each day to files and creates the ZIP file corresponding to that day. This way all data necessary to serve historical data requests is already prepared: if the team just requests one day, the ZIP file with the historical data is already prepared; if they request a date interval, it is only necessary to create the ZIP file from the plain text files already dumped; and just in case the current day is requested, it will be necessary to query the database.
- As stated previously, weather data is sent by devices as strings of comma separated values. Then these strings are parsed and their values stored separately. However, this strategy is not good enough in order to obtain a good performance when querying the database to deliver the data to teams in real-time or to dump the data to the historical files, because it requires rebuilding each string of comma separated values, which would be a waste of the system resources. To avoid this task, the full string of comma separated values received from a device is also stored in the database, after applying any necessary calibration. With this strategy, serving any request becomes almost as easy as executing a simple query on the database.

Design recommendation 6. If your system is delivering real-time data on demand, you should store the data in such a way that they could be retrieved and delivered as easily and quickly as possible, i.e., perform any required data modification (such as calibration or conversion) when data is received, and use any available feature that gives you a gain on performance (for example, cache on web servers, table indexes on databases, ...).

6 Implementation Details

As shown in Fig. 7 all data processing, storage and delivery tasks are done by the MDS cluster, which is a HP Proliant DL380 G4 packaged cluster with 2 Intel Xeon processors at 3.4 GHz and 2 GB of RAM. All data management applications of the MDS system run on this cluster and they were developed using Visual Basic .Net. One windows service was developed to receive and process the weather data sent by devices, and several web pages were created to attend the needs of teams and the MDS staff: one web page to provide teams with real-time weather data; another to allow them to download historical data; and another to provide the MDS staff with the necessary tools to manage the system.

Although the MDS cluster is powerful enough to run all the necessary applications, we decided to conduct some stress tests to verify whether the design decisions made to achieve a good performance have a real impact or not. In Fig. 8 we can observe

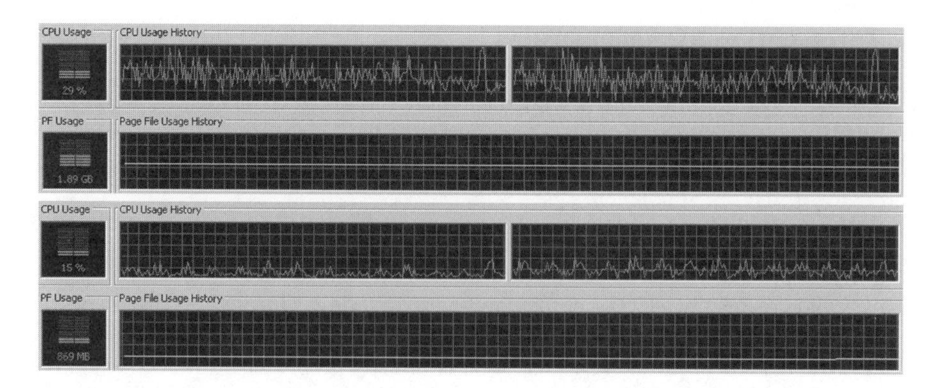

Fig. 8. Performance graphics of the system serving 100 users with (lower image) and without (upper image) design optimizations

how the system behaves when 100 users are connected to retrieve real-time data, while processing all the weather data sent by devices. The upper image belongs to the behaviour of the system without the performance related design decisions being executed. If we compare it with the lower image, which belongs to the results of the system with all the performance related design decisions running, we can see a great reduction of memory consumption (around 55%), and a more stable and lower processor load without peaks above the 50% mark.

7 Conclusions

We have presented in this paper an overview of the system that has being used to provide the participating teams of the 32nd America's Cup with the most accurate and up-to-date meteorological data possible. We have described the highly specialized hardware devices used to gather the weather data, and we have also presented the software system designed to manage and delivery those data with a high performance.

The design of the system allows having a continuous monitorization over 2 permanent course race (through buoys) and over another temporal course race (through buoys and boats). The use of specialized buoys have demonstrated to be a great advantage over the boats used in previous editions, as they are much cheaper and are continuously gathering data instead of just several hours on training or race days.

This system was also designed to work uninterruptedly and we can say that it has been up and running 24 hours a day and 7 days a week for 3.5 years, with only very minimal downtime due to remote software upgrades and hardware replacements or reconfiguration.

References

1. 32nd America's Cup official website (Accessed 17 June 2008), http://www.americascup.com
2. The New York Yacht Club website (Accessed 17 June 2008), http://www.nyyc.org

3. The Royal Yacht Squadron website (Accessed 17 June 2008), http://www.rys.org.uk
4. Elmenreich, W.: An Introduction to Sensor Fusion (2001) (Accessed 17 June 2008), http://www.vmars.tuwien.ac.at/php/pserver/extern/docdetail.php?DID=805&viewmode=paper
5. Hall, D.L., Llinas, J.: Multisensor Data Fusion. In: Hall, D.L., Llinas, J. (eds.) Handbook of Multisensor of Data Fusion. CRC Press, Boca Raton (2001)

A Proposal for Facilitating Privacy-Aware Applications in Active Environments

Abraham Esquivel, Pablo A. Haya, Manuel García-Herranz, and Xavier Alamán

Dpto. de Ingeniería Informática,
Universidad Autónoma de Madrid,
C. Francisco Tomás y Valiente, 11, 28049, Madrid, España
Abraham.esquivel@estudiante.uam.es,
{pablo.haya,manuel.garciaherranz,xavier.alaman}@uam.es

Abstract. This article analyses the nature of personal information according to privacy and proposes different privacy managing solutions for each type of information. Mainly focused on short-term information, the "Fair Trade" metaphor will be presented as a mean to control privacy in a balance between sharing and getting information. Combined with the appropriate monitoring and punitive tools we hypothesize that this balance will reduce system abuses and enrich the information services provided. Finally a context-aware Instant Messenger prototype, based on the "Fair Trade" metaphor, will be introduced to evaluate our hypothesis.

1 Introduction

Nowadays, social-based computing systems are a successfully experience, in particular, those based on the Web and communication technologies. In these systems, users play a central role in creating virtual communities connected through digital technologies. The success or failure of these systems depends mainly on its ability to encourage users' participation, being one of the most remarkable aspects of them the sharing of personal information between its members. Acquaintances, likes and dislikes, photos and videos are good examples of personal information. In addition, other types of context information, such as location or activity, can also be public. Nonetheless, this information is not always reliable since users have to update it by hand most of the times. Twitter[1] is an example of a social-networking platform where participants manually publish what they are doing.

Instant Messengers (IM), another example of social networking platforms, have grown as one of the most demanded communication applications. Commercial IMs include simple functionalities to manage context information. Location information, if any, must be established by hand, and activity information is limited to periodical updates of user's presence. On the contrary, there are several attempts in the research area that have obtained promising results IM [2, 6, 11, 12]. All of them require a sensor infrastructure, and retrieve context not only from raw data sensor, but also using inference mechanisms.

[1] http://twitter.com/

J.M. Corchado, D.I. Tapia, and J. Bravo (Eds.): UCAMI 2008, ASC 51, pp. 312–320, 2009.
springerlink.com

Context information handling is common to the so-called Active Spaces [10] as to refer to the above projects as social-aware applications in Active Spaces. These spaces are populated with a vast number of sensors and actuators and the extensive amount of information (directly achieved or inferred) they produce will, quite inevitably, compromise user's privacy. In consequence, if these new social-aware applications are to succeed, their users must feel in control of their personal data, being a key part of the system to win their confidence over the way it manages it. Regarding users' control over their information, a main drawback in the development of privacy-aware applications arises from the fact that users might be overwhelmed by managing privacy in a so changing and rich in information environment. Thus, in spite of all the attractive opportunities offered by the combination of social-based systems and active spaces, any attempt in that direction may fail if privacy management is not carefully considered.

According to these considerations, we propose three requirements to be fulfilled by any privacy control mechanism in Smart Spaces' social-aware applications [3]: **comprehensibleness**, **usefulness** and **flexibleness**. In this paper we present a proposal for privacy management in social-aware applications through the "Fair Trade" metaphor, which emerges from the application of the three previous principles to privacy control in Active Spaces. To test our assumptions, we have developed a context-aware instant messenger, exploiting not only the context information generated in the computer where is executed but also that of the Active Space.

This paper is organized as follows: We first provide background on the "Fair Trade" metaphor in section 2, establishing a clear distinction between when and how has to be applied. In section 3 we describe its implementation. Finally, section 4 and 5 summarizes our work and related work, offering a set of conclusions and several proposals for future works.

2 "Fair Trade" Metaphor

The "Fair Trade" metaphor [3] constitutes the keystone of our design. Privacy is a dynamic phenomenon depending on the user and its context: each individual has as many privacy issues as different situation he/she is involved in. In general, we can say that privacy management is a permanent negotiation where the private and public boundaries are constantly changing in response to the circumstances. Furthermore, we pose that a particular privacy configuration depends on the user sharing a service, and the service itself. In consequence, depending on the users and services involved, user's privacy needs will change.

One immediate problem arising from those considerations is privacy management. It seems reasonable that users would like to control their privacy boundaries. However, a very dynamic context can frustrate user's expectation due to an overwhelming amount of information. This is particularly true in ubiquitous environments where users must decide privacy configurations for every data gathered or deduced from sensors. We will show that two different kinds of management are required depending on the nature of the user's information.

According to the nature of the user's personal data, we classify information in two categories: **long-term information** and **short-term information**. The former comprises information with a low changing rate -e.g. telephone number, social security number, name, surname or postal direction- The latter, on the other hand, contains information of changing nature -e.g. location, activity or cohabitants. These two types of information have their own strengths and weakness. Thus, since static information is probably valid in the future, a single unwanted access will endanger the information in the long term. Consequently we can categorize this information as especially sensible in a time line. On the contrary, dynamic information may change with time, for what it could seem less sensible, on the other hand, is precisely this kind of information what really describes a person's way of living -e.g. where you are, doing what and with who. Therefore, even not especially sensible in a time line, this kind of information is more directly related to what we understand by privacy than the previous one. Summarizing, we pose that, in order to design privacy management solutions, we should be extremely aware of the idiosyncrasy of the information for which it is going to be designed.

With this purpose, we define dynamic information as what Dey identify as context [1]. This author split context information in four main variables: identity, location, activity and time. These four concepts will guide the implementation proposed in section 3.

2.1 Long-Term Personal Data

As we described before, privacy data could be split on two categories. One kind of information is named long-time personal data. This means that the value of this data remains constant with time.

In our environment model [4] each entity has a set of properties and relations associated. Regarding privacy issues, each property or relation must be included in one of the previous groups (i.e. long-term or short-term information). Depending on the assigned group, each property or relation will require a different privacy management policy. Long-term data are insensible to context changes, meaning that once the value is revealed to an unwanted receiver, the data is compromised forever. In this case, the main concern must be who has rights to access the information while when, where, or how is accessed are not relevant. Accordingly, the privacy mechanism associated to long-term personal data is based on a restrictive approach, in which the user must specify which individuals can access each piece of long-term information.

Thus, each property or relation has an associated privacy level, varying, from closeness to openness, in the following range.

- **Private.** A property or a relationship is tagged as private when the owner is the only one with access rights. Nobody, except the owner, can read the value and, if an entity is specified as private, all its properties and relations will be considered private too.
- **Protected.** The owner can specify who has permission to view the information by defining a list of receivers. Only members of the list can access the information. Any non-member will see the information as private.
- **Public.** There are no access restrictions to the information. Data is open to everyone for consulting. This policy can be applied to the whole entity or to a particular property or relation.

2.2 Short-Term Personal Data

Contrary to long term information, properties or relations more likely to change over time are classified as short-term information. Many examples can be found in an active environment in which sensors are constantly gathering information about the inhabitants and the environment itself. Similarly, information inferred from short-term raw data can be categorized as short-term information too. As a result, there is a huge amount of available information categorized as short-term, from inhabitant spatial-temporal coordinates such as location or activity information, to environment information such as temperature, humidity, lighting levels and appliances' status and so on. As a consequence of our previous research [3] we consider only three types of short-term information: identity, location and activity. Even though identity can be considered –as it is in fact- as a long-term information, it must be also kept within the short-term ones due to the indexing function it bears. Thus, without identity, nor location neither activity has any sense.

Short-term information is classified according to two discrete axes: privacy level and personal data. As illustrated in Table 1, the accuracy of the information varies with the privacy level. This accuracy is characterized by different degrees of granularity of each variable. Although different scales and values can be considered, our three levels approach is motivated by simplicity of use. We believe that more levels could lead to a non-feasible approach partly due to technical issues, but more importantly, because the model should remain simple enough to be comprehensible by non-technical user. Nevertheless, as we will suggest in future work, we will study if a four level approach is also reasonable.

Table 1. Context variables are filtered using different granularity depending on its privacy level

	Context variables		
Privacy Levels	Identity	Location	Activity
Restricted	Anonymous	Not available	Not available
Diffuse	Alias	Building Level	High level activity
Open	Real	Room level/GPS data	Raw data

The three privacy levels are:

1. **Restricted.** In this level a non-disclosure of information is achieved. Thereby, user's information is not distributed to the community.
2. **Diffuse**. Personal data can be retrieved with some level of distortion. Thus, for each variable, an intermediate value is established hinting at the real one without revealing it. This distorted value can be automatically obtained from the real one or hand-written in the user preferences.
3. **Open.** The information is revealed without any modification. Hence, the community receives the information as accurate as possible. Besides privacy configuration, the accuracy of the data will depend on the sensors' quality and deployed infrastructure.

The main challenge relies on how to assign these privacy levels to short-term information. One possible approach would be to apply the same strategy that for long-term information. However, due to the constantly changing nature of context, an approach that requires full-control over privacy will force the definition of particular privacy settings for each situation, meaning that users must anticipate every possible situation, or review their privacy preferences every time the situation changes. While the former is non-feasible the latter can be achieved by following an incremental approach, with the system asking for new privacy settings every time a new situation occurs and then remembering it. In this way, the system acquires, step by step, privacy settings close to the ideal. A real example can be found in firewall applications. On the other hand, the main problem with this kind of solutions is that users might find annoying –as they truly are in the case of firewalls-the repeated interruptions of the system, even more if we consider that notifications might be sent through ambient interfaces such as speakers. A completely opposite approach would be to consider totally automatic privacy control as an option. This implies that the system automatically infers the privacy configuration for each situation in behave of the user, whom stays aside. However, we believe that it will be hard for end-users to feel comfortable in delegating control over their privacy to an automatic system, unable to explain its behaviour [8]. Although automatic decision-control techniques can be very competent, user direct control is particularly relevant when the variables to be controlled are personal data.

Summarising, and following P. Maes [7] guide on trust and competence as main issues in developing software applications, we would like to emphasize the importance of choosing a comprehensible and useful solution. Where comprehensible means that the user has enough knowledge of the how, when, and why the system reacts to trust it; and useful means that the system is sufficiently easy to use and powerful to achieve a high degree of competence over the problem. If the solution is not trustable, users will end up setting their privacy so restricted that context-aware services would not have enough information to work properly. If the solution is not competent either the users will have problems with their settings or the services with their sources. This is the main reason to adopt an intermediate solution, simple enough for the common understanding, but rather flexible to fulfill users and services expectations.

The idea behind the "Fair Trade" metaphor is the stimulation of sharing information, hypothesizing that users will accept to harm their privacy if, on return, they receive valuable services. In addition to the services they obtain in exchange of their information, users must be able to track the flow of their disclosed data. Our approach emphasizes the "fair trade" metaphor in the following way: users configure their privacy settings according to the amount of information they would like to know from others, assuming that others would be able to access their own information in the same terms. For example, following the previous approach, if a user defines his privacy level as Diffuse, his personal data will be shown after being filtered but he would only be able to retrieve filtered information from others in the best case, even from those with an Open privacy level. At the end, the amount of shared information grows with information needs, making context-aware services more useful and valuable.

Despite all this, another issue of considerable importance for the success of the metaphor has been the optimistic access control included in the system [9]. This access control mechanism provides free information access, adding a logging mechanism of what data is visualized, who accessed it and when was accessed. As a result, this mechanism acts as a dissuasive measure to prevent abuses of the system combined with the punitive actions that can be taken in case ill-intentioned behaviour patterns are detected.

Summarizing, our privacy management proposal relays on trading quality of service for information. Users give their context information data in exchange of better services. A recording mechanism is provided to mitigate the risk of personal information disclosure. Users can revert their privacy preferences to a more restricted one if they believe somebody is misusing their context information.

3 AmIChat: A Context and Privacy Aware Instant Messaging Application

While developing our prototype, we focused particularly on supporting privacy control following the "Fair Trade" metaphor. In section 2 we classified personal data in two categories according to its durability, we motivated a different privacy management approach for each category and we explained both approaches. Even though AmiChat provides information access and privacy control to both long-term and short-term information this article will only explain in detail the latter, being the one in which the "Fair Trade" metaphor is applied.

In AmiChat short-term information is defined in terms of location and activity and the three privacy levels defined in section 2 are incorporated to the IM:

1. **Restricted.** Privacy has the highest level of closeness. Personal information remains secret; as a consequence the user looses IM quality of service. When the user connects to the IM his/her identity will be shown as "anonymous", as they will be all their buddies' identity too, regardless their privacy configuration. Albeit this lack of context information (identity, activity and location) the IM service remains active so the user can send and receive messages even though he does not know to whom or from whom.
2. **Diffuse.** In this privacy level, information is retrieved with certain granularity. User's real identity is replaced with an alias (see Figure 1). As a consequence, the user will only view the alias of his/her buddies with diffuse or open privacy levels. The activity to be shown is selected, as in many commercial IMs, from a list with three different options: "Write me!", "Away" and "Don't disturb". Regarding location, the result value is obtained from environment sensors after being filtered by the privacy middleware. Location information is gathered from two different kind of sensor. On one hand, working spaces such offices and laboratories are outfitted with RFID readers. Users getting in and out of environment should pass their RFID cards. Events on change locations are notified to the IM server. Additionally, a spatial model provides a hierarchy of environments including relationships between them, such inclusion or adjacency. On the other hand, IM monitors the user's personal computer status by means of a resident program that informs if the user is connected or is away. The previous spatial model also represents inclusion relationship between resources and environments. Thus, if a user is using her PC, IM can infer where she is.

3. **Open.** Personal information is completely open, hence the user is able to see others' information with the highest possible resolution: the one defined by the information owner. The identity shown is the real one and the location and activity values are directly inferred from sensors (e.g. "At the computer *abraxas*, but not working" or "At room *lab_b403*") Thus, personal information is traded for quality of service (see Figure 1).

Finally, since sensing technologies might not be available in every place, the IM client allows configuring manually an alternative version of the location and activity values. Nevertheless, information obtained directly from sensors will always be preferred to information stated manually, in order to avoid conscious cheating from malicious users.

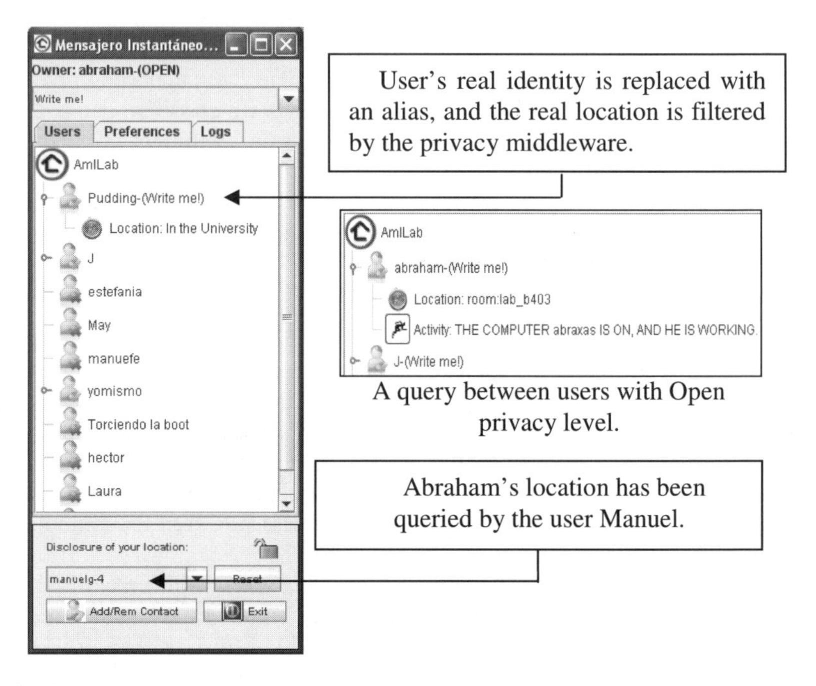

Fig. 1. Comparative between Diffuse (left side) and Open (right side) privacy levels

3.1 Supporting Mechanisms

As we already mentioned, the "Fair Trade" metaphor requires additionally considerations to achieve a satisfactory user experience. In this sense, our prototype uses two optimistic access control premises: free information access (whenever the privacy level allows it), and information access log. These premises, rather than preventing information access as the default policy, they complement the "Fair Trade" metaphor through data sharing: users can see others' information and, more importantly, they can check who is accessing their personal data. Due to the underneath "Fair Trade" policy, if user A access B's information with some granularity x, B will be able to check in the logs A's access with the same granularity x. This information helps B

regulating his/her privacy level at any moment if he/she feels uncomfortable. Additionally, *A* knows in advance that the logs and the "Fair Trade" policy combined will show *B* the same amount of *A*'s information as he/she is retrieving from *B*, explicitly entangling personal and alien privacies and deterring abuses such as requesting synchronously and repeatedly another user's location.

Considering these assumptions, we have included several logging mechanisms in the IM. Firstly, the number of location and activity accesses is summarized in the main window. A drop down list shows the requesters and the number of times each one has accessed the information. This list is ordered by the access number, as a measure of privacy risk. Secondly, the IM also keeps detailed record of each access in two logs (one for short-term and one for long-term information) where the user can view when each access has been produced. At that moment, the IM supports exclusion from the contact list as the only punitive measure against users considered to be doing an incorrect use of the application.

4 Related Work

In this section, we are going to review some of the more relevant privacy-aware projects that have adapted an instant messaging application for supporting context information. We will focus on how they manage the user privacy.

A project sharing a vision similar to the one that we propose in this paper is Babble [2]. This IM provides a graphical representation of the community social activity. They pose the idea of "Socially Translucent Systems". People have tendency to imitate others individual behaviour through observation. The key point is social conventions and rules supporting the cooperation will emerge, if systems facilitate the exposure of community activity. Therefore, these translucent systems aim to the creation of open communities.

Another context-aware instant messenger is Piazza [6]. The main goal is to achieve spontaneous interaction between users that shares the same activities. User's activity is published and announces to the community.

Project view IM [11] aims the improvement of collaborative tasks when the users are working simultaneously on several projects. Privacy is limited by providing information only to people on a shared project. Finally, Hsieh et al. [5] have designed privacy controls and feedback mechanisms for contextual IM. Their work is based on a group-level rule-based privacy control. Their results are very promising. We share a similar approach although we focus on facilitating the rules creation avoiding that the user has constantly worried about her privacy configuration.

5 Conclusions and Future Work

In this work, we have studied privacy issues in social and context aware applications. To address this challenge we propose the "Fair Trade" metaphor as the underlying policy of the privacy control mechanism. Categorizing personal information in two different classes: long-term and short-term information, we found the metaphor especially suitable to deal with short-term personal data, what is usually understood as context information. On the other hand, a traditional approach has been proposed to handle long-term information.

To validate our hypothesis we developed a context-aware privacy-sensitive instant messenger with which to evaluate how feasible is the “Fair Trade” metaphor as a privacy control mechanism. The prototype runs on top of the infrastructure deployed in the Universidad Autónoma de Madrid [4] and a working version is been used daily by the researcher staff of our group. We are planning to make an evaluation on a broad variety of end-users.

Finally, future evaluation will be necessary to determine whether the number of different privacy levels has to be grown or not and if user’s black lists should be public.

References

1. Dey, K.A.: Understanding and Using Context. Personal Ubiquitous Computing 5(1) (2001)
2. Erickson, T., Smith, D.N., Kellogg, W.A., Laff, M., Richards, J.T., Bradner, E.: Socially Translucent Systems: Social Proxies, Persistent Conversation, and the Design of “Babble”. In: Proceedings of CHI 1999, pp. 72–79 (1999)
3. Esquivel, A., Haya, P.A., García-Herranz, M., Alamán, X.: Managing Pervasive Environment Privacy Using the “fair trade” Metaphor. In: CoopIS OTM 2007 Workshops, Vilamoura, Portugal, November 25-30. LNCS, vol. 4806, pp. 804–813. Springer, Heidelberg (2007)
4. Haya, P.A., Montoro, G., Alamán, X.: A prototype of a context-based architecture for intelligent home environments. In: Meersman, R., Tari, Z. (eds.) OTM 2004. LNCS, vol. 3290, pp. 477–491. Springer, Heidelberg (2004)
5. Hsieh, G., Tang, K.P., Low, W.Y., Hong, J.I.: Field Deployment of IMBuddy: A Study of Privacy Control and Feedback Mechanisms for Contextual IM. In: Krumm, J., Abowd, G.D., Seneviratne, A., Strang, T. (eds.) UbiComp 2007. LNCS, vol. 4717, pp. 91–108. Springer, Heidelberg (2007)
6. Isaacs, E., Tang, J.C., Morris, T.: Piazza: A Desktop Environment Supporting Impromptu and Planned Interactions. In: Proceedings of CSCW 1996, pp. 315–324 (1996)
7. Maes, P.: Agents that Reduce Work and Information Overload. Communications of the ACM 37(7), 31–40 (1994)
8. Norman, D.: How might people interact with agents. Communications of the ACM 37(7), 68–71 (1994)
9. Povey, D.: Optimistic Security: A New Access Control Paradigm. In: WNSP: New Security Paradigms Workshop. National Information Systems Security Conference (1999)
10. Roman, M., Hess, C., Cerqueira, R., Ranganathan, A., Campbell, R.H., Nahrstedt, K.: A Middleware Infrastructure for Active Spaces. IEEE Pervasive Computing, 74–83 (2003)
11. Scupelli, P., Kiesler, S., Fussell, S.R., Chen, C.: Project View IM: A Tool for Juggling Multiple Projects and Teams. In: Proceedings of ACM CHI, vol. 2, pp. 1773–1776 (2005)
12. Tang, J.C., Yankelovich, N., Begole, J., Kleek, M.V., Li, F.C., Bhalodia, J.R.: ConNexus to Awarenex: Extending awareness to mobile users. In: Proceedings of CHI 2001, pp. 221–228 (2001)

People as Ants: A Multi Pheromone Ubiquitous Approach for Intelligent Transport Systems

Jorge Sousa and Carlos Bento

CISUC DEI/FCTUC University of Coimbra
Polo II, Coimbra, Portugal
pelaio@student.dei.uc.pt, bento@.dei.uc.pt

Summary. As our society develops, cities grow both in size and complexity. In order to find optimal paths in an urban environment, we need to consider several factors, such as traffic congestion or city rebuilding. It is, however, unpractical to centralize all this information in a single service. In this paper, we propose a multi pheromone architecture, depending mostly on the users and their potential collaborative ubiquitous computation in order to find optimal routes regarding several parameters.

Keywords: ACO, Intelligent Transport Systems, Multi Pheromones, Ubiquitous Advisor Systems.

1 Introduction

The popularization of PDAs and Smartphones, including a GPS receiver, popularized the use of navigation systems. These devices provide limited computational power, which makes necessary the use of routing algorithms that do not overestimate their capacities, in order to find the best route. Best routes are often assumed to be the shortest paths between two spots, as these would, empirically, save time when compared to routes with longer paths. However, most of these solutions lack the ability to assume that a city is not a static entity, but a rather complex one. Although this fact is getting into consideration [6, 1], the use of an optimal, deterministic algorithm regarding distance still far outweighs the risk on relying on more dynamic and stochastic approaches.

The characteristics that make a path optimal may vary from person to person and the standard algorithms are not capable of tackling these personalization issues lightly. The same can be said for multi variable and dynamics characteristics.

We propose an approach that makes use of the growingly ubiquitous environments to provide a computational ecosystem, a network of information and transformed knowledge. Regarding the problem of route finding in a city, we propose a solution that relies on the mimicry of ants by human beings, tackling the multi objective issue by using a multi pheromone approach.

J.M. Corchado, D.I. Tapia, and J. Bravo (Eds.): UCAMI 2008, ASC 51, pp. 321–325, 2009.
springerlink.com

2 Related Work

Although intelligent transport systems are a hot topic, the standard solutions are usually the most common, least we dont count the proprietary systems. A*[5] or Dijkstra [2] are the most used in the field, as they provide optimal solutions regarding discovery of the shortest path. There is a work by Kroon and Rothkrantz [4] from which our work took inspiration, although the problem considered in this work referred only to the traffic congestion problem.

3 People as Ants: The Proposed Evolutionary Approach

We start by introducing the ant based approach, with our own assumptions. We then regard the multi objective issue, by using the multi pheromone technique. Finally, we present a developed architecture.

3.1 Dynamics by Ants

In several works, ants [3] are used on a straight evolutionary style. Generaly, intelligence by swarms is used by releasing several artificial ants that organize their paths along the graph, discovering the optimal route. This, is unpractical on our problem, due to poor computational power in the user devices. We chose to take the metaphor one step further, as instead of releasing several artificial ants, we regard the each user as a single instance of an ant [4]. After going through the path, users send information to deploy pheromones on a virtual graph, in order to provide knowledge to other users (ants).

3.2 Multi Pheromones for Multi Objective Problems

As stated before, our system should provide its users the opportunity to choose the preferred aspect of their travel. As such, we consider multi pheromones on a graph. This approach is not new [7] but it provides for a simple way of having multiple goal information on the same graph.

As the users insert their information in a diffuse way (0-3 in the 3 categories), we compute the quality of a path by applying formula 1.

$$PathQuality = w1 * eco + w2 * speed + w3 * dist \tag{1}$$

With $w1$ = weight of ecological path; $w2$ = weight of speed preference; $w3$ = weight of distance preference; eco = amount of ecological pheromones on the path; $speed$ = amount of speed pheromones on the path; and $dist$ = amount of distance pheromones on the path.

3.3 General Architecture

Given a graph, each node has a destination table with entries for every neighbor (connected node) and a destination, along with the corresponding pheromone concentrations for each entry. An example is provided in table 1.

Table 1. Portion of a destinations table (for Node 0)

Neighbours / Destinations	5	4	7
1	$PheroSet(1,5)$	$PheroSet(1,4)$	$NotAssigned$
2	$PheroSet(2,5)$	$NotAssigned$	$PheroSet(2,7)$
3	$PheroSet(3,5)$	$NotAssigned$	$PheroSet(3,7)$

Note: PheroSet(X,Y) is the group of (3) pheromones assigned to the entry

In this example, node 0 is connected to node 1, 2 and 3, being possible to use them to reach destinations 5, 4, and 7. As the user requests a path, providing the initial node, the final node and the trip preferences, the following algorithm ensues:

1. Check if current node equals target node. If so, go to step 7;
2. Get all connections (from current node) that lead to the destination;
3. For each connection, assert its value according to user defined parameters;
4. Chose the best future node, stochastically;
5. Increase and register costs (for simulation only);
6. Assign current node as the chosen one and go back to step 1;
7. Assert costs of voyage (for simulation only);
8. Return the chosen path to the user.

When the user receives the advised path, s/he has the opportunity to follow or ignore it, if s/he thinks the system is wrong. Either the user followed the advice or not, the feedback of its track will refine the pheromone amount on the graph located on the server, by updating accordingly to the performance of the path taken, as seen in formula 2. The formula is used for each objective (type of pheromone) and may possess different fitness functions. All the paths and sub paths with the same destination will suffer evaporation, according to formula 3.

$$Pher_{new} = \frac{Pher_{old} + Fitness}{1 + Fitness} \tag{2}$$

$$Pher_{new} = \frac{Pher_{old}}{1 + A} \tag{3}$$

With $Pher_{new}$ = new pheromone concentration; $Pher_{old}$ = current pheromone concentration; $Fitness$ = fitness value; and A = evaporation constant.

4 Preliminary Observations and Results

The first attempt to test the algorithm was by observing its performance in a multi-variable environment. Figure 1a shows a graph (a segment of a possible map) with different values for distance, average speed and CO2 concentration. After a run with 1000 potential users, all with a preference concerning speed factor and route (0 to 5), advised path to the 1001th user has the probabilities

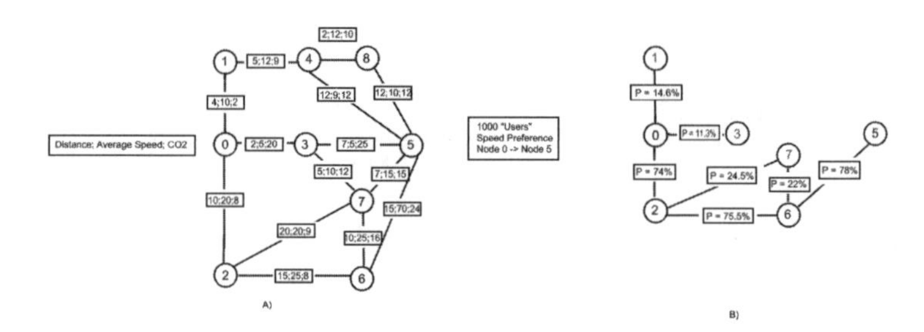

Fig. 1. a) The initial virtual graph; b) the probabilities graph over 1000 user runs

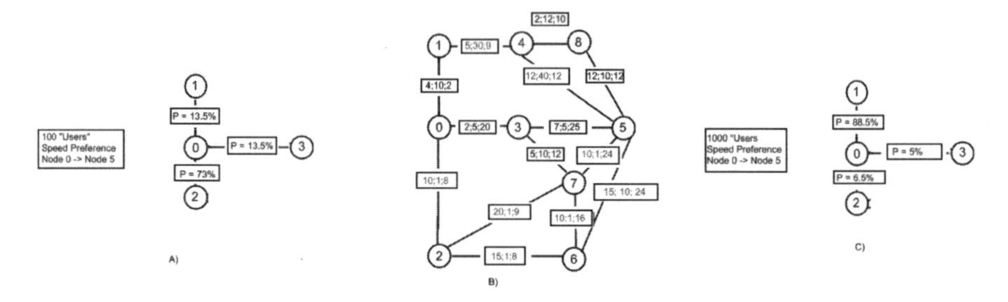

Fig. 2. a) The probabilities after 100 user runs; b) the graph with modifications regarding graph from figure 1a); c) the new probabilities, after 1000 runs

presented in figure 1b. Route (0-2-6-5) has the best value in average speed, being the optimal path, and it has a high probability to be recommended.

To test the algorithm under dynamic conditions we performed the test shown in figure 2. Figure 2a has the values obtained after a run of 100 users (with same preferences on characteristics and trip). In figure 2b, are a few changes made after those 100 users went by. Notice the abrupt decrease in average speed in the previous optimal path, (0-2-6-5), possibly representing, for example, a traffic congestion. On the other hand, the traffic on (1-4) and (4-5) has become much faster. As such, after another 1000 users pass by considering these new conditions, the next user would be presented with probabilities like shown in 2c. Notice how the probabilities changed drastically. Now the optimal path is clearly (0-1-4-5), and the algorithm adopted it as the most probable route to advise users who desire speed above it all.

5 Future Work

Further tests are needed in order to refine some of the parameters of the algorithm. Testing the algorithm against other approaches is also relevant. We can in addition study the inclusion of heuristics in the choice of the path, regarding certain variables. In order to provide the solution as imagined, a mechanism that transforms the graph and its topology in real time will have to be implemented, in order to make a completely collaborative application. Last but not least, the application is relying on some techniques that are still a matter of research, such as GPS track analysis.

6 Conclusions

Although this is a preliminary work, we believe this is a direction worth to pursue. The approach is inspired in several different works, but provides for an innovative tackling of a rather new problem, where we look at individual transportation as a multi objective ant like problem. The approach taken in this work is another step on change of paradigm from static to dynamic. The collaborative aspects taken into account have a very interesting potential. The solution proposed has the potential to be integrated in a network of ubiquitous computing advisors that mimic the ant approach and are embedded in an ecosystem of intelligent transport system devices.

Acknowledgments

To Miguel Silva (who actually pointed out of the multi pheromone approach) and Rui Craveirinha, for the insightful conversations.

References

1. Berezhnoy, A., Chudinov, K.: Approach for use of intelligent transport systems for urban trip choice (2007)
2. Dijkstra, E.W.: A note on two problems in connexion with graphs (1959)
3. Dorigo, M.: Swarm Intelligence: From Natural to Artificial Systems (1999)
4. Kroon, R., Rothkrantz, L.J.M.: Dynamic vehicle routing using an ABC-algorithm (2003)
5. Nilsson, N.J.: Principles of Artificial Intelligence (1980)
6. Taylor, M.A.P.: Intelligent transport systems: emerging technologies and methods in transportation and traffic (2004)
7. Xia, Y., Chen, J., Meng, X.: On the dynamic ant colony algorithm optimization based on multi-pheromones (2008)

A System of Cooperation Based on Ubiquitous Environments for Protection against Fires in Buildings

Clarissa López and Juan Botia

Departamento de Ingeniería de la Información y las Comunicaciones
Universidad de Murcia
Campus Espinardo, Murcia, Spain
clarissa.lopez@alu.um.es, juanbot@um.es

Abstract. The constant danger in our daily lives caused by natural means or by the same hand of man has led the field of research to gather knowledge about how to react to these disasters in the best possible way, trying to avoid deaths. This research work suggests modeling an ontology that will be implemented on a system of intelligent actuators agents inside an automated building, which will take decisions in a ubiquitous atmosphere, to save lives with the resources they have available.

Keywords: ubiquitous, ontology, intelligent agents, automation, semantic web.

1 Introduction

The building automation becomes a reality not only within the reach of corporations; in addition the requirements of installation and space for this type of automated systems have declined with the new ubiquitous technologies. For this reason is no longer surprising to include capabilities to control smart appliances: lights, different types of alarm and response to emergencies, etc. which also allows the sharing and links the functions of the various components that form them.

"An intelligent building is one that by itself can create personal, environmental and technological conditions to increase productivity and satisfaction of its occupants, within an environment of maximum comfort and safety, adding saving energy resources from monitoring and Control of common systems of the building" [1]. Another advantage of building's automation are the continuously monitoring of fire systems, automating the pressurization of evacuation staircases; smoke sensors that prevent the flames from spreading activating regulatory doors or hermetic closure of floodgates firewall, they can also control the level of carbon monoxide in the affected areas, checking entrances and exits in the evacuation, registration of movement and location of persons within and so on.

The proposal for this research is a decentralized cooperative system of intelligent agents in a ubiquitous environment, capable to act in emergency situations, designing a strategy for action within the building and a plan of action outside of it coordinating and providing information, as far as possible to help the teams involved in the emergency. The first step in the development of the research has been modeling the Ofepac ontology, which will form the basis of knowledge shared by the intelligent system in decision-making.

J.M. Corchado, D.I. Tapia, and J. Bravo (Eds.): UCAMI 2008, ASC 51, pp. 326–334, 2009.
springerlink.com

1.1 The Fires in Great Buildings

The theme of disaster prevention has taken an important place on investigations, has acknowledged that it is imperative to develop strategies and programs aimed at preventing far-reaching and reduce their effects, as well as act in emergencies and disasters. Certainly much has been achieved on the issue in recent years. In most cases the time between the alarm and the actions of firefighters has been an important factor for the early spread of fire; ignore the structural information of the building and information about people who find in it difficult immediate action, for these reasons it is necessary to design programs and actions to mitigate and reduce these risks prior to the occurrence of the phenomenon, this can be achieved through the strengthening and adequacy of the technological infrastructure that allows efficient data collection, planning and communication of knowledge generated.

1.2 Our Approach

Drawing on the definition Weiser [2] posed for ubiquitous computing, if we have a ubiquitous device in every part of our world for which navigates information that should always be available and understandable to both, humans and computers, is necessary to codify semantics Web documents through languages and metadata ontologies (a hierarchical structure of a set of terms to describe a domain that can be used as a skeletal foundation for a knowledge base) [3]. The key of the functionality of the ubiquitous in addition to its presence and invisibility is the managing of standard ontologies that enable opportune management and not obsolete information. The use of ontologies ensures a standard model for all ubiquitous entities software on a system, the main advantages include a common information model in the context of users and the possibility of re-using previously defined ontologies in this system type [4]. The model of cooperation that aims to raise, is based on the participation of a multi-system operators who will share knowledge through ontologies, it is necessary that all information is accessible at all times, for this, the model will consist of the participation of entities (actuators agents) in one or more processes each partially solving the problem, will define the way in which these players will participate on stage (Role) and how to be part of the logic processes. We propose the use of agents surrounding the system [5] (*wrapper agent*: encapsulates the software and rises to the level of society agents) which would stabilize at a high level autonomy for processing and delivering of planned information. The information will be accessible via the Internet by *wrapper agents*, as a result, all information will be available to all planners in a standard format and so states in planning and decision-making may be amended at any time according to the circumstances as they arise. The Ontologies contain all information concerning to the building; rules and methods that agents need to decide and act. Fig. 1 represents the approach of the stage in the cooperation model where agents are related in their work, communication between them and the outside world and ontologies as the knowledge base. Actuators agents or *Fire Protection Agents* must have 3 types of models of information for proper planning: Rules of Fire Protection (allow the proper processing model protection in a network of cooperative planning), model of fire protection (guidance and regulations for proper fire protection) and a model of the building.

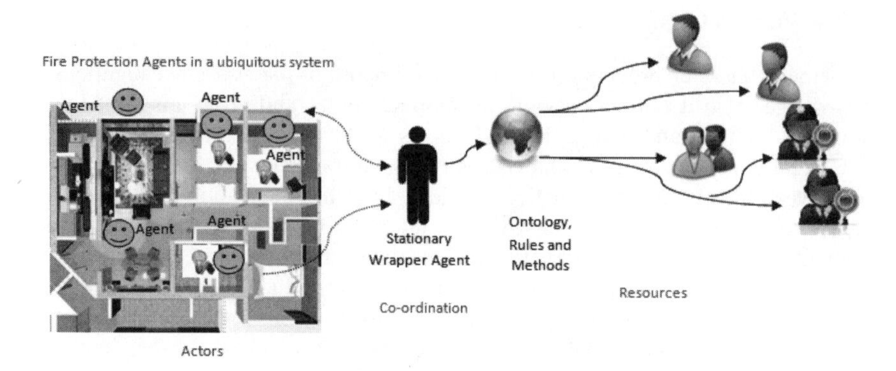

Fig. 1. Stage approach in the model of cooperation

2 Ofepac (Ontology Fire for Emergency Plans and Contingency)

The first step in design this system of cooperative fire protection in buildings has been modeling the ontology of the building, which also contain concepts concerning the structural part of it, contains too, an alarm security system modeling in the building, this security system is an important part in shaping the ontology because is subsumed into the building to detect and control the fire through actuators agents.

2.1 Modeling Ofepac

The process of building ontology can be simplified by using an appropriate methodology. There are several methodologies applicable, after analyzing most of them; Ofepac was built following the methodology Methontology [6] for its simplicity and effectiveness. According to its composition Ofepac will have a multiple structure according to the classification proposed by Wache [7], because it consists of various local ontologies which will serve the actors to make decisions. In addition, concepts have been modeled on a generic domain in the fire of buildings and therefore give the quality of being a standard extensible ontology.

Ofepac has been created with the editor of Ontologies Protégé [8]; Protégé is one of the most widely used editors and stable for the construction of Ontologies.

As already mentioned, one of the objectives of ontologies is to achieve a standard conceptualization of knowledge, therefore for the construction of Ofepac ontologies have been reused in the domain, for it has been used a tool called PROMPT proposed by [9].

One of the integrated ontologies is the ontology of e-response [10] for modeling the building. This focuses on the classification of types of buildings habitable and not habitable. With this classification we can know some characteristics of the buildings like types of materials into the spaces or if there are permanently people into the building. It has been integrated a second ontology developed at the University of Murcia [11] who modeled the building after classes of SOUPA [12] which is an ontology for the development of ubiquitous computing systems used in many projects; ontology modeled from the University of Murcia, part of the classes *Fixed Structure* and *Space In A Fixed Structure*, so it can be considered as an extension of the *SPACE* ontology (This ontology is designed to support reasoning about the spatial relations between various types of geographical regions, mapping from the geo-spatial coordinates to the symbolic representation of space and vice versa, and the representation of geographical measurements of space), the terms of this ontology has inclusions of spatial relations (*SpatiallySubsumes* and *SpatiallySubsume By*); properties very important in the creation of the building ontology, since they allow infer, for example, in which floor of the building is located a specific room. In Fig. 2 we see the concepts that form the ontology.

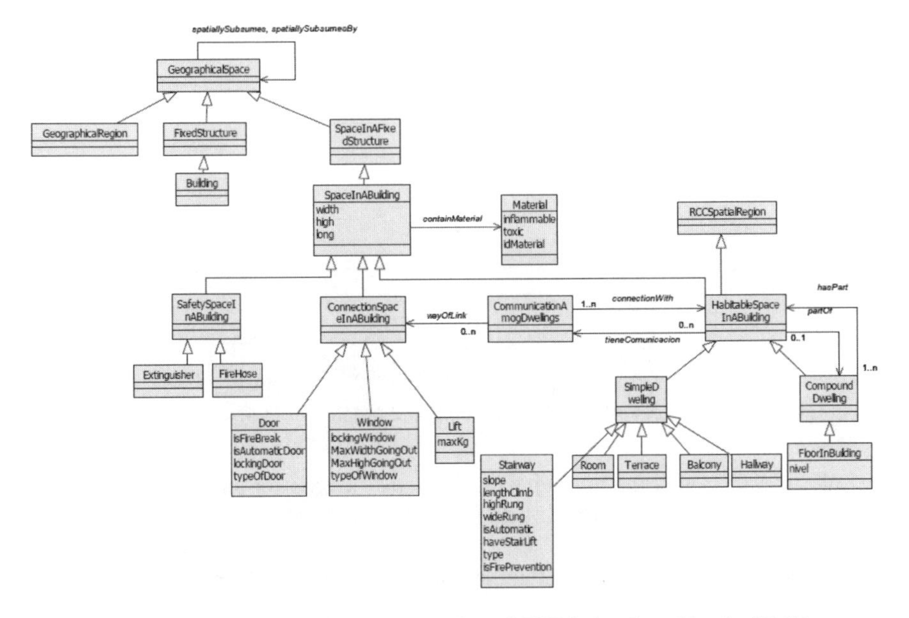

Fig. 2. Ontology from the Space ontology SOUPA developed by the UMU

Fig. 3 shows the result of the integration of the two aforementioned ontologies with the ontology Ofepac. The new properties *ControlledFor* and *ToControl* are therefore crucial to infer in the security system which points in the building can be controlled or manipulated by agents at the time of a fire. Principally *SecuritySystem* ontology of OFEPAC is designed to support a mapping information about all the points in the building that can be manipulate for the agent for evacuate people or thwart the fire effects; for this mission the ontology needs information about the plan´s building to make evacuation routes, this is the reason of the integration of the UM ontology, with this ontology the agents can knows the safety spaces (with extinguisher, for example), connection spaces (window, door), habitable spaces (like rooms), etc. The e-response ontology provides information about the buildings classification; this ontology divides the buildings in habitable and not habitable and all subtypes of them (hospital, school, house, etc.) and has a complement of types of construction materials for know what the resistance at fire of the materials buildings is. We propose, this ontologies integration to make a complete ontology for responses a contingency fire, because we try to

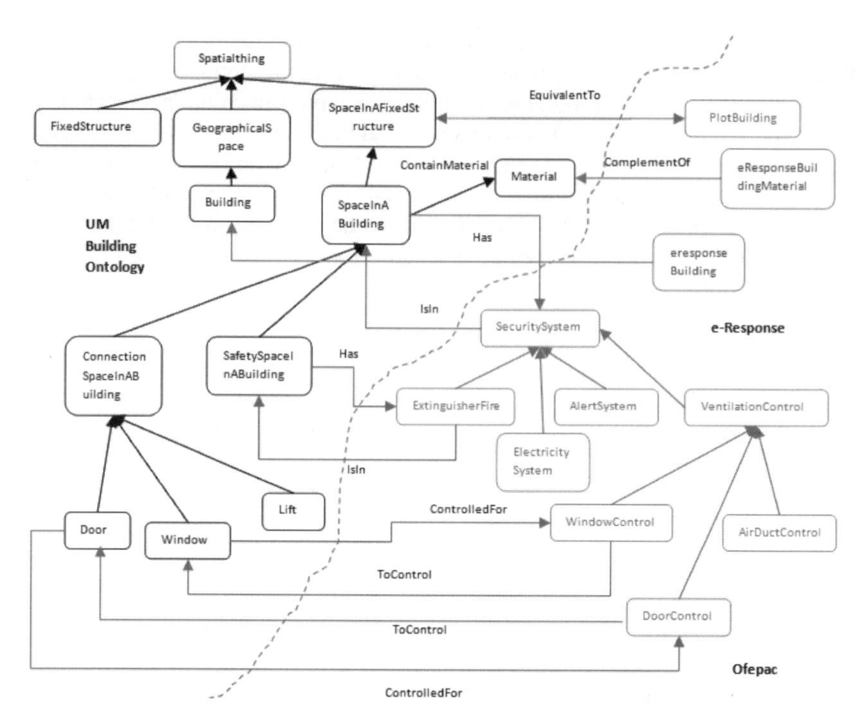

Fig. 3. Ontologies integration and ontology of Ofepac

cover all aspects around the buildings in conflicts, the exterior and interior spaces and how manipulate this information to help people and mitigate fire.

2.2 Taxonomic Evaluation and Performance Analysis to Ofepac

Being Ofepac the engine of knowledge for the system operators, which will be the fundamental basis in decision-making at the time to act before an emergency; it is necessary that the ontology has a high margin of confidence. To be sure, the ontology must be evaluated on structural and rational aspects. We have grouped the assessments into two types:

1. **Structural assessment.** That will be those tests to verify that the ontology has an efficient taxonomic structure, avoiding inconsistencies, redundancies, etc.
2. **Rational assessment.** That will be those tests showing that the ontology generates satisfactory results before processes inference.

These assessments as well provide own information of behavior ontology before an emergency, also include measurements of response time, it is important to know these response times, because before an emergency time is a predominant factor that limits the decision-making. In this document, we present the structural assessment of the ontology, but also include an assessment of performance in which small inferences are made to the ontology. According to [13] [14] the evaluation of ontology includes the inspection of its taxonomy, which should evaluate the inconsistency, incompleteness and redundancy. For this assessment was used pellet's reasoning and a model created with Jena, where the results of all tests were satisfactory.

After these tests, the following step was to evaluate performance of the ontology, there were made Query Type researches using a query engine of SPARQL. The queries were done so much for a single client up to an estimated population of 50 clients, this sample was taken of Customers on the basis of a standard number of 50 rooms per floor in a building; reference was pointed out that in every room there will be an actuator agent which will consult anything to the ontology to make decisions for action, it has been designed in this way so that the system is not centralized but independent; as well avoid the saturation of a centralized network and those already known that creates problems in time and avoid delays in action. The model will be automatically instanced with Jena and queries will be defined from less to more complexes, in order to assess the response time of the ontology as an increasing number of instances and customers are added.

Queries of selection. In a real scenario, to initiate an emergency, the model of knowledge must be settled in a taxonomic structure that allows the optimal intelligent agent, for example, quickly find the elements needs, give the location of the start of fire inside the building, giving the location of safe places for the inhabitants, etc. it is therefore important that the ontology has its relations (concept-concept, concept-relation) correctly defined. This category will use simple queries to search elements in the ontology. It is important to measure the responsiveness of the ontology before finding certain graph to see if their paths are optimal. In this type of queries has been observed that the ontological model is capable of responding optimally in up to 20 clients with up to 800 instances; a maximum of 10 clients with 2000 instances, a maximum of 5 clients with up to 6000 instances and 1 Client with a maximum of up to 20 000 instances in a time less than 2 seconds. If the measurement is done according to the number of instances in the ontology we can see that the ontology is able to respond within an optimal time under 2 seconds (mark in the graph with a horizontal line) in the following manner (See *Fig. 4*). The results indicate that 10 agents can consult the ontology in parallel in a building containing until 2000 places, namely that in a building compound from 1 to 2000 places will be able to count on the collaboration of up to 10 agents at the time a fire, can search through ontology; these searches can be evacuation routes, safe places in the building to shelter people, location of elements of extinguishing fire, and so on.

Queries of Restructuring. In a real scene of a fire behavior, more often than not, it is unstable and as a result knowledge changes, to make decisions the agent can go changing knowledge according to situations that are submitted, is therefore important to know whether the ontology is sufficiently flexible and adaptable to changes and new knowledge inferred. This is will make queries that allow deductions on the ontological model. It uses the CONSTRUCT clause that allows return information of new possible relations since a number of conditions. In the queries of restructuring a maximum of 20 customers can consult with the ontology and up to 400 instances populating it; if ontology were populated with a larger number of instances the number of customers incurred should decrease, with a number of up to 10 clients and until 1600 instances of population; also decreasing the number of customers to 5 and increasing the number of instances in 6000, the ontology optimally responds to an event

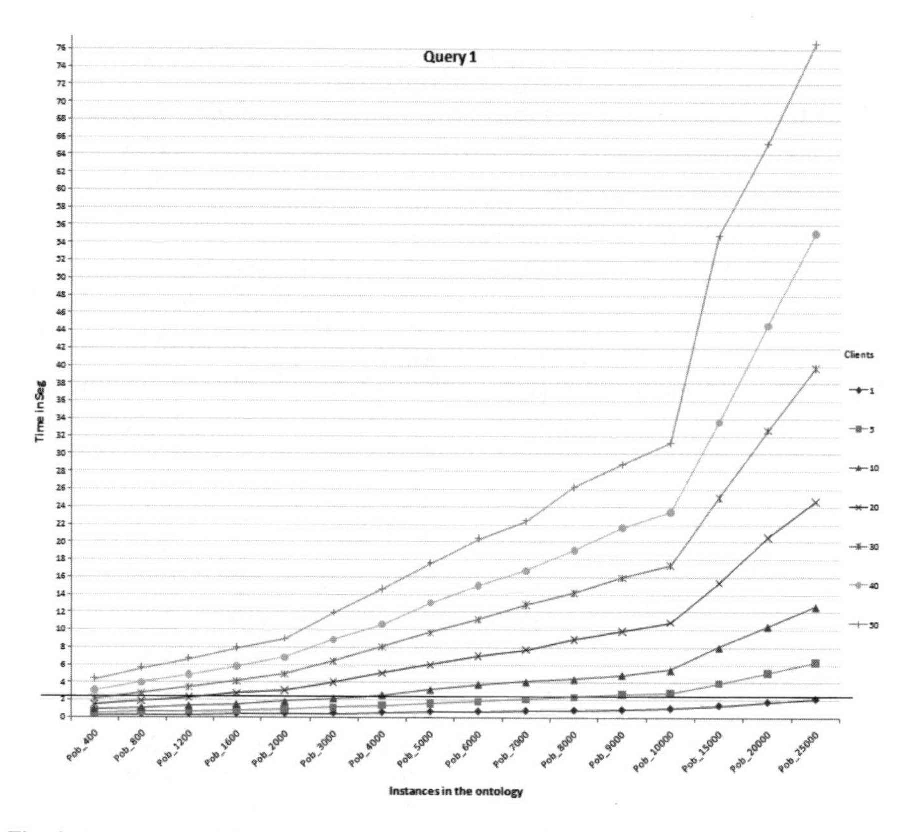

Fig. 4. Assessments of the time in simple queries, according to the number of instances populating the ontology

of contingency, and 1 customer with a population of up to 20000 instances in a time less than 2 seconds to respond to queries (mark in the graph with a horizontal line).

On the contrary, if the analysis is done according to a population of instances in the ontology we see the following results (see Fig. 5). Regarding the performance in response to restructuring queries in a building that is composed with places until 1600, 10 agents can infer new knowledge in Ofepac parallel. It is also possible to have 20 agents acting at the time of the fire, if the building had a maximum of 400 places in its structure, these agents may be inferring new knowledge in the ontology and taking decisions according to what is present at the time of the fire. This new knowledge may be tor example, drawing alternative route of new roads for evacuation. Both queries show that the ontology is not responding optimally to a larger number of 20000 instances or more than 20 clients consulting at once. The time to run queries is dependent on the number of instances of the knowledge base; time reveals an increased exponential tendency relative to the population in the ontology.

In General, we can see that queries of restructuring doing a small inference in the ontology, get better response times to simple queries evaluated, with this, we can say that the ontology is sufficiently extensive to create queries with a greater response time, but its semantic structure is optimum to generate minor response times in the

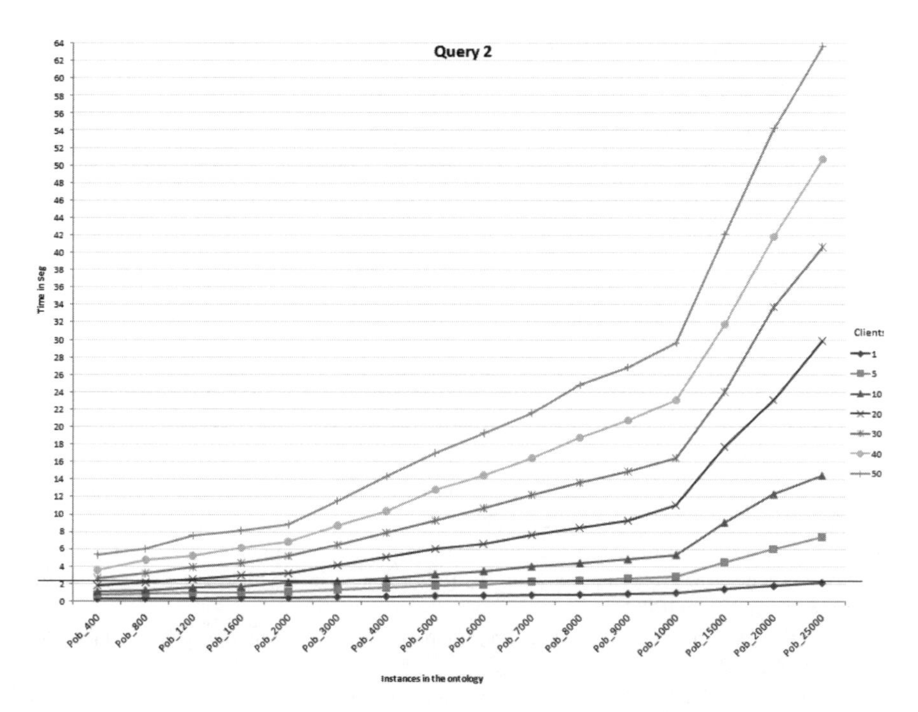

Fig. 5. Assessments of time in queries of Restructuring, according to the number of instances populating the ontology

inferences. It has not been ruled out the possibility of using declarative methods to optimize the assessments made previously to the ontology, which will surely yield results more precise about the behavior of Ofepac in a real fire scene. The rational validation of the ontology is in its development phase, which has been raised through the use of rules of inference and the insertion of new facts inferred. The rules of inference will be established analyzing critical situations that occur in a fire.

3 Conclusions and Future Works

The integration of ontologies in ubiquitous systems is an extensive issue that requires much research in the field, it is important to get appropriate knowledge so that intelligent agents can take correct decisions, it is true that automation in buildings is an issue that already exists, but we need to not only the building acting through sensors, but to have other intelligence agents to take appropriate decisions not only taking into account the conditions of the fire but also the people inside it and the resources it has to mitigate the contingency.

Having assessed the ontology is intended to carry it to a virtual scene of intelligent agents under the system of INGENIAS. One scenario is aimed to achieve a goal within the intelligent system. So in a scenario is essential to determine how and in what order is to establish communication between the entities, what possibilities there are to organize the flow of control between processes and what is the flow of information that is sent and received between entities, all this being achieved with the integration of a planning module in the ontology. For all this, will also be necessary to model scenarios taking into account the behavior of intelligent systems on runtime. Once this is done, it will be possible to obtain formal specifications of the various scenarios, which transcribed to a process specification language can be implemented on an automatically or semi-automatic form.

Bibliography

1. Redacción SicaNews, Edificios inteligentes (2002), http://paginadigital.com.ar/articulos/2002rest/2002terc/tecnologia/sica98.html
2. Mark, W.: Ubiquitous Computing "Hot Topics", vol. 26, pp. 71–72. IEEE, Los Alamitos (1993)
3. Bill, S., Ramesh, P., Kevin, K., et al.: Toward Distributed use of large-scale ontologies. USC/Information Sciences Institute (1996)
4. Javier, B., Molina José, M., Corchado Juan, M., et al.: Ubiquitous computing for mobile environments. In: Antonio, M., Juan, P. (eds.) Issues in Multi-Agent Systems. A Birkhäuser book, Basel (2008)
5. Kinshuk, T.L.: Improving mobile learning environments by applying mobile. Massey University, New Zealand (2004)
6. Gómez-Pérez, A., Corcho, O.: Ontology specification languages for the semantic web, vol. 17, pp. 54–60. IEEE, Los Alamitos (2002)
7. Wache, H.: Ontology-Based integration of information –A survey of existing approaches. In: Proceedings of the International Joint Conference on Artificial Intelligence (2001)
8. Stanford University Protégé, http://protege.stanford.edu
9. Natalya, N.F., Musen Mark, A.: PROMPT: Algorithm and Tool for automated ontology merging and alignment. In: Proceedings of AAAI 2000. MIT Press/AAAI Press, Austin (2000)
10. e-response intelligence virtual organisation to responde to highly dinamics events, http://e-response.org
11. Victoria, B., Fernando, T.: Middleware para la gestión de información contextual. Universidad de Murcia (2005)
12. Harry, C., Filip, P., Tim, F., et al.: SOUPA: Standard Ontology for Ubiquitous anda Pervasive Applications. University of Maryland, Baltimore Country (2004)
13. Gómez-Pérez, A., Fernández-López, M., Corcho, O.: Ontological Engineering. Advanced Information and Knowledge Processing series. Springer, Heidelberg (2004)
14. Gómez-Pérez: Evaluation of Taxonomic Knowledge on Ontologies and Knowledge-Based system. In: Proc. North American Workshop on Knowledge Acquisition, Modeling, and Management, KAW (1999)

CARM: Composable, Adaptive Resource Management System in Ubiquitous Computing Environments

Roberto Morales and Marisa Gil

Computer Architecture Department
Universitat Politècnica de Catalunya
Jordi Girona 1-3, 08034
Barcelona, Spain
{rmorales,marisa}@ac.upc.edu

Summary. Presently heterogeneous devices provided with several communications interfaces are everywhere, with this, we are increasingly coming in contact with "shared" computer-enhanced devices such as cars, portable media players, or home appliances, commonly called ubiquitous computing environments. These environments require special properties that traditional computing does not support, such as a proper resource management which plays an important role in pervasive computing where adaptation and dynamic reconfiguration of resources take place. This work presents a new adaptive resource management approach that supports adaptation for the required resources. We use a component-based model to abstract system's ubiquitous resources in a transparent and uniform way to the applications.

Keywords: Resource management, component, ubiquitous, heterogeneous.

1 Introduction

Nowadays pervasive computing environments are increasing and covering from small areas like home, transports, and museums to large open spaces like shopping centers, technology parks, or universities. These smart spaces are populated with several mobile or fixed heterogeneous devices provided with different resource capacities: from large servers (lot of memory, CPU, I/O, fast connectivity, etc.) to small sensors with limited and scarce resources. Such devices are provided with one or several communication technologies like WiFi, Bluetooth, IrDA, or Zigbee which enable interactive collaboration among them. Mobile phones, MP3 players, PDA's, and Laptops are an example of these devices.

Mobility, spontaneous networking, collaboration and other important factors are leading research to explore new forms of interaction, management of devices or designing new applications in order to fit in these environments [2, 7], as depicted in Figure 1a. We also need to consider additional characteristics to manage this hardware diversity like operating system (OS) heterogeneity. Devices comes with a wide variety of system software (close or open) working at different protection levels, different programming models, or customized for

J.M. Corchado, D.I. Tapia, and J. Bravo (Eds.): UCAMI 2008, ASC 51, pp. 335–342, 2009.
springerlink.com

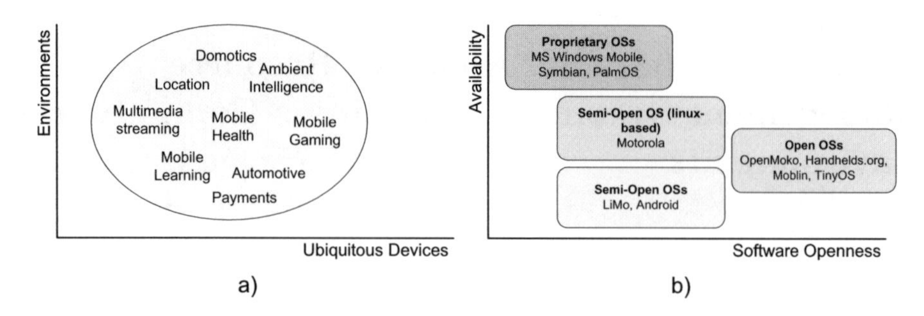

Fig. 1. The left chart shows the new pervasive applications according to the heterogeneity in environments and ubiquitous devices. The right chart gives an example of software heterogeneity and openness.

specific platforms as shown in Figure 1b. The inherent complexity in the diversity of devices and network connectivity that pervasive environments entail opens the door to novel resource management paradigms, where devices are dynamically combined in different ways to use and exploit each of their resources.

The main contribution of this work is a composable adaptive resource management (CARM) approach, which supports dynamic adaptation for the required resources. In the context of our proposal, a "resource" means any shared entity that an application relies on in their execution. Typical resources are memory, processor, and communications.

The proposed new adaptive resource management offers important benefits, we point out some of them:

- *Efficiency:* Effective and efficient management of the system's pool of resources within a pervasive computing environment such as memory and storage.
- *Composition:* Combine resources into a seamless whole when required.
- *Integration:* Incorporation of devices into pervasive computing environments with the objective of resources availability without any manual intervention.
- *Optimization:* Better resource utilization.

For these benefits, we have identified the following set of features in our system:

- *Extensibility:* Pervasive resources are shown as an extensible virtual hardware architecture.
- *Transparency:* Applications should be unaware that they are using remote resources without application modification.
- *Uniformity:* The management of pervasive resources is independent from different system types. Resource access could be at different system levels (e.g. user/kernel level).
- *Adaptability:* The resources should be dynamically reconfigurable and capable of adapting automatically to the addition or removal of resources at runtime to handle resource variation due to user mobility and because of the changing environment.

- *Portability:* The system should be easily portable to multiple architectures and operating systems.

The rest of the paper is organized as follows. In Section 2, we present the adaptive resource management methodology. Section 3 discusses the ongoing work. Section 4 compares our work to related work and Section 5 summarizes the main conclusions of this work.

2 Composable, Adaptive Resource Management Methodology

The trend in current and future pervasive applications (e.g. games) is to consume much more resources such as memory or computing processing. As devices are spread and embedded in pervasive environments, applications are able to use its resources by opportunistically annexing friendly devices they encounter, improving the application life-cycle using its available resources in a transparent and uniform manner. To cope with these needs we propose a composable adaptive resource management (CARM) infrastructure. The CARM infrastructure deals with the resource-hungry applications which usually need more resources than what is currently available; by working at different system levels (user/kernel) depending on the OS openness, and specifically on the ubiquitous computing resources management.

CARM interacts directly with ubiquitous devices to provision shared hardware resources among applications, providing high-level management of memory, processor, and other resources. This requires a layer to be present on user devices that will control its shared resources in the pervasive computing environment. Such layer is responsible for keeping the CARM up to date about the resources status available at each device.

2.1 Resources Classification

A common feature in pervasive environments today is that they consist of multiple devices with different capacities, functionalities and objectives. We can differentiate at least three resource levels. The first level corresponds to self-contained resources (i.e. embedded) in its own personal device such as memory, CPU, or storage. The second level corresponds to those resources which are surrounded to its personal device (e.g. earphones, digital pen, or camera). Finally, at the third level there are several resources provided by the environment itself like printers, displays, or wireless hotspots.

In our work we will refer to resources at level 1. At this level, devices can share and borrow available resources to other devices in its environment, which in consequence will use these resources to meet its goals.

2.2 Resource as a Component

To model ubiquitous resources in a uniform way, we use a componentization approach [10]. Resources are treated as individual components, which are placed along the host system. Each component encapsulates specific behavior (depending of resource type) and provides a virtual view of the specified resource and a secure runtime environment to support applications. For example, when a new resource enters into the system, a new component instance is created to manage remote memory segments or remote code execution. When the resource leaves the system or the task is finished, the component is destroyed. Each component can operate as required (server or client) and does not depend on the operating system or device models; the structure of a component is depicted in Figure 2.

Next we define three different types of components:

1. *Server component:* Its main function is to publish and manage available host resources.
2. *Client components:* These components requests for available ubiquitous resources (when needed). When a new resource is localized, encapsulates the required resource and presents a virtualized instance to applications.
3. *Host components:* These components work internally (client/server) and have other specific tasks such as optimizing communication paths, monitor battery level or CPU load, etc.

All components share the same properties:

- *Dynamic un/loading*: components are un/loading dynamically when required by the applications or system itself and according to the changing pervasive environment.
- *System placement*: due to the different protection levels (user/kernel mode), each component can be placed at different software layers hiding heterogeneity and complexity to the host system and to the applications.
- *Communication channels*: each component has communication mechanisms to send/receive messages to other local components (e.g. IPC) or to remote components (e.g. sockets).

The component mechanism allows changes to the underlying implementation without requiring changes to the applications, i.e. transparency. Additionally, the platform will facilitate the creation and integration of new components (component discovery and downloading) through a modular plug-in architecture. The basic component architecture is composed of three modules named: runtime execution, component abstraction layer and messaging. The runtime execution

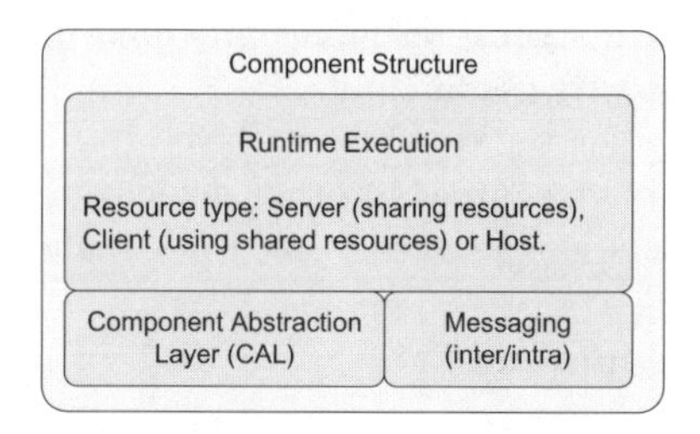

Fig. 2. Internal component architecture design

module is the core system functionality (e.g. virtual memory management). The component abstraction layer hides implementation and specific hardware details. The messaging module provides the communication mechanisms to send and receive data from other components.

To deal with hardware and software diversity (different architectures and operating systems), components will use virtualization techniques to abstract away resource heterogeneity and complexity [8]. In particular, we will apply the idea of virtualization to the management of the availability of pervasive resources. Virtualization introduces a layer of abstraction, which means that components have to snoop out what resources are available and try to adapt an application to use them.

2.3 Component Management Architecture

It is important to mention that components functionality need not to primarily come in the form of software architecture, a lot of components can be deployed in hardware, firmware or downloadable drivers. It will be up to individual OSs the most suitable form of deployment. That's why in this component-based environment, we need additional software infrastructure to control and support components in the system.

So far, we have identified seven different modules, which are necessary to manage all components in the system:

- *Component Repository (CR):* This module keeps all runtime information regarding to all available components. The runtime information includes instance id, component state, access locking, etc.
- *Activity Manager (AM):* This module monitors component activity and informs to the CH the current changes which in turn take the appropriate action.
- *Component Handler (CH):* This module acts as a Component Loader and is responsible for component launch (configuration and registration), maintenance of component runtime information and maintenance of component life cycle. When the Loader launches and component instance, it adds the information of the component instance to the runtime information table. At initialization, all components are recognized and available to the client.
- *Messaging (inter/intra):* The messaging module provides services for components to send and receive messages over arbitrary message delivery mechanisms such as IPC or sockets. The messaging module is built on a modular, plug-in architecture for the flexibility and extendability of the messaging systems: i.e. consider several interfaces (WiFi, Bluetooth, etc.) and several protocols (TCP, etc.).
- *Component Discovery (CD):* The discovering of new components in the environment is the main task in this module. CD also is built on a modular, plug-in architecture for the flexibility and extendability of the discovery system: i.e. considering different mechanisms (Bluetooth inquiry, UPnP, OSGi, etc.).

- *Resources Monitoring (RM):* This module keeps track of the current devices status, informing to the CH when a new device appear or disappear.
- *Publish/Subscribe (PS):* Publish the available components in the system or it subscribes to required components in remote nodes.

Because there is no strict line between components running in kernel and user space mode, each component can run in a separate or shared address space with either kernel or user-space privileges. The component manage architecture will provide a transparent mechanism to allow this interoperability among local or remote components dealing with the protection modes and system layers.

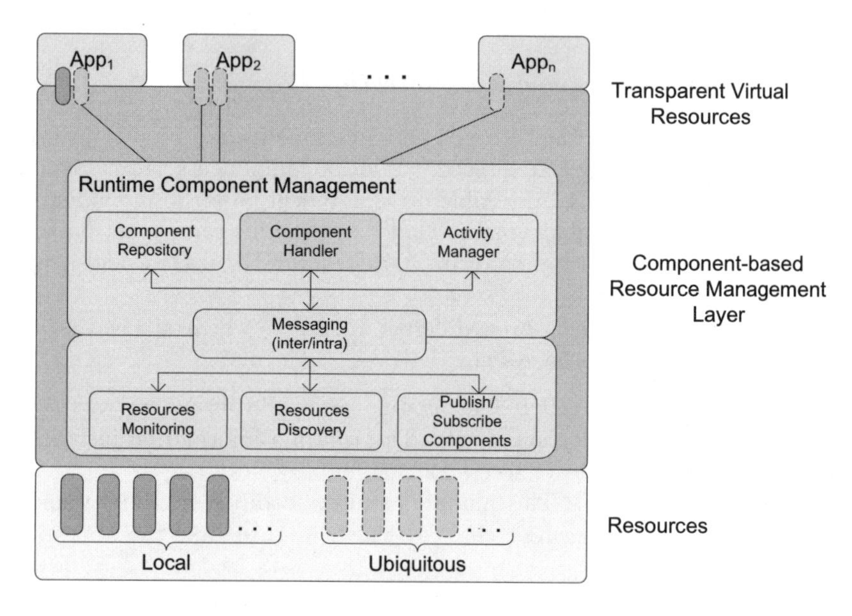

Fig. 3. General component management architecture design

3 Ongoing Work

The complexity of resource management over dynamic and heterogeneous environments dictates novel solutions integrated with high- and low-level data, providing management configurability while hiding low-level mechanisms and implementation details. We are defining the methodology to build and manage components and how they can be matched to pervasive resources.

Currently we are tuning and extending the methodology proposed and developing mobile components on top of it. As an example we are looking into components which are able to optimize resource usage like power management by distributing computation and networking. Additionally, we are seeking for new applications to show the utility of our proposal in a pervasive computing environment.

4 Related Work

The interest in pervasive computing environments has given rise to a proliferation of systems that allow resources to be dynamically discovered and utilized. Recent research shows how to adapt component-based applications in mobile environments but not tackle directly the resource management considering different protection levels, heterogeneous platforms and operating systems.

D. Preuveneers and Y. Berbers [1] proposed a self-adapting middleware to optimize resource usage in a component-based application. The main difference with our work is that they do not consider heterogeneous platforms and their middleware runs inside a Java Virtual Machine. Also, in the middleware community, there exist many technologies, which are developer oriented. UPnP[9], OSGi[6], or Jini[5] are an example of this kind of middleware capable of managing pervasive resources from applications. The main difference with our proposal is that we do not modified or build an application depending on a specific programming model. Instead, we present resources to the existing applications in a transparent manner.

In the operating systems community, THINK [4] project offers a component library, namely Kortex that contains some elementary components for building operating systems (e.g. memory allocators, schedulers, file systems, etc). The TinyOS [3] is another example of existing component-based OS, which is targeted on small-embedded devices (sensors). We share similarities in the way that we see each resource as a component but we do not build an operating system, instead we help the operating system to manage pervasive resources (when needed) efficiently and transparent to the applications.

Finally, we took some ideas from the virtualization community, the μDenali [11] project virtualizes each hardware resource as a component to share with each guest operating system. In our proposal we look for pervasive resources and apply similar concepts of virtualization so the system and the applications use virtual resources without noticing if there are remote or local. We also implement each resource as a component but we improve resource utilization through optimizations or efficient usage of shared resources.

5 Conclusions

In this work we presented an adaptive resource management infrastructure for pervasive computing environments, which enables resources efficient usage and the improvement application life-cycle. Our CARM proposal is designed to run in different levels of operating systems, which provides the features and mechanisms necessary to develop components. Our proposal permits the interoperability among heterogeneous platforms in a transparent and uniform manner to the applications and to the user itself.

We have only given a preliminary sketch of our work and until now we have offered the basis for significant improvements in resource management. Resource management in a pervasive computing environment is a challenging area and we believe that our methodology of componentization is a promising approach. Furthermore, we are actively examining and developing suitable mechanisms to support and evaluate its use and practice.

In general, the work described here can be viewed as part of widespread research efforts to make OSs more adaptive and extensible in pervasive computing environments. Our work is complementary to the above-mentioned related work but which in addition is primarily focused on a mechanism for managing generic resources in a highly dynamic and heterogeneous system.

Acknowledgements

This work has been supported by the HiPEAC European Network of Excellence and the Ministry of Science and Technology of Spain and the European Union (FEDER) under contract TIN2007-60625. Additionally, and a PhD scholarship from Fundación Carolina-CONACyT. All products or company names mentioned herein are the trademarks or registered trademarks of their respective owners.

References

1. Davy, P., Yolande, B.: Towards context-aware and resource-driven self-adaptation for mobile handheld applications 1244255, 1165–1170 (2007)
2. George, H.F., John, Z.: The challenges of mobile computing. Computer 27(4), 38–47 (1994)
3. Jason, H., Robert, S., Alec, W., Seth, H., David, C., Kristofer, P.: System architecture directions for networked sensors. SIGOPS Oper. Syst. Rev. 34(5), 93–104 (2000)
4. Jean-Philippe, F., Jean-Bernard, S., Julia, L.L., Gilles, M.: Think: A software framework for component-based operating system kernels 713860, 73–86 (2002)
5. Jini, http://www.jini.org
6. OSGi, http://www.osgi.org
7. Satyanarayanan, M.: Fundamental challenges in mobile computing 248053 1–7 (1996)
8. Smith, J.E., Nair, R.: The architecture of virtual machines. Computer 38(5), 32 (2005)
9. UPnP, http://www.upnp.org
10. van Ommering, R., van der Linden, F., Kramer, J., Magee, J.A.: The koala component model for consumer electronics software. Computer 33(3), 78 (2000)
11. Whitaker, A., Cox, R., Shaw, M., Gribble, S.: Constructing services with interposable virtual hardware. In: 1st Symposium on Networked Systems Design and Implementation, pp. 169–182 (2004)

Mobile Habits: Inferring and Predicting User Activities with a Location-Aware Smartphone

Andrei Papliatseyeu[1] and Oscar Mayora[2]

[1] University of Trento
papliats@disi.unitn.it
[2] Create-Net
oscar.mayora@create-net.org

Summary. In this paper we present a work in progress dedicated to the recognition and prediction of mobile user activities. In contrast with related projects that generally use GPS for localization, we employ a fusion of wireless positioning methods available in current smartphones (GPS, GSM, Wi-Fi). Our positioning system offers high availability and accuracy without dedicated calibration. We demonstrate how such a positioning information can improve place extraction algorithms and enable the recognition of the new types of user activities both indoors and outdoors. Besides that, the project addresses a number of open challenges in activity and place prediction, such as detection of behaviour changes, prediction of unseen places.

Keywords: activity recognition, behaviour prediction, location awareness.

1 Introduction

The knowledge of user activities and habits is a crucial factor for the development of highly personalized applications, that can be beneficial in many areas of daily life. First of all, the activity recognition is important in assisted living and health care systems. These methods can be used to support memory and planning capabilities of elderly [1], enable early detection of possible health problems [2, 3]. Mobile content providers can adapt the information delivery accordingly to the user preferences and the current context. There also are more general applications, like adaptation of mobile phone interface accordingly to the user's current activity [4]. Moreover, the prediction of user activity (and its location) can be used to improve routing and overall performance of wireless networks [5, 6, 7].

This project will explore and develop novel methods for recognition and prediction of user activities. Mobile smart devices present an ideal platform for this task; they usually possess considerable computational resources, a reach set of wireless communication and multimedia features. Assuming that people tend to always carry their phones along, we employ a modern smartphone as the main sensory and processing unit for learning user's behaviour.

A large number of research works have been dedicated to smartphone-based context recognition [8, 9, 10]. The authors focused primarily on detection of

J.M. Corchado, D.I. Tapia, and J. Bravo (Eds.): UCAMI 2008, ASC 51, pp. 343–352, 2009.
springerlink.com

user's activity, paying little attention to the location; in most cases, only a single positioning technology was used, e.g. GPS [11, 12] or Wi-Fi [13]. However, any individual localization method has its limitations (GPS works only outdoors, Wi-Fi is rarely available outside of big cities) that impose certain constraints on the obtained results. Context prediction, in turn, enables the development of proactive features, such as taking preventive measures in health care, effective roaming in networks [5, 7], in-time traffic notifications [14], and others [15, 16]. The area of context prediction is not mature yet and offers many open challenges [15, 17].

In this project we propose to use a fusion of multiple positioning techniques available in modern smartphones (GPS, GSM, Wi-Fi, Bluetooth) for ubiquitous coverage and high accuracy of localization. We expect that such a positioning system will enable detection of new types of user activities and improve overall recognition accuracy. Another part of the project addresses the open issues of activity prediction, such as detection of changes in habitual behaviour, recognition of activities with complex periodicity and data uncertainty handling.

The paper proceeds as follows. Section 2 describes related work and some of the current challenges. Section 3 presents our objectives and approach. We conclude with a brief description of the current state and future work on the project.

2 Related Work

2.1 Positioning Frameworks

There is no ideal positioning technology. GPS provides an accuracy better than 10m, but only outdoors and with non-obstructed view of sky. Median positioning error of a state-of-the-art Wi-Fi positioning system can be less than 2m, but Wi-Fi coverage is very limited in less populated areas and developing countries. GSM-based solutions provide high coverage but have low accuracy (hundreds of meters). Bluetooth can be used for sub-room-level positioning, but stationary Bluetooth-enabled devices are not widely spread yet.

The current approach to this problem is to combine the data from multiple positioning sensors in order to increase the overall coverage of the system and improve its accuracy. This process is called *sensor fusion* and defined as

> ...the use of multiple technologies or location systems simultaneously to form hierarchical and overlapping levels of sensing... [It] can provide aggregate properties unavailable when using location systems individually. [18]

In the last decade, a number of location frameworks featuring sensor fusion methods have been proposed [19, 20, 21, 22]. Some of these works were rather generic [19, 22], while others were targeted specifically for mobile devices with limited resources [20, 21]. PlaceLab [20] uses multiple wireless technologies for positioning (GPS, GSM Cell-ID, Wi-Fi). It is implemented in Java with some native code, and is available for different platforms. All the measurements are converted to a common physical coordinate system. Data fusion (called by the authors as "conflict solving") can be done using a simple Venn diagram-like method or more resource-demanding particle filtering approach. However, the training of PlaceLab is difficult, as it requires extensive mapping of the RSSI and beacon-ID data to the ground-truth position provided by GPS; this becomes a particularly problematic task indoors because GPS is not available there. Also, the evaluation has shown that the conflict solving method does not increase the positioning accuracy, but instead improves the coverage at the cost of lower accuracy [20, p. 129]. Another localization framework, Location Stack, developed by Hightower et al. [19] represents a layered OSI-like architecture of a positioning system. The data from multiple positioning sensors are first converted to an internal representation in Cartesian coordinates; Bayesian particle filtering and motion modelling are then used for probabilistic data fusion. Although we do not use Location Stack as a positioning system, our work is closely related to its top layers, namely, Activities and Intentions, that were not considered by the authors.

Depending on their type, positioning systems provide either *physical* or *symbolic* location [18]. The systems of the first type report the location as coordinates (absolute or relative), while the second type systems output a label or a name, associated with the place. Most of the current positioning systems, except GPS, are symbolic; in order to obtain physical coordinates from them one needs a database which maps symbolic locations to their coordinates. However, this approach requires considerable calibration efforts and is not well-scalable. The inverse conversion from physical coordinates to a meaningful place name (*place extraction*) is not a straightforward process; some of its aspects remain a challenge.

2.2 Place Extraction and Prediction

Kang et al. define *place* as

> ...a locale that is important to an individual user and carries important semantic meanings such as being a place where one works, live, plays, meets socially with others, etc.[23]

A number of algorithms have been developed for recognition of personally important (or frequent) places. Ashbrook and Starner [11] made use of poor indoors GPS reception: the loss of GPS signal for more than 10 minutes was treated as being in an important place; hierarchical clustering was then used to identify subspaces. The method inherits the drawbacks of GPS, namely, limited availability in urban environment, and is not suitable for place recognition indoors. Laasonen et al. [24] offered a method for place extraction basing on GSM cell transitions; the accuracy was limited due to low resolution of GSM Cell-ID positioning. An adaptive clustering approach proposed in [23] used time and distance thresholds for computationally effective place extraction; the method was tested on Wi-Fi positioning traces obtained via PlaceLab. Similarly, Rekimoto et al. [13] used a custom Wi-Fi keychain logger and offline k-means clustering. A different view of place extraction was presented by Hightower et al. [25]. In order to avoid the limitations of single-sensor positioning and the calibration requirement of PlaceLab, they offered a BeaconPrint clustering algorithm that simultaneously uses Wi-Fi and GSM fingerprints for place recognition.

To our best knowledge, all current place extraction methods apply a binary decision rule to classify a place either as important or not. We argue that such an approach considerably reduces the flexibility of personalized location-aware applications. Instead, we propose to use a more gradual classification, which preserves the "importance" rank of the place (see Section 3.1).

Prediction of the user's future location is often an integral part of the place recognition works. Ashbrook and Starner [11] used a Markov model to predict user movements between places. However, the model was updated offline and the time of the move was not considered. A similar work, but using Bayes predictor, was conducted by Krumm and Horvitz [26]. Their approach addresses the problem of prediction of "never-seen-before" locations, but is not applicable for a mobile device due to the high computational complexity. Song et al. [27] have compared four prediction algorithms in a Wi-Fi network, and shown that a simple Markov model performed almost as good as other, more complicated methods. Mayrhofer in his PhD thesis [16] addressed a generalized problem of context prediction on a mobile device. The consideration of only partial history (using sliding window) and prediction of abstract context instead of its particular features can be seen as limitations of the work.

The places that have never been visited before represent significant difficulties for prediction. This problem has been addressed by open-world modelling approach [26] and by shortening the context until the algorithm is able to provide a prediction [27]. Another challenge is to relax the *stability assumption* [27], which implies that the user behaviour does not change through the time. For example, a college student's behaviour alters rapidly at the beginning of a new semester as the class schedule changes, and the predictor can take the whole new semester to adapt [11]. A typical approach is to assign recent events more weight (however, with little effect [27]). In this paper we present novel methods addressing both of these challenges.

2.3 Activity Recognition

A number of projects have addressed the problem of inferring user's activity from position. A two-step hidden Markov model (HMM) and the variance of Wi-Fi signal strength were used in LOCADIO project [28] to detect whether the user is in movement. Patterson et al. [12] were able to recognize the transportation mode of the user (by bus, by car or by feet) using an online unsupervised Bayesian classifier. A similar task has been done in [29] for GSM network; they have demonstrated that unsupervised HMM and a supervised ANN have similar recognition accuracy (about 80%). Sohn [30], in turn, achieved 85% accuracy by applying a boosted logistic regression with one-node decision tree. A large-scale study of user activity patterns has been performed by Reality Mining project [8]. One hundred users with ContextPhone-running smartphones [10] were building the biggest known publicly available DB of contextual data, over the period of 9 month. The project enabled the researchers to analyze many different perspectives of stand-alone and cooperative user behaviour, activity patterns, etc. Although one of the key ideas of the project was to use both Cell-ID and Bluetooth proximity to other devices, so that the two techniques could "augment each other for location and activity inference", the user location data were very coarse-grained due to the limitations of the ContextPhone platform [10].

3 Our Objectives and Approach

The aim of the project is the development of methods for recognition and prediction of mobile user activities by means of sensors and resources available in a modern smartphone. In particular, we focus on utilization of wireless positioning techniques and user movement history obtained from their fusion. The project objectives are the following.

- Analyse and, if necessary, develop place extraction methods for multi-sensor positioning system, using the resources available in modern smartphones. Address the issues of localization uncertainty, privacy and minimization of user distraction.
- Analyse the types of activities that can possibly be inferred from user's current location and movement history; develop corresponding classification methods, considering the probabilistic nature of location estimates and their varying accuracy. Identify auxiliary sources that can improve classification accuracy (e.g. time, day of the week, recent calls, etc).
- Analyse the applicability of machine learning techniques for recognition of periodic patterns in the user behaviour and prediction of future user activities (intentions).
- Evaluate the performance of the developed methods via user study.

3.1 Hybrid Positioning and Place Recognition

As it has been shown in the Section 2.3, almost all of activity recognition projects employ a single positioning technology, usually GPS. The novelty of our approach is in the utilization of a fusion of positioning methods that provides higher coverage and accuracy. These factors are expected to enable recognition of new types of activities and improve recognition performance. In order to avoid expensive and non-scalable calibration, we adapt a hybrid positioning approach: places are recognized by their raw fingerprints; geographic coordinates are assigned only when available. Our approach is similar to that of Hightower et al. [25]; however, while they have used purely symbolic positioning, we augment the fingerprints with GPS coordinates when possible. This enables the use of coordinates-based inference, like reverse geocoding, obtaining a list of nearby businesses, etc.

The increased availability of positioning information enables us to recognize personally important places both outdoors and indoors with high accuracy. Consideration of places at different scale is a challenging task, because inter-place

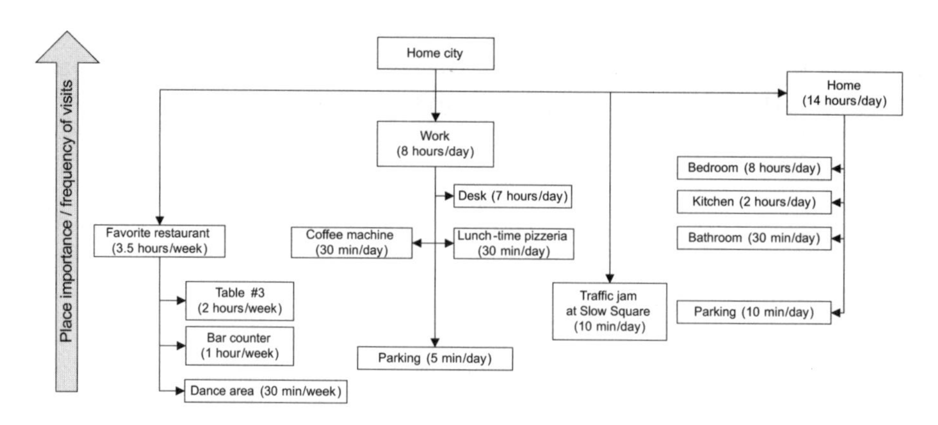

Fig. 1. An example of a user's weekly movement pattern taking into account places "importance"

distances indoors and outdoors vary by orders of magnitude, approaching positioning error when indoors [31]. We propose to augment the recognized places with the "importance" rank. Although most of the current place extraction algorithms already use some kind of ranking in their intermediate stages, the final step is usually a threshold-based binary decision on whether a place is significant or not. We believe that consideration of "Work" place being more important than "Post box" might be more beneficial than the traditional assumption they are equally significant (see Figure 1). This approach alleviates the choice of proper threshold value, because in our approach the latter only defines the lower bound of places to be considered.

3.2 Activity Recognition

An important issue for a symbolic positioning system we employ is the labelling of discovered places with personally significant names. For example, the user would probably prefer to label his/her house as "Home" instead of address or geographic coordinates. One possible approach can be to request the user to input place's label manually. However, this is distracting and can easily become irritating. Another approach for automatic place labelling can be beneficial for the users who schedule their meetings using phone's agenda software. Meeting description usually includes some kind of "location" field, where the user enters personalized name of the meeting place. Observing the user's place at the meeting times for a certain period, the labelling algorithm can assign the corresponding name to the place, along with some degree of confidence that this name is appropriate. Unfortunately, sometimes it is impossible to infer the name of the place without direct user input. For such cases we propose to separate place labels into two categories: private and public. The latter can be uploaded to the Internet and made available for other users, so that they can avoid repetitive manual labelling of public places.

Accurate positioning information augmented by basic inference rules can substantially widen the range of recognizable activities. For example, given GPS coordinates of a place, the system may apply reverse geocoding followed by white pages lookup and retrieve the name of the business located at this place and its category (e.g., "pub") [32]. Then, relying on timing and the observations of internal movements at this place, the system can infer additional information about the user. In the "pub" example, being there at the same time period longer than 4-6 hours, more than 4-5 times a week for some weeks, means that the user is either a *barman* (if staying at the same place during the whole period) or a *waiter/waitress* (moving between a number of places, i.e. tables); in other cases the user can be considered as a *visitor*.

Hybrid positioning also enables the graceful degradation of system performance: if Wi-Fi is not available, the system is unable to detect internal movements but still knows the GPS coordinates of the place (and, possibly, its category); if the GPS signal is poor, the system can still recognise the place by its GSM and/or Wi-Fi fingerprints.

3.3 Place and Activity Prediction

Existing methods for the prediction of user movements, described in Section 2.2, usually model user behaviour as a graph, where nodes represent significant places and edges represent routes. However, this model misses such important factors as time and periodicity. Indeed, the probability of going from "Home" to "Work" during weekend is arguably lower than during working days. Moreover, some activities have long (e.g., visiting a dentist) or complex periods (e.g., visiting friends' house when anyone of them has birthday). For the recognition of periodic activities, we plan to further explore the Fourier transform approach demonstrated in the Reality Mining project [8].

The activity prediction methods presented in related research are also affected by the so-called *"schedule change" problem*. For example, the behaviour of a university student is to a large extent defined by the timetable of classes [11]. During the semester the system learns the schedule and reaches reasonable prediction accuracy. However, when a new semester begins and the schedule changes, the system might take the whole semester to retrain. Typical workarounds are either to shorten the training history or to weight it, assigning more importance to the recent data [27]. Unfortunately, both of these approaches reduce the system capability of prediction of rare events, and should be used only when the need for retraining is detected. We propose a self-evaluation approach for detection of behaviour change. For an "always on" system it is possible to constantly check whether its forecasts come true within certain time interval. If the number of errors increases over its usual value and remains high for some period, it might be an indication of a schedule change. In this case the system should facilitate quicker retraining by application of the weighting method described above.

Availability of positioning history enhanced with categorical information (see Section 3.2) can offer a step towards the prediction of previously unseen places. Let us consider a user whose activity profile exhibits a tendency to spend lunchtime in places categorized as "pizzeria". In this case, even if the user is in a different city, the system may utilize GPS coordinates and web search to get a list of nearby pizzerias and treat them as possible location at lunchtime.

3.4 Project Plan and Current State

Currently, we have finished the literature review and are implementing the data collection application based on POLS framework [33] with a GSM smartphone and an external GPS receiver. Ground truth is to be provided by the user via graphical interface. The project will proceed with a data collection phase: 9 month by the author and, after ensuring the robustness of the logging system, 2 weeks by lab members. For the place extraction, we will adapt the Beacon-Print algorithm [25] augmented with the place importance measure described in Section 3.1.

We have planned a number of experiments for testing our hypotheses and performance of learning methods. First of all, the positioning traces will be used to compare stand-alone and joint availability of different positioning technologies. Then, we plan to analyse the performance of activity recognition methods (including k-means clustering, k-nearest neighbour, naïve Bayes, hidden Markov models and simple neural networks). For the detection of the behavioural patterns, we will evaluate clustering algorithms, fast Fourier transform and self-organizing Kohonen maps. Finally, the prediction methods that take into account the time will be additionally evaluated with a separate dataset collected on city bus routes.

4 Conclusion

This paper presents our approach to recognition and prediction of user's activities using a location-aware smartphone. We propose a multi-sensor localization system which offers almost ubiquitous coverage without prior calibration. The system uses hybrid position estimates based on GSM and Wi-Fi fingerprinting, optionally augmented with GPS coordinates when they are available. We argue that such a system can increase the range of detectable activities. We also demonstrate how basic common-sense reasoning can further improve recognition. Besides that, the article describes our approach to the recognition and automatic labelling of personally important places, prediction of periodic activities and behaviour change detection.

References

[1] Kidd, C.D., Orr, R., Abowd, G.D., Atkeson, C.G., Essa, I.A., MacIntyre, B., Mynatt, E., Starner, T.E., Newstetter, W.: The Aware Home: A Living Laboratory for Ubiquitous Computing Research. In: Streitz, N.A., Hartkopf, V. (eds.) CoBuild 1999. LNCS, vol. 1670, pp. 191–198. Springer, Heidelberg (1999)

[2] Tacconi, D., Mayora, O., Lukowicz, P., Arnrich, B., Setz, C., Troester, G., Haring, C.: Activity and Emotion Recognition in the P-cube Framework to Support Early Diagnosis of Psychiatric Diseases. In: Proc. 2nd Int'l Conf. on Pervasive Computing Technologies for Healthcare (to appear, 2008)
[3] Kautz, H., Arnstein, L., Borriello, G., Etzioni, O., Fox, D.: An overview of the assisted cognition project. In: Proc. AAAI 2002 Worskhop on Automation as Caregiver: The Role of Intelligent Technology in Elder Care, pp. 60–65 (2002)
[4] Weld, D.S., Anderson, C., Domingos, P., Etzioni, O., Gajos, K., Lau, T., Wolfman, S.: Automatically personalizing user interfaces. In: Proc. IJCAI 2003 (2003)
[5] Liu, G., Maguire, G.: A class of mobile motion prediction algorithms for wireless mobile computing and communications. Mobile Networks and Applications 1(2), 113–121 (1996)
[6] Lee, J.-K., Hou, J.C.: Modeling steady-state and transient behaviors of user mobility: formulation, analysis, and application. In: Proc. 7th ACM Int'l Symposium on Mobile ad hoc networking and computing, pp. 85–96. ACM, New York (2006)
[7] Su, W., Lee, S.J., Gerla, M.: Mobility prediction and routing in ad hoc wireless networks. Int. J. Network Management 11(1), 3–30 (2001)
[8] Eagle, N., Pentland, A.: Reality mining: sensing complex social systems. Personal and Ubiquitous Computing 10(4), 255–268 (2006)
[9] Mulder, I., ter Hofte, H., Otte, R., Ebben, P.: Collecting user experience in context with Xensor for Smartphone. In: Proc. Int'l Symposium on Intelligent Environments (2006)
[10] Raento, M., Oulasvirta, A., Petit, R., Toivonen, H.: ContextPhone: A Prototyping Platform for Context-Aware Mobile Applications. IEEE Pervasive Computing 4(2), 51–59 (2005)
[11] Ashbrook, D., Starner, T.: Using GPS to learn significant locations and predict movement across multiple users. Personal and Ubiquitous Computing 7(5), 275–286 (2003)
[12] Patterson, D., Liao, L., Fox, D., Kautz, H.: Inferring high-level behavior from low-level sensors. In: Dey, A.K., Schmidt, A., McCarthy, J.F. (eds.) UbiComp 2003. LNCS, vol. 2864, pp. 73–89. Springer, Heidelberg (2003)
[13] Rekimoto, J., Miyaki, T., Ishizawa, T.: LifeTag: WiFi-Based Continuous Location Logging for Life Pattern Analysis. In: Hightower, J., Schiele, B., Strang, T. (eds.) LoCA 2007. LNCS, vol. 4718, pp. 35–49. Springer, Heidelberg (2007)
[14] Zhang, K., Torkkola, K., Li, H., Schreiner, C., Zhang, H., Gardner, M., Zhao, Z.: A Context Aware Automatic Traffic Notification System for Cell Phones. In: Proc. 27th Int'l Conf. on Distributed Computing Systems Workshops. IEEE, Los Alamitos (2007)
[15] Mynatt, E., Tullio, J.: Inferring calendar event attendance. In: Proc. IUI 2001, pp. 121–128 (2001)
[16] Mayrhofer, R.: An Architecture for Context Prediction. PhD thesis, Johannes Kepler University of Linz, Austria (2004)
[17] Mayrhofer, R.: Context Prediction based on Context Histories: Expected Benefits, Issues and Current State-of-the-Art. In: Proc. ECHISE 2005 (2005)
[18] Hightower, J., Borriello, G.: Location systems for ubiquitous computing. Computer 34(8), 57–66 (2001)
[19] Hightower, J., Brumitt, B., Borriello, G.: The location stack: a layered model for location in ubiquitous computing. In: Proc. 4th IEEE Workshop on Mobile Computing Systems and Applications, pp. 22–28 (2002)

[20] LaMarca, A., Chawathe, Y., Consolvo, S., Hightower, J., Smith, I., Scott, J., Sohn, T., Howard, J., Hughes, J., Potter, F., Tabert, J., Poweldge, P., Borriello, G., Schilit, B.: Place Lab: Device Positioning Using Radio Beacons in the Wild. In: Gellersen, H.-W., Want, R., Schmidt, A. (eds.) PERVASIVE 2005. LNCS, vol. 3468, pp. 116–133. Springer, Heidelberg (2005)
[21] Nord, J., Synnes, K., Parnes, P.: An architecture for location aware applications. In: Proc. HICSS 2002, pp. 3805–3810 (2002)
[22] Kargl, F., Bernauer, A.: The COMPASS Location System. In: Strang, T., Linnhoff-Popien, C. (eds.) LoCA 2005. LNCS, vol. 3479, pp. 105–112. Springer, Heidelberg (2005)
[23] Kang, J.H., Welbourne, W., Stewart, B., Borriello, G.: Extracting places from traces of locations. ACM SIGMOBILE Mobile Computing and Communications Review 9(3), 58–68 (2005)
[24] Laasonen, K., Raento, M., Toivonen, H.: Adaptive On-Device Location Recognition. In: Proc. Pervasive 2004, vol. 4, pp. 287–304. Springer, Heidelberg (2004)
[25] Hightower, J., Consolvo, S., LaMarca, A., Smith, I., Hughes, J.: Learning and Recognizing the Places We Go. In: Beigl, M., Intille, S.S., Rekimoto, J., Tokuda, H. (eds.) UbiComp 2005. LNCS, vol. 3660, pp. 159–176. Springer, Heidelberg (2005)
[26] Krumm, J., Horvitz, E.: Predestination: Inferring Destinations from Partial Trajectories. In: Dourish, P., Friday, A. (eds.) UbiComp 2006. LNCS, vol. 4206, pp. 243–260. Springer, Heidelberg (2006)
[27] Song, L., Kotz, D., Jain, R., He, X.: Evaluating next-cell predictors with extensive Wi-Fi mobility data. IEEE Transactions on Mobile Computing 5(12), 1633–1649 (2006)
[28] Krumm, J., Horvitz, E.: LOCADIO: Inferring Motion and Location from Wi-Fi Signal Strengths. In: Proc. MobiQuitous 2004, pp. 4–13. IEEE, Los Alamitos (2004)
[29] Anderson, I., Muller, H.: Practical Activity Recognition using GSM Data. Technical Report CSTR-06-016, Department of Computer Science, University of Bristol (2006)
[30] Sohn, T., Varshavsky, A., LaMarca, A., Chen, M.Y., Choudhury, T., Smith, I., Consolvo, S., Hightower, J., Griswold, W.G., de Lara, E.: Mobility Detection Using Everyday GSM Traces. In: Dourish, P., Friday, A. (eds.) UbiComp 2006. LNCS, vol. 4206, pp. 212–224. Springer, Heidelberg (2006)
[31] Aipperspach, R., Rattenbury, T., Woodruff, A., Canny, J.: A Quantitative Method for Revealing and Comparing Places in the Home. In: Dourish, P., Friday, A. (eds.) UbiComp 2006. LNCS, vol. 4206, pp. 1–18. Springer, Heidelberg (2006)
[32] Castelli, G., Mamei, M., Rosi, A.: The Whereabouts Diary. In: Hightower, J., Schiele, B., Strang, T. (eds.) LoCA 2007. LNCS, vol. 4718, pp. 175–192. Springer, Heidelberg (2007)
[33] Privacy-Observant Location System (Accessed 2008.05.07), `http://pols.sourceforge.net/`

Author Index

Printing: Krips bv, Meppel, The Netherlands
Binding: Stürtz, Würzburg, Germany